ADVANCES IN PLANT AND ANIMAL BORON NUTRITION

Advances in Plant and Animal Boron Nutrition

Proceedings of the 3rd International Symposium on all Aspects of Plant and Animal Boron Nutrition

Edited by

FANGSEN XU
Huazhong Agricultural University, Wuhan, China

HEINER E. GOLDBACH
University of Bonn, Bonn, Germany

PATRICK H. BROWN
University of California, Davis, California, U.S.A.

RICHARD W. BELL
Murdoch University, Perth, Australia

TORU FUJIWARA
The University of Tokyo, Tokyo, Japan

CURTISS D. HUNT
USDA-ARS Grand Forks Human Nutrition Research Center, Grand Forks, ND, U.S.A.

SABINE GOLDBERG
USDA-ARS George E. Brown, Jr., Salinity Laboratory, California, U.S.A.

and

LEI SHI
Huazhong Agricultural University, Wuhan, China

Springer

A C.I.P. Catalogue record for this book is available from the Library of Congress.

ISBN-10 1-4020-5381-9 (HB)
ISBN-13 978-1-4020-5381-8 (HB)
ISBN-10 1-4020-5382-7 (e-book)
ISBN-13 978-1-4020-5382-5 (e-book)

Published by Springer,
P.O. Box 17, 3300 AA Dordrecht, The Netherlands.

www.springer.com

Cover illustration courtesy of Yunhua Wang, unpublished (the 'disorders' of 'multi-apex' in cotton, a typical boron deficiency symptom. In 1970s, the discovery of cotton and oilseed rape (Brassica napus) suffering from B deficiency initiated a new research area for plant B nutrition and B fertilization in China (See this book p.83-91).

Printed on acid-free paper

Sponsors of the 3rd International Symposium on all Aspects of Plant and Animal Boron Nutrition

September 10~13, 2005

Huazhong Agricultural University

Wuhan, China

National Natural Science Foundation of China
Ministry of Education, China
U. S. Borax. Inc.
Borochemie International Pte. Ltd
College of Resources and Environment, China Agricultural University
Key Laboratory of Subtropical Agricultural Resources and Environment, Ministry of Agriculture, China
Wuhan Qianfang Biochemistry Ltd
Huazhong Agricultural University

1986
NATIONAL NATURAL SCIENCE FOUNDATION OF CHINA

BORAX

BOROCHEMIE
International Pte. Ltd. - Singapore

International Scientific Committee

Benjavan Rerkasem	Thailand
Curtiss Hunt	USA
Fusuo Zhang	China
Fangsen Xu	China
Heiner Goldbach	Germany
Michel Thellier	France
Patrick Brown	USA
Richard Bell	Australia
Sabine Goldberg	USA
Toru Fujiwara	Japan
Toru Matoh	Japan

Local Organizing Committee

Chengxiao Hu
Chongfa Cai
Duanwei Zhu
Fan Liu
Fangsen Xu
Mingjian Geng
Jianwei Lu
Lishu Wu
Nengwen Song
Qiling Tan
Shi Lei
Shuxin Tu
Youliang Yuan
Yunhua Wang
Zheng Liu
Zhuqing Zhao

The 3rd International Symposium on All Aspects of Plant and Animal Boron Nutrition
September 9~13, 2005 Huazhong Agricultural University Wuhan, China

Contents

Part II Boron in Animals and Humans

Part III Boron in Soils

Preface

Boron has been known as an essential micronutrient for higher plants since 1923 by the work of Katherine Warington, but the physiological role of boron in plants and its molecular basis have not been known for a long time. This lack of knowledge left ample room for conflicting views and hypotheses. Even today, many questions are left open. However, considerable progress has been made in understanding the role of boron with the development of new techniques for boron analysis, as well as those for the examination of pertinent aspects of physiology and cellular and molecular biology. Significant progress has been made not only in understanding the uptake, translocation, and physiological role of boron in plants, but also in establishing boron as an essential element in animals and humans. Thus, the **Third International Symposium on all Aspects of Plant and Animal Boron Nutrition** was held from September 10 to 13 in 2005, Wuhan, P.R. China, and provided a forum for scientists focusing on boron research to exchange their latest achievements and organize and collaborate in planning future research activities.

The Boron 2005 symposium focused on all aspects of B research in soils, plants, animals, and humans, similar to its predecessors held in Chiangmai, Thailand (1997) and in Bonn, Germany (2001), despite being a satellite meeting to the XV International Plant Nutrition Colloquium. So, a total of one hundred and four representatives from eighteen nations of the world gathered at Wuhan, and enjoyed six sessions of reports on recent developments regarding B sorption mechanisms in soils, deficiency and toxicity of B, B fertilizer application and basic research on the physiology and molecular biology of plant B nutrition, and nutritional function of B in animals and humans, and exchanged their views and experiences. Some recent key findings were reported, such as new information about gene expression and control of B transporters, the continuum of B re-translocation and the cold tolerance of low B plants, and many more.

Based on the aim of this symposium and the submission of manuscripts from the participants, the conference book of the symposium Boron 2005, **Advances in Plant and Animal Boron Nutrition,** consists of four parts: Plenary Review, Boron in Plants, Boron in Animals and Humans, and Boron in Soils. The second part "Boron in Plants" is again divided into three sections: Physiology and Metabolism of Boron in Plants, Boron Nutrition and Boron Application in Crops, Genotypic Differences of Boron Nutrition in Plants. In this book, readers will find thorough coverage of all recent developments in boron nutrition research as well as suggestions for future research focus.

We would like to thank all participants for their contributions to the symposium Boron 2005 and conference book. We also sincerely thank our sponsors on behalf of the International Scientific Committee and the Local Organizing Committee. Without their generous funding support, it would have been impossible to hold the symposium and publish this volume. Meanwhile, we would also like to express our sincere thanks to the members of the Local Organizing Committee and the large number of Ph.D. and Masters students for their dedicated efforts that made the symposium an outstanding success. We would especially like to thank Professor Yunhua Wang, well known for his 30 years of research on plant boron nutrition and boron fertilizer application in China. He enthusiastically engaged himself with many aspects of symposium planning including design of the scientific program, the logistical needs of participants, and the important task of raising funds.

Finally, we would like to thank Springer for its willingness to again publish the conference book in the series of international symposia on boron research. Hopefully, this summary of recent progress in all aspects of boron nutrition research will prove valuable during future implementation of relevant improvements in agricultural and nutritional practices, policy development, and planning of boron nutrition research around the globe. Due to the time constraints, some errors in manuscript presentation might have escaped the attention of the editorial staff. We therefore apologize in advance to authors and readers for any such possible errors. We are confident, though, that the readers of this volume will find the information found herein to be interesting and a motivation for making further advances in boron nutrition research.

Fangsen Xu
In the name of the organizers and
the members of the editorial board

Plenary Review

Boron Functions in Plants and Animals: Recent Advances in Boron Research and Open Questions

Heiner E. Goldbach[1], Longbin Huang[2] and Monika A. Wimmer[1]

[1]INRES-Plant Nutrition, University of Bonn, Karlrobert-Kreiten-Str. 13, D 53115 Bonn, Germany

[2]The University of Queensland, Hartley Teakle Building (83), Q 4072 Adelaide, Australia

Introduction

Boron deficiency is a widespread problem for field crop production where large losses of yield occur annually both quantitatively (e.g. in southeast China over 40% yield reductions may occur in oilseed rape: Wei et al. 1998), as well as qualitatively (Stephenson and Gallagher 1987; Ram et al. 1989; Bell et al. 1990; Nyomora et al. 1997). Significant losses of yield or quality resulting from boron deficiency may occur as well in vegetable crops (e.g. Kotur 1991). Even eucalyptus trees in large areas of southern China (Dell and Malajczuk 1994), and pine trees in southeast Australia (Hopmans and Flinn 1984) may be severely affected by boron deficiency in both growth and quality.

Although reports from agricultural practice have well established that adequate boron supply is imperative for obtaining high yields and good quality, and increasing evidence suggests a metabolic function or at least beneficial effects of boron in animal metabolism (Park et al. 2002, Hunt, Nielsen et al. and Spears, this volume), knowledge about metabolic functions of boron is yet incomplete. Nevertheless, recent research findings have greatly improved our understanding for boron uptake and transport processes (Brown and Shelp 1997; Hu and Brown 1997; Brown et al. 2002; Frommer and von Wiren 2002; Takano et al. 2002), and roles of boron in cell wall formation (Matoh 1997; O'Neill et al. 2004), cellular membrane functions (Goldbach et al. 2001), and anti-oxidative defence systems have been suggested (Cakmak and Römheld 1997). A beneficial or even essential role of boron in animal metabolism is supported by the findings that low boron concentrations induce the MAPK pathway in cultured animal cells and that cell lines with a knockout of the boron transporter NaBC1, the mammalian homolog of AtBor1, stop to develop and proliferate (Park et al. 2004). The finding of boron being an essential part of a signal molecule (AI2) in bacteria (see below) highlights the possibility that, besides being an indispensable factor for RGII cross-linking (see review by O'Neill et al. 2004), boron might play further roles in both, animal and plant metabolism.

This paper summarizes recent advances and major achievements in boron research, mostly since the last boron meeting in 2001, and highlights open questions for

F. Xu et al. (eds.), Advances in Plant and Animal Boron Nutrition, 3–25.

further elucidating the role(s) of boron in plant and animal metabolism.

Boron binding bio-molecules and stability of boron complexes

There are already well known boron containing biomolecules such as the boron containing macrolides (aplasmomycine, boromycine, and tartrolon B: Moore and Hertweck 2002).

The first boron-containing compound identified in the plant kingdom, which is stable under physiological conditions, is the pectic polysaccharide rhamnogalacturonan II (RGII), where boron cross-links two RGII monomers and thus provides stability to the cell wall matrix (O'Neill et al. 2004). As can be seen in RGII, the steric arrangement of molecules can make a large difference in the stability of boron complexes: although the pectic polysaccharide contains two apioses on side-chains A and B, only the one in chain A is responsible for forming the stable borate bridge, whereas apiose from chain B does not participate in the formation of the RGII dimer (Reuhs et al. 2004). Furthermore, replacing L-fucose by L-galactose in the GDP-D-mannose-4,6-dehydratase deficient *Arabidopsis mur1* mutant, significantly reduces growth and leads to malformation, which can be compensated by higher boron concentrations (Reuhs et al. 2004) or by supply of fucose (O'Neill et al. 2001). The stability of the altered RGII complex is thus lowered through a comparatively small change, and for obtaining a regular degree of cross-linking, the ratio of boron to RG II has to be increased. The glycosyl sequence and the three-dimensional conformation of RG-II are therefore important in regulating the interaction of this pectic polysaccharide with borate (Reuhs et al. 2004).

Other putative boron binding biomolecules have also been identified. Using capillary electrophoresis, Ralston and Hunt (2001) compared adenosine and molecules with adenosine moieties including S-adenosylmethionine (SAM) and diadenosine polyphosphates (Ap_nA). S-adenosylmethionine (SAM) proved to be the compound forming the most stable borate complexes next to apiose. The stability of boron complexes decreases in the order: SAM $\cong$ Ap_6A $\cong$ Ap_5A > Ap_4A > Ap_3A $\cong$ NAD^+ > Ap_2A > NADH $\cong$ 5'ATP > 5'ADP > 5'AMP > adenosine > 3'AMP $\cong$ 2'AMP $\cong$ cAMP $\cong$ adenine. Species with vicinal *cis*-diols bind boron, species without those moieties do not. Boron binding affinity increases when proximal cationic moieties are present, and anionic moieties remote from the *cis*-hydroxyl binding site also seem to positively influence boron binding affinity. In the Ap_nA species, cooperative complexing of boron between the terminal ribose moieties apparently occurs, as boron affinity is greater than expected for two monocomplexes and binding affinities increase as more phosphate groups (beyond three) are present separating the terminal moieties, probably by reducing the strain on the bonds (Hunt 2002). At physiological pH, the adenine moieties of Ap_nA are driven together by hydrophobic forces and stack interfacially (Kolodny and Collins 1986, Hunt this volume). The work of these authors thus clearly showed the ability of different biomolecules to form borate complexes. It has to be emphasised, though, that the capillary electrophoresis used by Hunt and co-workers to show the complex stability of these borate esters has been carried out at pH 8.4, which is higher than the pH usually found in living cells. This favours complex formation due to a higher ratio of borate to undissociated boric acid (about 12% *vs.* less than 2% in natural systems). Thus, complex formation might be overestimated. This should not, however, question the results cited before as the observations made by Hunt's group show the *relative* complex stability of a number of relevant biomolecules. It is, however, extremely difficult to assess the stability of the respective borate-complexes *in vivo*, as the cellular fluids are a complex multi-solute system where water activity may be lower than in simplified *in vitro* systems. Furthermore, the complex stability of the borate complexes depends highly on steric conditions, charge distribution and the molecules environment (e.g. hydrophobic interactions). Borate complexes could be stabilized in a way similar to the inhibition of bacterial enoyl-reductase by diazaborines (Baldock et al. 1996).

Considering compartmentation within cell organelles and differences in the steric arrangement of potentially B-binding molecules, it is yet hard to decide which of those ligands besides RGII monomers do play a major role *in vivo.*

An interesting hypothesis has been launched by Ricardo et al. (2004), suggesting that the formation of di-pentose-borate complexes might have stabilized ribose/ribulose (besides other cyclic pentoses such as arabinose, xylose, and lyxose) in pre-biotic phases in interstellar dust or early during earth history. When formed from glycolaldehyde or formaldehyde, borate prevents the hydrolytic decay of pentoses under alkaline conditions. In a reaction mixture at pH values around 12 and in the presence of borate, ribose was the main reaction product, whereas in the absence of borate, pentoses rapidly degraded to ill-defined brown polymers. Formation of borate complexes might thus have enabled the accumulation of ribose as a pre-condition for early development of life on earth. The fact that boric acid resp. borate react with hydroxyl groups is considered as the key for understanding boron functions (Bolaños et al. 2004). The authors stated that "the primary role of boron in biological systems is stabilization of molecules with *cis*-diol groups, independently of their function", and "boron chemistry makes it a perfect candidate for atomic diester bridging". Whether the formation of one- or two-sided complexes is related to boron's function(s) is still under debate. According to Hu et al. (personal communication), the chemical conditions in living systems do not favour monoester formation, and only di-esters could achieve stabilities high enough to be of physiological relevance.

Using phenylboronic acids as a probe for boron binding ligands and the fact that these are forming exclusively one-sided esters makes it thus possible to search for those functions where boron is required for (cross-) linking ligands with *cis*-hydroxyl groups as in RGII (Bassil et al. 2004).

Boron in the animal metabolism

In the past years, a wealth of new evidence has been gathered for boron being an essential or beneficial element in animals/humans (Hunt 2003, see as well contributions by Hunt, Nielsen et al. and Spears, this volume).

Earlier reports showed that boron is essential for embryo development, at least for vertebrates. Boron deprivation disrupted embryonic development resulting in a high percentage of necrotic eggs and abnormal development of the gut in *Xenopus laevis* (Fort et al. 1999). It seems as if at least the early stage of development is especially sensitive to boron deficiency such as described for mated zebrafish (*Danio rerio*) (Rowe and Eckhert 1999). Although the target molecules are likely to be different, there is a coincidence in animal and plant metabolism for boron to be especially required at initial phases of differentiation, as Behrendt and Zoglauer (1996) have shown for *Larix decidua*.

At least beneficial effects of boron were suggested in a number of nutritional studies, with many of the effects related to bone metabolism. For example, boron supplementation of a low-boron diet reduced gross bone abnormalities in the vitamin D-deficient chick (Hunt and Nielsen 1981; Bai and Hunt 1996). In vitamin D-deficient rats fed a low-boron diet, supplemental dietary boron enhanced the apparent absorption and retention of calcium and phosphorus and increased femur magnesium concentrations (Hegsted et al. 1991). In male pigs, bone lipid was lower and the bending moment higher when boron was supplemented to a low-boron diet (Armstrong et al. 2000). It might even be envisaged that the effect of boron on bone metabolism could be at least one of the essential functions of boron in humans. It has been found as well that physiologic concentrations of boron reduce the amount of insulin required to maintain plasma glucose (Bakken and Hunt 2003).

Recently, Park et al. (2004) identified the mammalian homologue NaBC1 of the AtBOR1 transporter (see below), which suggests that there is possibly a need for a relatively close control of boron levels in animal cells, too. The finding that low concentrations of borate activate the MAPK pathway and that the knockdown of NaBC1 halted cell growth and proliferation, point to a possibly essential functional role of boron in animal metabolism.

There is evidence from several laboratories that dietary boron plays a role in immune function in a variety of organisms. The boron-containing antibiotic boromycin from *Streptomyces sp.* strain A-3376 was recently found to be a potent anti-human immunodeficiency virus (HIV) antibiotic (Kohno et al. 1996), acting by (in part) still unknown mechanisms. Generally, boron-containing antibiotics such as boromycin, tartrolon B and aplasmomycin are borodiesters and act as ionophores. When boron is removed from at least one of these antibiotics (aplasmomycin) by mild acid hydrolysis (pH 3), the resulting desboroaplasmomycin loses its functionality as K^+ ionophore (Sato et al. 1978), which can be re-constituted by treatment with boric acid at pH 6 and 8 (Chen et al. 1980).

Interleukin-6 (IL-6) is a systemic "proinflammatory" and as such an important regulator of the immune system by inducing several processes (Lowik 1992). When using rhamnogalacturonan IIs isolated from *Panax ginseng* leaves, monomerization of the RG-II dimer significantly decreased its IL-6 production-enhancing activity. Boron may affect production of TNF, a proinflammatory cytokine, in chicks and humans, since TNF concentrations were elevated in the culture medium of pelvic cartilage isolated from chick embryos after they were incubated with boron as a 3% boric acid solution (Benderdour et al. 1997). Further examples are presented by Hunt (2003); and Nielsen et al. and Hunt, this volume). The treatments, though, were carried out with elevated, non-physiological amounts of boron, and there is a need to determine whether the effects of boron on TNF production are of physiological importance.

As mentioned above, diadenosine phosphates (Ap_nA), which function as signal nucleotides associated with platelet aggregation and neuronal response and which are putative "alarmones" reportedly regulating cell proliferation, stress response, and DNA repair (McLennan 1992), have higher affinities for boron than any other currently recognized boron ligand present in animal tissues including NAD^+ (Ralston and Hunt 2001). A close interaction between boron and serine proteases has also been suggested (Hunt 2003). Serine proteases are major proteolytic enzymes and have, in addition to degrading structural proteins, regulatory roles in normal inflammation processes (Kettner et al. 1988). Boron is able to form covalent bonds with the nitrogen atom of amine groups, e.g. in hemerythrin (a nonheme iron-containing oxygen transport protein of *Golfingia gouldii*), where it binds near the coordination iron site (Garbett et al. 1971). Whether the stability of this type of boron complex is high enough and comparable to polyols to be of biological relevance remains to be shown. In serine proteases, the boron atom is thought to inhibit the formation of a tetrahedral boron adduct (the transition-state analogue) that mimics the tetrahedral adduct formed during normal substrate hydrolysis (Berry et al. 1988). The adduct includes a covalent bond between boron and a specific nitrogen at the active site of these enzymes. Nanomolar concentrations of certain synthetic peptide boronic acids effectively inhibit chymotrypsin, cathepsin G, and both leukocyte and pancreatic elastase *in vitro* (Kettner and Shenvi 1984). The serine protease thermitase (E.C. 3.4.21.66) was partially inactivated by hydrogen peroxide in the presence of 50 mM sodium borate (Hausdorf et al. 1987). This concentration, however, is rather high and well beyond concentrations usually found in organisms *in vivo*. To favour the formation of such a complex, compartmentation and/or hydrophobic interactions would be needed, if boron would not reach similar concentrations in biological systems. Whether this effect is part of the anti-inflammatory pharmaceutical properties of elevated doses of boric acid (e.g. Newnham 2002) still remains to be elucidated.

Boron uptake and translocation

Boron uptake has earlier been considered as a merely passive process following mass flow of water (Hu and Brown 1997). These authors pointed out, though, that the wide variation of boron uptake under field conditions indicates that there are more processes involved in boron uptake and distribution. Since then, there has been a tremendous growth of information in this special field, ranging from the discovery of complex formation with polyols as a mechanism for boron phloem mobility (Brown and Shelp 1997), the finding of several transport mechanisms (Dordas et al. 2000), to the recent identification of boron transporters (Takano et al. 2002). Below, we will only briefly address the development in this area during the past years.

For agricultural purposes, it is important to find cultivars which thrive well under boron–limited conditions. In rape, Xu et al. (2001) found by QTL analysis, that there are one major and three minor gene loci for boron efficiency. In this species, boron efficiency was also more closely related to the (free) sugar contents than to tightly bound or "free" boron. Boron use efficiency is further related to phenology: the boron efficient cultivar shows earlier bolting and flowering (Du et al. 2002), thus probably reducing the amount of boron needed for vegetative biomass development.

In wheat, boron efficiency was likely conferred by two major genes *Bod1* and *Bod* (Jamjod et al. 2004). Enhanced boron transport into the ear was identified as the mechanism for boron efficiency rather than re-translocation of boron (Huang et al. 2001). Breeding for boron efficiency may boost cereal productivity in SE Asia (Jamjod et al. 2004; Nachiangmai et al. 2004).

Boron and water channels

Water channels (or aquaporins) in the plasma membrane of root cells play an important role in root hydraulic conductivity and plant water relations. In plants, 75-95% of the water moving across the plasma membranes pass through water channels (Steudle 2000). The contribution of aquaporins to water uptake increases in response to changing environmental factors (e.g. drought, nutrient deficiency, low temperature: Tyerman et al. 1999; 2002). Water channels also seem to play an important role in the process of boron uptake across the plasma membrane (Dordas and Brown 2000; Dordas et al. 2000). This was demonstrated in purified plasma membrane vesicles from squash roots, where the boron permeability coefficient (3×10^{-7} cm s^{-1}) was six times higher than that of microsomal vesicles, and where boron permeation across the plasma membrane vesicles was reduced by 30-39% by the addition of the non-specific channel-blocking agent $HgCl_2$ (Dordas et al. 2000). The contribution of channel-mediated boron uptake to total root boron uptake may be particularly significant when external boron concentration is more than adequate, whereas active transporter-mediated uptake seems to be necessary to account for rates of boron uptake required to avoid boron deficiency when boron concentration at the root surface is low.

Severe boron deficiency (e.g. interruption of boron supply to the root tips for one hour) caused a rapid decrease in the amount of the plasma membrane water channel proteins (ZmPIP1 aquaporins) in transgenic tobacco (Goldbach et al. 2001).

Boron transporters

Boron loading into the xylem after uptake into root cells requires the function of boron transporters located in the plasma membrane of the root pericycle, particularly when external boron levels are low (Takano et al. 2002). The recently suggested mechanism of active boron uptake and transport in roots at low boron supply (Dannel et al. 2000; 2002) may be related to the functions of the BOR1 (Takano et al. 2002), and possibly further boron transporters. Wild type *Arabidopsis* plants show an enhanced transport of boron to the shoot under low boron supply, which can be suppressed upon resupply of high levels of boron within a few hours (Takano et al. 2005b). This enhanced transport was not observed in the *bor1-1 Arabidopsis* mutant. Takano et al. (2002; 2005b) were able to show the accumulation in the plasmalemma of a constitutively expressed BOR1-GFP fusion protein under conditions of limited boron supply. Upon resupply of high levels of boron, BOR1-GFP was degraded within several hours. Posttranscriptional mechanisms seem to play a major role in the regulation of BOR1 accumulation. Endocytosis and degradation of BOR1 seem to be regulated by boron availability, possibly to avoid accumulation of toxic levels of boron in shoots under high boron supply, while protecting the shoot from boron deficiency under boron limitation.

Recent reports of a boron transporter as mammalian homologue of the AtBOR1 borate transporter, occurring ubiquitously in animal cell membranes (Park et al. 2004), raises questions about whether related transporters are generally found in plant and animal cell membranes. This NaBC1 transporter conducts Na^+ and OH^- (H^+) ions in the absence of borate, whereas in the presence of borate, NaBC1 behaves as an electrogenic, voltage-regulated, Na^+-coupled $B(OH)_4^-$ transporter (Park et al. 2004). Knockdown of NaBC1 halted cell growth and proliferation.

Boron's function(s) in plants

Structural role(s) in cell walls: RG II

The structural role of boron in the cell wall of higher plants has been reviewed extensively (Brown et al. 2002). It seems as if enhanced formation of RGII (and cross-linking by borate) is closely related to the conquering of land during evolution. Avascular bryophytes contain only about 1% of the amount of RGII of vascular plant species, and the amount of RGII in cell walls increased during the evolution of vascular plants (Matsunaga et al. 2004). Development of boron-dependency during evolution may thus correlate as well with upright growth and lignified secondary walls. The highly conserved structure of RGII and the fact that its genes appeared early during land plant evolution, point to RGII as a fundamental molecule for wall structure (Matsunaga et al. 2004).

The dimerisation of RGII by a borate cross-link has already been addressed above as well as the influence of small molecular changes on its structure (see above). Partial replacement of L-fucosyl by L-galactosyl residues in xyloglucans and in RGII of the dwarf mutant *Arabidopsis thaliana mur1* resulted in reduced growth and malformation of the plants (Reuhs et al. 2004). It was also shown that tensile strength was reduced in the *mur1* mutant compared to the wild type (Ryden et al. 2003). It could, however, be completely rescued with higher boron levels in the hypocotyl and the stem, demonstrating that the lack of fucose in RG II rather than in xyloglucan is important for the mechanical phenotype. Experiments with the *nolac-H18* (non-organogenic-loosely attached cells) tobacco callus mutant also demonstrate the structural importance of dB-RGII for normal growth because the mutant lacks glucuronyltransferase 1, needed for glucuronic acid addition to RGII, causing reduced formation of dB-RGII and consequently reduced intercellular attachment in meristems and tissues as well as the inability to form shoots (Iwai et al. 2002). This highlights the importance of pectin as an adhesion molecule (Lord and Mollet, 2002). This agrees with our earlier observations of a reduced cell wall elasticity modulus under boron deprivation (Findeklee and Goldbach 1996) and observations by Fleischer et al. (1999), showing that RGII is a key component of structural integrity of cell walls. The observations reported by Ryden et al. (2003) also suggest that B–RGII complexation plays a role in the expanded primary wall and is important for secondary wall structure or assembly. Tensile properties of the cell wall depend both on a xyloglucan cross-linked microfibrillar network and RGII-borate complexes (Ryden et al. 2003). Rapid loosening of the cell wall under boron deficiency could also affect the aperture of specific aquaporins, which are responsive to short pressure pulses in a dose dependent manner (Lee et al. 2005; see also below).

The cell walls of the *Arabidopsis* bor1-1 mutant show a lower degree of cross-linking under limited boron supply, which is likely a consequence of lower boron shoot levels in the mutant compared to the WT (Noguchi et al. 2003), highlighting again the importance of boron for maintaining cell wall integrity.

Boron and membranes

Several earlier studies have demonstrated roles of boron in the functioning of enzymes and other proteins of the plasma membrane, transport processes across the membrane and membrane integrity (Cakmak and Römheld 1997; Goldbach et al. 2001; Brown et al. 2002). For example, boron deficiency altered the membrane potential (Blaser-Grill et al. 1989; Schon et al. 1990; Goldbach et al. 1991), and reduced the activity of proton-pumping ATPase and thus the proton gradient across the plasma membrane (Heyes et al. 1991; Ferrol and Donaire 1992; Lawrence et al. 1995; Obermeyer et al. 1996), and of Fe-reductase (Goldbach et al. 1991; Ferrol and Donaire 1992). Some of these changes are observed within minutes after changing the boron supply, which is in line with the assumption of a direct interaction between boron and membranes. Although the results are sometimes contradictory and not always attributable to an early effect of boron deprivation, the inhibition of PL-bound oxidoreductase activity within minutes of boron deprivation has been confirmed (Barr et al. 1983; Wimmer 2000).

Effects of boron deprivation may exert multiple direct and indirect effects on membrane-bound processes. A direct role of boron in maintaining membrane structure is likely through *cis-diol* complexation with glycoproteins, which are structural constituents of the plasma membrane (see below) (Goldbach et al. 2001; Brown et al. 2002). Effects of boron deficiency pointing to a structural role of boron in membrane stabilization are an altered permeability for K and sugars (Pollard et al. 1977; Parr and Loughman 1983; Goldbach 1985; Cakmak et al. 1995; Wang et al. 1999), a damage of the peribacteroid membrane in nodules (Bolanos et al. 1994; 2001) or a change in membrane-bound Ca levels (Mühling et al. 1998; Wimmer and Goldbach 1999). Also, boron deficiency reactions can be compensated by an enhanced Ca supply in cyanobacteria (Bolanos et al. 1993; 2002) Boron seems to be essential for the full functioning of N_2 fixing rhizobial as well as actinomycal symbioses and for heterocyst formation of free living *Cyanophyceae* (see review by Bolanos et al. 2004). The authors hypothesize that the primary role of boron in biological systems is the stabilization of molecules with *cis*-diol groups, independently of their function.

Verstraeten et al. (2005) suggested that boron preferentially interacts with negatively charged phospholipids, or with those containing sugar moieties in their headgroup. Their work showed that boron, at concentrations as low as 0.5 µM, interacts with the lipid bi-layer affecting membrane rheology. Lipid composition determined the magnitude and direction of the boron effects. Boron may thus play a role in the maintenance of membrane rheology by modulating hydration and fluidity of lipid bilayers. This could be a modulating function equally distributed in animal and plant kingdom. More evidence, though, is still required for a special mode of action, and whether the respective function is essential or just "beneficial".

The fact that PL-bound enzymatic activities respond remarkably fast to changes in boron supply (within minutes to one hour) point to at least in part post-transcriptional and post-translational control by the boron level. Another evidence for a post-translational control of plasmalemma-bound proteins comes from the observation that the boron transporter AtBor1-1 is regulated by boron levels (see above, (Takano et al. 2005a).

Membrane function could also be affected by the accumulation of oxidative free radicals (including *OH) in cells, which is one of the consequences of boron deficiency in root and leaf cells (Cakmak and Römheld 1997). Even though this might be rather one of the later secondary responses, water channels in the plasma membrane are reversibly closed by hydroxyl radicals (*OH) (Henzler et al. 2004), which is in line with our findings that GFP-transformed ZMPiP water channel activity almost disappeared in tobacco root tips within one hour of boron deficiency (Yu et al. 2002).

Boron deficiency has been shown to affect leaf photosynthesis, although the existing evidence has been mostly obtained from *in vivo* experiments with far too long (10 days or longer) treatments of plants with deficient boron supply (Kastori et al. 1995; Plesnicar et al. 1997; El-Shintinawy 1999). The primary mechanisms of boron's roles in photosynthesis are unknown, but boron could affect the functions of chloroplastic membranes by disrupting thylakoid electron transport, resulting in photoinhibition. In own preliminary experiments (unpublished), we obtained only weak effects, if at all, of boron deficiency with isolated spinach chloroplasts. It is quite feasible that the effects observed in chloroplasts are secondary and caused by growth inhibition in root and shoot tips, i.e. a reduced sink activity, finally leading to an over-saturation of the electron acceptors of PSII or PSI. These possible effects may increase the rate of photo-oxidative damage in response to further stresses.

Boron and stress responses with special emphasis to chilling

It is challenging to follow more closely possible interactions between boron supply and further stresses such as chilling (Ye et al. 2000; 2003) and salinity (Wimmer et al. 2005). Below, we will focus especially on interactions between boron supply and chilling tolerance as this is discussed since long as a special interaction (Cooling and Jones 1970; Hanson and Breen 1985) and there seems to be an additive or even multiplicatory effect of both stresses (Ye et al. 2000; 2003). So far, however, underlying processes and reactions still remain largely obscure.

The primary interaction site between boron and low temperature may lie with plant cellular membranes, including plasmalemma and chloroplastic membranes. The interaction between boron and low temperature in warm season species has been recently reviewed, particularly in relation to root functions, shoot water use and boron uptake/utilisation in plants (Huang et al. 2005).

Biochemical and physical properties seem to affect both low temperature tolerance and boron permeation at the cellular level. In chilling sensitive species, cellular membrane alterations precede other cellular changes and adverse effects on different cellular organelles are dependent on the duration of chilling and associated growth conditions (e.g. light intensity and relative humidity: Lyons et al. 1979). Dordas and Brown (2000) demonstrated that different proportions of sterols and longer chain fatty acids in the plasma membrane of root cells significantly changed boron uptake in *Arabidopsis thaliana* mutants, and related these changes to different permeability coefficients for boric acid across plasma membranes containing different groups of lipids and fatty acids. The decrease in sterol contents in the plasma membrane may increase membrane fluidity and permeability to water and ions (Lyons et al. 1979), which is correlated with plant chilling tolerance (Hugly et al. 1990). Increased membrane rigidity is a common response to chilling temperature in chilling sensitive species such as *Coffea arabica* L. (Queiroz et al. 1998). As a result, chilling-induced reduction in membrane fluidity and permeability of root cells may have also contributed to the inhibition of boron uptake in chilling sensitive species (Ye et al. 2000; 2003).

These interactions at the membrane level may be translated into root functions at organ levels, particularly in relation to boron uptake and external boron requirements of plants. Chilling stresses in roots may result in an increased external boron requirement, leading to a higher risk of boron deficiency during the period of seasonal transition from cold to warm climate when air warms up much faster than the soil, such as late winter and early spring. On the other hand, the pre-existence of boron deficiency in young roots may enhance the sensitivity of roots to chilling. Recent work has shown that low root temperature has a consistent set of effects on temperate and tropical species, but at different threshold temperatures (oilseed rape: 5-10°C, sunflower: 12-17°C; e.g. Ye et al. 2000; 2003).

Boron internal requirements in leaf cells may also be altered at low temperatures in warm climate species, which is largely related to the roles of boron in anti-oxidative systems and possibly in the photoinhibition responses of the photosystems after chilling stress. Boron deficiency can decrease the levels of antioxidants in leaves (Cakmak and Römheld 1997). In leaves of a subtropical *Eucalyptus grandis* × *E. urophylla* hybrid, the production rate of superoxide (O_2^-) and polyphenol oxidase activity increased significantly under boron deficiency treatments (0 and 5 μM boron for up to 96 hours at 5°C), but not in plants with adequate (15μM) boron supply at the same temperature (Lu and Huang 2003). In addition, the activities of anti-oxidative enzymes (superoxide dismutase, peroxidase, catalase and ascorbate peroxidase) in leaves were decreased at 5°C with boron deficiency (< 10 μM B), but not at 15 μM B. As a result, boron deficiency may increase the sensitivity of leaf cells to chilling, through enhanced generation of oxidative free radicals and weakened anti-oxidative capacity.

At the whole plant level, root hydraulic conductance is important in both boron uptake and post-chilling recovery of shoots. In species sensitive to root chilling such as cucumber and sunflower, chill-induced water loss is one of the most significant physiological consequences, resulting from decreased root hydraulic conductance and excessive transpiration due to loss of stomatal control (delayed closure or closure failure) (Allen and Ort 2001). At adequate boron levels, root boron uptake is mostly a passive process of permeation of undissociated boric acid across the plasma membrane, which is largely determined by the rate of water uptake through the plasma membrane of root cells, in addition to boron concentration around the root surface (Hu and Brown 1997). The decrease in root hydraulic conductance caused by root chilling would thus have a negative impact on boron supply to new shoot growth due to limited boron uptake and transport from root to shoot. On the other side, boron deficiency rapidly decreased the amount of ZmPIP1 aquaporins in tobacco (Goldbach et al. 2001) and enhanced the accumulation of oxidative free radicals (e.g. *OH) (Cakmak and Römheld 1997), which are also known to (reversibly) close certain aquaporins (Henzler et al. 2004). Boron deficiency-induced reduced water flow through aquaporins may then lower the ability of roots to maintain hydraulic conductivity in response to chilling stress, even in chilling-tolerant species or genotypes. Experimental evidence is required to test this hypothesis, especially at low boron concentrations, which may be maintained by using the boron-buffered solution culture with realistically low solution boron concentrations (Huang et al. 1999).

Boron and quorum sensing

One of the most exciting findings in the past years was the identification of the bacterial quorum sensor autoinducer 2 (AI2) as a boron containing stabile complex (Chen et al. 2002). This molecule is synthesized by bacteria from S-adenosylmethionine (which by itself shows a rather high complex stability with borate at higher pH values; see Hunt 2002) and activates luminescence when it has accumulated to high enough levels. Bacteria thus sense their cell density *via* the autoinducer concentration (Miller and Bassler 2001; Schauder and Bassler 2001). AI2 is recognized by a large number of different bacteria. In a concerted way, two autoinducers (AI1, a non boron bearing member of the family of AHL signals, and the boron-containing AI2) are required to control a signal chain (Cao and Meighen 1989; Bassler et al. 1993; 1994). Detection of AI-2 requires a periplasmic protein that resembles the ribose binding protein (LuxP) and a second two-component kinase (LuxQ) (Bassler et al. 1994). Both signals are needed for bioluminescence, because in the absence of one signal the cognate receptor acts as a potent kinase to block *lux* gene expression (Mok et al. 2003).

Outlook: how could boron's functional roles be described and assessed?

Until quite recently, the role of boron has only been attributed to an apoplastic function (Kobayashi et al. 1996; O'Neill et al. 1996). Although cross-linking rhamno-galacturonan II (RGII) and hence stabilizing the cell wall is one main and meanwhile well documented structural function (see above), there is increasing evidence for an essentiality in organisms without cell walls such as yeasts (Bennett et al. 1999) and animals (Eckhert 1998; Fort et al. 1998; Rowe and Eckhert 1999; Lanoue et al. 2000). In the latter, boron seems to play a major role in membrane-bound processes or where large amounts of membrane material are required (e.g. dysplasia in zebra fish during embryogenesis: Eckhert and Rowe 1999). In humans, boron shortage resulted in decreased activities of several membrane-bound hormones (Nielsen 2000). In plants, growth (O'Neill et al. 2001) as well as cell differentiation (e.g. xylem differentiation: Lovatt 1985, embryogenesis in *Larix*: Behrendt and Zoglauer 1996) are affected.

Functions of boron at the plasma membrane have been postulated on the basis of quite a number of observations, even though the underlying mechanisms are still a matter of speculation (see reviews by Goldbach 1997; Blevins and Lukaszewski 1998; Brown et al. 2002). The presence of specific acceptor molecules, which bear ligands able to form complexes with boric acid/borate, is indispensable for any boron function to occur. Sugar moieties (esp. in their furanose form) such as mannose, apiose or galactose, but as well other hydroxylated ligands such as serine or threonine, may form ester-like complexes with boron (Ralston and Hunt 2000). In membranes, glycoproteins and glycolipids are good candidates for a possible boron function. A number of membrane-bound proteins and membrane structures are specifically interesting, as they seem to be related to still not well understood processes in cell growth, differentiation and perception (Kohorn 2000), which are also reported to be boron dependent. Most prominent among these putative B-binding membrane structures are surface proteins attached to the membrane *via* a glycosyl-phosphatidyl-inositol anchor (GPI) (Ferguson and Williams 1988; Thompson and Okuyama 2000). They typically contain three mannose-residues as well as phosphatidylinositol, which are all possible ligands for boron with a strong binding capacity (van Duin et al. 1984; Ralston and Hunt 2000). A modification of the GPI by complexing with boron may alter the access of phospholipases, which loosen this anchor. Thus the ratio of free and bound membrane proteins may be changed under boron deficiency with consequences such as lower contents of hydroxyproline-rich proteins in cell walls of *Phaseolus vulgaris* (Bonilla et al. 1997) or reduced incorporation of proteins into *Petunia* pollen tube walls (Jackson 1989). In both cases, the control was exerted at the post-transcriptional or post-translational level, which is in line with the assumption that boron exerts posttranslational control (Goldbach et al. 2001). A subgroup of GPI proteins is represented by classical arabinogalactanproteins (AGP) (Sherrier et al. 1999; Majewska-Sawka and Nothnagel 2000). There is a striking coincidence between many AGP-dependent processes and their dependency on boron supply (e.g. xylem differentiation: Stacey et al. 1995, pollen tube growth: Jackson 1989; Cheng and Rerkasem 1993; Majewska-Sawka and Nothnagel 2000). GPI-anchored proteins are components of "membrane rafts", micro-domains of membranes with specific functions (Brown and Rose 1992), rich in sphingolipid and cholesterol and insoluble in non-ionic detergents (Brown and London 2000). A specific function of boron in the formation and stability of membrane rafts *via* the formation of two-sided borate-complexes with e.g. mannose residues, seems to be plausible (Brown et al. 2002). Finally, changes in boron concentrations may lead to a mechanical cascade of signals starting by an altered conformation of membrane-bound proteins (Watson 1991; Morris and Homann 2001) and extending into the cytoplasm (Ligterink and Hirt 2001) *via* the cell wall-PL-cytoskeleton-continuum. An altered membrane tension may as well directly influence exo- and endocytosis (Fricker et al. 2000). The accumulation of vesicles at the cytosolic side of the plasma membrane is in line with this assumption (Kouchi and Kumazawa 1976; Hirsch and Torrey 1980), pointing to a possibly inhibited exocytosis. We have shown that actin and tubulin levels increased within 20-40 min of withholding boron in *Arabidopsis* and maize (Yu et al. 2001), but not in zucchini roots. The polimerization rate of these proteins was altered as well (Yu et al. 2003), likely at the translational or post-translational level. Boron deficiency led to an increased level of JIM5-reactive pectins with low (<40%) methyl esterification and dimeric B-RGII-complexes in the cell walls, whereas their internalization in BrefeldinA induced compartments was reduced or even completely inhibited (Yu et al. 2002).

One or more functions of boron at the plasma membrane are thus highly likely. The challenge is now to identify the relevant components, boron-binding ligands as well as their function. Promising tools for this attempt may include the use of mutants either being deprived off or over-expressing possible target molecules. Also the availability of GFP-fusion proteins or fluorescing markers should greatly improve our understanding of boron's functions. It will be especially challenging to determine the stability of potential boron complexes *in vivo* where the complex stability could be highly influenced by molecules directly surrounding the boron-binding moiety. A very useful tool became available recently with phenylboronic acids such as 3-naphtyl-boronic acid (Bassil et al. 2004), as they bind strongly to *cis*-diols but prevent the formation of cross-links. Boronic acids could be used to test whether the essential function of boron is restricted to the formation of cross-links, i.e. 1:2 boron complexes, or whether it may also occur *via* formation of 1:1 complexes. The latter would suffice for e.g. preventing the oxidation of phenolic acids and alcohols (although this is less likely to occur under physiological conditions).

In any case, it has to be seen, that boron possibly acts only *via* post-translational control. If a boron–containing compound such as AI2 is involved in a signal chain in bacteria, boron might be needed to form or modify signalling complexes in a way similar to AI2 in higher organisms, too. The stability of such complexes can be predicted to vary according to the physicochemical environment (see above). It could be possible that complex formation with boron reduces or modifies the polarity of the complex surface, altering its binding behaviour to apolar sites (from enzymes such as MAPK). To exert a significant effect, concentrations needed could be rather low. It is envisaged that significant progresses will be made in boron research in the next years, especially when considering that evidence accumulates for a more general importance of boron in a wide array of organisms, ranging from bacteria to higher plants, animals and humans.

References

Allen D.J. and Ort D.R., 2001 Impacts of chilling temperatures on photosynthesis in warm-climate plants Trends in Plant Science **6**(1): 36–42.

Armstrong T.A., Spears J.W., Crenshaw T.D., and Nielsen F.H., 2000 Boron supplementation of a semipurified diet for weanling pigs improves feed efficiency and bone strength characteristics and alters plasma lipid metabolites Journal of Nutrition **130**: 2575–2581.

Bai Y.S. and Hunt C.D., 1996 Dietary boron enhances efficacy of cholecalciferol in broiler chicks Journal of Trace Elements in Experimental Medicine **9**(3): 117–132.

Bakken N.A. and Hunt C.D., 2003 Dietary boron decreases peak pancreatic in situ insulin release in chicks and plasma insulin concentrations in rats regardless of vitamin D or magnesium status Journal of Nutrition **133**(11): 3577–3583.

Baldock C., Rafferty J.B., Sedelnikova S.E., Baker P.J., Stuitje A.R., Slabas A.R., Hawkes T.R., and Rice D.W., 1996 A mechanism of drug action revealed by structural studies of enoyl reductase. Science **274**(5295): 2107–2110.

Barr R., Böttger M., and Crane F.L., 1993 The effect of boron on plasma membrane electron transport and associated proton secretion by cultured carrot cells. Biochemistry and Molecular Biology International **31**(1): 31–39.

Barr R., Crane F.L., and Craig T.A., 1983 Transmembrane ferricyanide reduction in tobacco callus cells J Plant Growth Regul. 243–249.

Bassil E., Hu H.N., and Brown P.H., 2004 Use of phenylboronic acids to investigate boron function in plants. Possible role of boron in transvacuolar cytoplasmic strands and cell-to-wall adhesion. Plant Physiology **136**(2): 3383–3395.

Bassler B.L., Wright M., Showalter R.E., and Silverman M.R., 1993 Intercellular Signaling in Vibrio-Harveyi Sequence and Function of Genes Regulating Expression of Luminescence. Molecular Microbiology **9**(4): 773–786.

Bassler B.L., Wright M., and Silverman M.R., 1994 Sequence and Function of Luxo, a Negative Regulator of Luminescence in Vibrio-Harveyi. Molecular Microbiology **12**(3): 403–412.

Behrendt U. and Zoglauer K., 1996 Boron controls suspensor development in embryogenic cultures of Larix decidua Physiologia Plantarum **97**: 321–326.

Bell R.E., Rerkasem B., Keerati-Kasikorn P., Phetchawee S., Hiranburana N., Ratanarat S., Pongsakul P., and Loneragan J.F., 1990 Mineral Nutrition of Food Legumes in Thailand with Particular Reference to Micronutrients. ACIAR Technical Report 19.

Benderdour M., Hess K., Gadet M.D., Dousset B., Nabet P., and Belleville F., 1997 Effect of boric acid solution on cartilage metabolism. Biochemical and Biophysical Research Communications **234**(1): 263–268.

Bennett A., Rowe R.I., Soch N., and Eckhert C.D., 1999 Boron stimulates yeast (Saccharomyces cerevisiae) growth. Journal of Nutrition **129**: 2236–2238.

Berry S.C., Fink A.L., Shenvi A.B., and Kettner C.A., 1988 Interaction of Peptide Boronic Acids with Elastase - Circular-Dichroism Studies. Proteins-Structure Function and Genetics **4**(3): 205–210.

Blaser-Grill J., Knoppik D., Amberger A., and Goldbach H.E., 1989 Influence of boron on the membrane potential in Elodea densa and Helianthus annuus roots and H^+ extrusion of suspension cultured Daucus carota cells. Plant Physiology **90**: 280–284.

Blevins D.G. and Lukaszewski K.M., 1998 Boron in plant structure and function. Annual Reviews in Plant Physiology and Plant Molecular Biology **49**: 481–500.

Bolanos L., Cebrian A., Redondo-Nieto M., Rivilla R., and Bonilla I., 2001 Lectin-like glycoprotein PsNLEC-1 is not correctly glycosylated and targeted in boron deficient pea nodules. Molecular Plant-Microbe Interactions **14**: 663–670.

Bolanos L., Esteban E., Delorenzo C., Fernandezpascual M., Defelipe M.R., Garate A., and Bonilla I., 1994 Essentiality of boron for symbiotic dinitrogen fixation in pea (Pisum sativum) rhizobium nodules. Plant Physiology **104**(1): 85–90.

Bolanos L., Mateo P., and Bonilla I., 1993 Calcium-mediated recovery of boron deficient Anabaena Sp PPCC 7119 grown under nitrogen fixing conditions. Journal of Plant Physiology **142**(5): 513–517.

Bolanos L., Redondo-Nieto M., El-Hamdaoui A., and Bonilla I., 2002 Interaction of boron and calcium in the Rhizobium-legume N_2-fixing symbiosis. Boron Nutrition in Plants and Animals. H.E. Goldbach, et al. New York USA, Kluwer Academic/Plenum Publishers: 255–260.

Bolanos L., Redondo-Nieto M., Rivilla R., Brewin N.J., and Bonilla I., 2004 Cell surface interactions of Rhizobium bacteroids and other bacterial strains with symbiosomal and peribacteroid membrane components from pea nodules. Molecular Plant-Microbe Interactions **17**(2): 216–223.

Bolaños L., Lukaszewski K., Bonilla I., and Blevins D., 2004 Why boron? Plant Physiol. Biochem. **42**(11): 907–912.

Bonilla I., Mergold-Villasenor C., Campos M.E., Sánchez N., Pérez H., López L., Castrejón L., Sánchez F., and Cassab G.I., 1997 The aberrant cell walls of boron-deficient bean root nodules have no covalently bound hydroxyproline-/ proline-rich proteins. Plant Physiology **115**: 1329–1340.

Brown D.A. and London E. 2000 Structure and function of sphingolipid- and cholesterol-rich membrane rafts. The Journal of Biological Chemistry **275**(23): 17221–17224.

Brown D.A. and Rose J.K., 1992 Sorting of GPI-anchored proteins to glycolipid-enriched membrane subdomains during transport to the apical cell surface. Cell **68**: 533–544.

Brown P.H., Bellaloui N., Wimmer M.A., Bassil E.S., Ruiz J., Hu H., Pfeffer H., Dannel F., and Römheld V., 2002 Boron in plant biology. Plant Biology **4**: 205–223.

Brown P.H. and Shelp B.J., 1997 Boron mobility in plants. Plant and Soil **193**: 85–101.

Cakmak I., Kurz H., and Marschner H., 1995 Short-term effects of boron, germanium and high light intensity on membrane permeability in boron deficient leaves of sunflower. Physiologia Plantarum **95**: 11–18.

Cakmak I. and Römheld V., 1997 Boron deficiency-induced impairments of cellular functions in plants. Plant and Soil **193**: 71–83.

Cao J.G. and Meighen E.A., 1989 Purification and Structural Identification of an Autoinducer for the Luminescence System of Vibrio-Harveyi. Journal of Biological Chemistry **264**(36): 21670–21676.

Chen T.S.S., Chang C., and Floss H.G., 1980 Biosynthesis of the Boron-Containing Antibiotic Aplasmomycin Nuclear Magnetic-Resonance Analysis of Aplasmomycin and Desboroaplasmomycin. Journal of Antibiotics **33**(11): 1316–1322.

Chen X., Schauder S., Potier N., Van Dorsselaer A., Pelczer I., Bassler B.L., and Hughson F.M., 2002 Structural identification of a bacterial quorum-sensing signal containing boron. Nature 415.

Cheng C. and Rerkasem B., 1993 Effects of Boron on Pollen Viability in Wheat. Plant and Soil **156**: 313–315.

Cooling E.N. and Jones B.E., 1970 The importance of boron and NPK fertilisers to eucalypts in the southern Province, Zambia East African Agricultural and Forestry Journal **36**: 185–194.

Dannel F., Pfeffer H., and Römheld V., 2000 Characterization of root boron pools, boron uptake and boron translocation in sunflower using stable isotopes ^{10}B and ^{11}B. Australian Journal of Plant Physiology **27**: 397–405.

Dannel F., Pfeffer H., and Römheld V., 2002 Update on boron in higher plants - Uptake, primary translocation and compartmentation. Plant Biology **4**: 193–204.

Dell B. and Malajczuk N., 1994 Boron Deficiency in Eucalypt Plantations in China. Canadian Journal of Forest Research-Revue Canadienne De Recherche Forestiere **24**(12): 2409–2416.

Dordas C. and Brown P.H., 2000 Permeability of boric acid across lipid bilayers and factors affecting it. Journal of Membrane Biology **175**: 95–105.

Dordas C., Chrispeels M.J., and Brown P.H., 2000 Permeability and channel-mediated transport of boric acid across membrane vesicles isolated from squash roots. Plant Physiology **124**(3): 1349–1361.

Du C.W., Wang Y.H., Xu F.S., Yang Y.H., and Wang H.Y., 2002 Study on the physiological mechanism of boron utilization efficiency in rape cultivars. Journal of Plant Nutrition **25**(2): 231–244.

Eckhert C.D., 1998 Boron stimulates embryonic trout growth. J Nutr. **128**: 2488–2493.

Eckhert C.D. and Rowe R.I., 1999 Embryonic dysplasia and adult retinal dystrophy in boron-deficient zebrafish. J Trace Elem Exp Med **12**: 213–219.

El-Shintinawy F., 1999 Structural and functional damage caused by boron deficiency in sunflower leaves. Photosynthetica **36**: 565–573.

Ferguson M.A.J. and Williams A.F., 1988 Cell-Surface Anchoring of Proteins Via Glycosyl-Phosphatidylinositol Structures Annual Review of Biochemistry **57**: 285–320.

Ferrol N. and Donaire J.P., 1992 Effect of boron on plasma membrane proton extrusion and redox activity in sunflower cells. Plant Science **86**: 41–47.

Findeklee P. and Goldbach H.E., 1996 Rapid effects of boron deficiency on cell wall elasticity modulus in Cucurbita pepo roots. Bot Acta **109**: 463–465.

Fleischer A., O'Neill M.A., and Ehwald R., 1999 The pore size of non-graminaceaous plant cell walls is rapidly decreased by borate ester cross-linking of the pectic polysaccharide rhamnogalacturonan II. Plant Physiology **121**: 829-838.

Fort D.J., Propst T.L., Stover E.L., and Strong P.L., 1998 Adverse reproductive and developmental effects in Xenopus from insufficient boron. Biological Trace Element Research **66**: 237–259.

Fort D.J., Stover E.L., Strong P.L., Murray F.J., and Keen C.L., 1999 Chronic feeding of a low boron diet adversely affects reproduction and development in Xenopus laevis. Journal of Nutrition **129**(11): 2055–2060.

Fricker W., Jarvis M.C., and Brett C.T., 2000 Turgor pressure, membrane tension and the control of exocytosis in higher plants. Plant, Cell and Environment **23**: 999–1003.

Frommer W.B. and von Wiren N., 2002 Plant biology Ping-pong with boron. Nature **420**(6913): 282–283.

Garbett K., Darnall D.W., and Klotz I.M., 1971 The effects of bound anions on the reactivity of residues in hemerythrin. Arch. Biochem. Biophys. **142**: 455–470.

Goldbach H.E., 1985 Influence of boron nutrition on net uptake and efflux of ^{32}P and ^{14}C-glucose in Helianthus annuus roots and cell cultures of Daucus carota. Journal of Plant Physiology **118**: 431–438.

Goldbach H.E., 1997 A critical review on current hypotheses concerning the role of boron in higher plants: suggestions for further research and methodological requirements. J Trace and Microprobe Techniques **15**(1): 51–91.

Goldbach H.E., Blaser-Grill J., Lindemann N., Porzelt M., Hörrmann C., Lupp B., and Gessner B., 1991 Influence of boron on net proton release and its relation to other metabolic processes. Current Topics in Plant Biochemistry and Physiology. D.D. Randall et al. University of Missouri-Columbia. **10**: 195–220.

Goldbach H.E., Yu Q., Wingender R., Schulz M., Wimmer M., Findeklee P., and Baluska F., 2001 Rapid response reactions of roots to boron deprivation. Journal of Plant Nutrition and Soil Science **164**: 173–181.

Hanson E. and Breen P. 1985 Effects of fall boron sprays and environmental factors on fruit set and boron accumulation in Italian prune flowers. J. Amer. Soc. Hort. Sci. **110**: 389–392.

Hausdorf G., Krüger K., Küttner G., Holzhütter H-G., Frömmel C., and Höhne W.E., 1987 Oxidation of a methionine residue in subtilisin-type proteinases by the hydrogen peroxide/borate system an active site-directed reaction. Biochimica et Biophysica Acta **952**: 20–26.

Hegsted M., Keenan M.J., Siver F., and Wozniak P., 1991 Effect of boron on vitamin D deficient rats. Biological Trace Element Research **28**: 243–255.

Henzler T., Ye Q., and Steudle E., 2004 Oxidative gating of water channels (aquaporins) in Chara by hydroxyl radicals. Plant Cell and Environment **27**(9): 1184–1195.

Heyes J.A., White P.J., and Loughman B.C., 1991 The role of boron in some membrane characteristics of plant cells and protoplasts. Current Topics in Plant Biochemistry and Physiology. D D Randall et al. (Eds.), University of Missouri-Columbia. **10**: 179–194.

Hirsch A.M. and Torrey J.G., 1980 Ultrastructural changes in sunflower root cells in relation to boron deficiency and added auxin. Canadian Journal of Botany **58**: 856–866.

Hopmans P. and Flinn D.W., 1984 Boron Deficiency in Pinus-Radiata D Don and the Effect of Applied Boron on Height Growth and Nutrient-Uptake. Plant and Soil **79**(2): 295–298.

Hu H. and Brown P.H., 1997 Absorption of boron by plant roots. Plant and Soil **193**: 49–58.

Huang L., Bell R.W., and Dell B., 1999 Factors controlling equilibrium boron (B) concentration in nutrient solution buffered with B-specific resin (Amberlite IRA-743). Plant and Soil **208**(2): 233–241.

Huang L., Bell R.W., and Dell B., 2001 Boron supply into wheat (Triticum aestivum L cv Wilgozne) ears whilst still enclosed within leaf sheaths Journal of Experimental Botany **52**: 1731–1738.

Huang, L.B., Ye Z.Q., Bell R.W., and Dell B., 2005 Boron nutrition and chilling tolerance of warm climate crop species. Annals of Botany **96**(5): 755–767.

Hugly S., McCourt P., Browse J., Patterson G.W., and Somerville C., 1990 A Chilling Sensitive Mutant of Arabidopsis with Altered Steryl-Ester Metabolism. Plant Physiology **93**(3): 1053–1062.

Hunt C.D., 2002 Boron-binding biomolecules: A key to understanding the beneficial physiological effects of dietary boron from prokaryotes to humans. Boron in plant and animal nutrition. H.E. Goldbach, et al. (Eds). New York, Kluwer Academic/Plenum Publ. 21–36.

Hunt C.D., 2003 Dietary boron: An overview of the evidence for its roles in immune function. Trace Elem Exp Med. **16**(4): 291–306.

Hunt C.D. and Nielsen F.H., 1981 Interaction between boron and cholecalciferol in the chick. Trace Element Metabolism in Man and Animals-4 J Gawthorne et al. (Eds). Canberra, Australian Academy of Science: 597–600.

Iwai H., Masaoka N., Ishii T., and Satoh S., 2002 A pectin glucuronyltransferase gene is essential for intercellular attachment in the plant meristem. Proc. Natl. Acad. Sci. **99**: 16319–24

Jackson J.F., 1989 Borate control of protein secretion from Petunia pollen exhibits critical temperature discontinuities. Sex. Plant Reprod. **2**: 11–14.

Jamjod S., Niruntrayagul S., and Rerkasem B., 2004 Genetic control of boron efficiency in wheat (*Triticum aestivum L.*). Euphytica **135**(1): 21–27.

Kastori R., Plesnicar M., Pankovic D., and Sakac Z., 1995 Photosynthesis, chlorophyll fluorescence and soluble carbohydrates in sunflower leaves as affected by boron deficiency. Journal of Plant Nutrition **18**(9): 1751–1763.

Kettner C.A., Bone R., Agard D.A., and Bachovchin W.W., 1988 Kinetic-Properties of the Binding of Alpha-Lytic Protease to Peptide Boronic Acids. Biochemistry **27**(20): 7682–7688.

Kettner C.A. and Shenvi A.B., 1984 Inhibition of the Serine Proteases Leukocyte Elastase, Pancreatic Elastase, Cathepsin-G, and Chymotrypsin by Peptide Boronic Acids. Journal of Biological Chemistry **259**(24): 5106–5114.

Kobayashi M., Matoh T., and Azuma J.-I., 1996 Two chains of rhamnogalacturonan II are cross-linked by borate-diol ester bonds in higher plant cell walls. Plant Physiology **110**: 1017–1020.

Kohno J., Kawahata T., Otake T., Morimoto M., Mori H., Ueba N., Nishio M., Kinumaki A., Komatsubara S., and Kawashima K., 1996 Boromycin an anti-HIV antibiotic. Bioscience Biotechnology and Biochemistry **60**(6): 1036–1037.

Kohorn B.D., (2000): Plasma membrane-cell wall contacts. Plant Physiology **124**: 31–38.

Kolodny N.H. and Collins L.J., 1986 Proton and P-31 Nmr-Study of the Dependence of Diadenosine Tetraphosphate Conformation on Metal-Ions. Journal of Biological Chemistry **261**(31): 4571–4575.

Kouchi H. and Kumazawa K., 1976 Anatomical responses of root tips to boron deficiency. III Effect of boron dificiency on sub-cellular structure of root tips, particularly on morphology of cell wall and its related organelles. Soil Science and Plant Nutrition **22**: 53–71.

Lanoue L., Trollinger D.R., Strong P.L., and Keen C.L., 2000 Functional impairments in preimplantation mouse embryos following boron deficiency. FASEB J 14 A: 539.

Lawrence K., Bhalla P., and Misra P.C., 1995 Changes in NAD(P)H-dependent redox activities in plasmalemma-enriched vesicles isolated from boron- and zinc-deficient chick pea roots. Journal of Plant Physiology **146**: 652–657.

Lee S.H., Chung G.C., and Steudle E., 2005 Low temperature and mechanical stresses differently gate aquaporins of root cortical cells of chilling-sensitive cucumber and -resistant figleaf gourd. Plant Cell and Environment **28**(9): 1191-1202.

Ligterink W. and Hirt H., 2001 Mitogen-activated protein (MAP) kinase pathways in plants: versatile signalling tools International Review of Cytology **201**: 209-275.

Lord E.M. and Mollet J.C., 2002 Plant cell adhesion: A bioassay facilitates discovery of the first pectin biosynthetic gene. PNAS **99**(25): 15843–15845.

Lovatt C.J., 1985 Evolution of xylem resulted in a requirement for boron in the apical meristems of vascular plants. The New Phytologist **99**: 509–522.

Lowik C., 1992 Differentiation inducing factors: leukemia inhibitory factor and interleukin-6. Cytokines and Bone Metabolism. M Gowen Boca Raton, CRC Press: 299–324.

Lu C. and Huang B., 2003 Effects of boron on membrane lipid peroxidation and endogenousprotective systems in leaves of Eucalyptus grandis X Eucalyptus urophylla under low temperature. Journal of Tropical and Subtropical Botany **11**: 217–222.

Lyons J.M., Graham D., and Raison J.K., 1979 Low temperature stress in crop plants: the role of the membrane. New York, Academic Press.

Majewska-Sawka A. and Nothnagel E.A., 2000 The multiple roles of arabinogalactan proteins in plant development. Plant Physiology **122**: 3–9.

Matoh T., 1997 Boron in plant cell walls. Plant and Soil **193**: 59–70.

Matsunaga T., Ishii T., Matsumoto S., Higuchi M., Darvill A., Albersheim P., and O'Neill M.A., 2004 Occurrence of the primary cell wall polysaccharide rhamnogalacturonan II in pteridophytes, lycophytes, and bryophytes. Implications for the evolution of vascular plants. Plant Physiology **134**(1): 339–351.

McLennan A.G., 1992 Dinucleoside phosphates an introduction. Ap_4A and other dinucleoside polyphosphates. A.G. McLennan. Boca Raton, CRC Press: 1–9.

Miller M.B. and Bassler B.L., 2001 Quorum sensing in bacteria. Annual Review of Microbiology **55**: 165–199.

Mok K.C., Wingreen N.S., and Bassler B.L., 2003 Vibrio harveyi quorum sensing: a coincidence detector for two autoinducers controls gene expression. Embo Journal **22**(4): 870–881.

Moore B.S. and Hertweck C., 2002 Biosynthesis and attachment of novel bacterial polyketide synthase starter units. Natural Product Reports **19**(1): 70–99.

Morris C.E. and Homann U., 2001 Cell surface regulation and membrane tension. Journal of Membrane Biology **179**: 79–102.

Mühling K.H., Wimmer M., and Goldbach H.E., 1998 Apoplastic and membrane-associated Ca^{2+} in leaves and roots as affected by boron deficiency. Physiol Plantarum **102**: 179–184.

Nachiangmai D., Dell B., Bell R., Huang L.B., and Rerkasem B., 2004 Enhanced boron transport into the ear of wheat as a mechanism for boron efficiency. Plant and Soil **264**(1-2): 141–147.

Newnham R.E., 2002 How boron is being used in medical practice. Boron in Plant and Animal Nutrition. H.E. Goldbach, et al. New York, Kluwer Academic Publishers: 59–62.

Nielsen F.H., 2000: The emergence of boron as nutritionally important throughout the life cycle. Nutrition **16**: 512–514.

Noguchi K., Ishii T., Matsunaga T., Kakegawa K., Hayashi H., and Fujiwara T., 2003: Biochemical properties of the cell wall in the Arabidopsis mutant bor1-1 in relation to boron nutrition. Journal of Plant Nutrition and Soil Science-Zeitschrift Fur Pflanzenernahrung und Bodenkunde **166**(2): 175–178.

Nyomora A.M.S., Brown P.H., and Freeman M., 1997 Fall foliar-applied boron increases tissue boron concentration and nut set of almond. J. Amer. Soc. Hort. Sci. **122**: 405–410.

Obermeyer G., Kriechbaumer R., Strasser D., Maschessnig A., and Bentrup F-W., 1996 Boric acid stimulates the plasma membrane H^+-ATPase of ungerminated lily pollen grains. Physiologia plantarum **98**: 281–290.

O'Neill M., Eberhard S., Albersheim P., and Darvill A., 2001 Requirement of borate cross-linking of cell wall rhamnogalacturonan II for Arabidopsis growth. Science **294**(5543): 846–849.

O'Neill M.A., Ishii T., Albersheim P., and Darvill A.G., 2004 Rhamnogalacturonan II: Structure and function of a borate cross-linked cell wall pectic polysaccharide. Annual Review of Plant Biology **55**: 109–139.

O'Neill M.A., Warrenfeltz D., Kates K., Pellerin P., Doco T., Darvill A.G., and Albersheim P., 1996 Rhamnogalacturonan-II, a pectic polysaccharide in the walls of growing plant cell, forms a dimer that is covalently cross-linked by aborate ester. The Journal of Biological Chemistry **271**(37): 22923–22930.

Park M., Ko S.B.H., Choi J.Y., Muallem G., Thomas P.J., Pushkin A., Lee M.S., Kim J.Y., Lee M.G., Muallem S., and Kurtz I., 2002 The cystic fibrosis transmembrane conductance regulator interacts with and regulates the activity of the HCO_3^- salvage transporter human Na^+-HCO_3^- cotransport isoform 3 Journal of Biological Chemistry **277**(52): 50503–50509.

Park M., Li Q., Shcheynikov N., Zeng W.Z., and Muallem S., 2004 NaBC1 is a ubiquitous electrogenic Na^+-coupled borate transporter essential for cellular boron homeostasis and cell growth and proliferation. Molecular Cell **16**(3): 331–341.

Parr A.J. and Loughman B.C., 1983 Boron and membrane function in plants. Metals and micronutrients: uptake and utilization by plants. D.A. Robb, et al. New York, Academic Press: 87–107.

Plesnicar M., Kastori R., Sakac Z., Pankovic D., and Petrovic N., 1997 Boron as limiting factor in photosynthesis and growth of sunflower plants in relation to phosphate supply. Agrochimica **41**(3-4): 144–154.

Pollard A.S., Parr A.J., and Loughman B.C., 1977 Boron in relation to membrane function in higher plants. Journal of Experimental Botany **28**: 831–841.

Queiroz C.G.S., Alonso A., Mares-Guia M., and Magalhaes A.C., 1998 Chilling-induced changes in membrane fluidity and antioxidant enzyme activities in Coffea arabica L roots Biologia Plantarum **41**(3): 403–413.

Ralston N.V.C. and Hunt C.D., 2000 Biological boron interactions: Charge and structure characteristics required for boroester formation with biomolecules. FASEB J. **14**: A538.

Ralston N.V.C. and Hunt C.D., 2001 Diadenosine phosphates and S-adenosylmethionine: novel boron binding biomolecules detected by capillary electrophoresis. Biochimica et Biophysica Acta **1527**: 20–30.

Ram S., Bist L.D., and Sirohi S.C., 1989 Internal fruit necrosis of mango and its control. Acta Hort. **231**: 805–813.

Reuhs B.L., Glenn J., Stephens S.B., Kim J.S., Christie D.B., Glushka J.G., Zablackiş E., Albersheim P., Darvill A.G., and O'Neill M.A., 2004 L-Galactose replaces L-fucose in the pectic polysaccharide rhamnogalacturonan II synthesized by the L-fucose-deficient mur1 Arabidopsis mutant. Planta **219**(1): 147–157.

Ricardo A., Carrigan M.A., Olcott A.N., and Benner S.A., 2004 Borate minerals stabilize ribose. Science **303**(5655): 196–196.

Rowe R.I. and Eckhert C.D., 1999 Boron is required for zebrafish embryogenesis. Journal of Experimental Biology **202**: 1649–1654.

Ryden P, Sugimoto-Shirasu K., Smith A.C., Findlay K., Reiter W.D., and McCann M.C., 2003 Tensile properties of Arabidopsis cell walls depend on both axyloglucan cross-linked microfibrillar network and rhamnogalacturonan II-borate complexes. Plant Physiology **132**(2): 1033–1040.

Sato K., Okazaki T., Maeda K., and Okami Y., 1978 New Antibiotics, Aplasmomycins-B and Aplasmomycins-C. Journal of Antibiotics **31**(6): 632–635.

Schauder S. and Bassler B.L., 2001 The languages of bacteria. Genes & Development **15**(12): 1468–1480.

Schon M.K., Novacky A., and Blevins D.G., 1990 Boron induces hyperpolarization of sunflower root cell membranes and increases membrane permeability to K^+. Plant Physiology **93**: 566–571.

Sherrier J.D, Prime T.A., and Dupree P., 1999 Glycosylphosphatidylinositol-anchored cell-surface proteins from Arabidopsis. Electrophoresis **20**: 2027–2035.

Stacey N.J., Roberts K., Carpita N.C., Wells B., and McCann M.C., 1995 Dynamic changes in cell surface molecules are very early events in the differentiation of mesophyll cells from Zinnia elegans into tracheary elements. Plant Journal **8**: 891–906.

Stephenson R.A. and Gallagher E.C., 1987 Effects of Foliar Boron Sprays on Yield and Quality of Macadamia Nuts. Scientia Horticulturae **32**(1-2): 97–103.

Steudle E., 2000 Water uptake by roots: effects of water deficit. Journal of Experimental Botany **51**(350): 1531–1542.

Takano J., Miwa K., Wiren von N., and Fujiwara T., 2005a Boron dependent endocytosis and degradation of boron transporter AtBOR1. Plant and Cell Physiology **46**: S151–S151.

Takano J., Miwa K., Yuan L.X., Wiren von N., and Fujiwara T., 2005b Endocytosis and degradation of BOR1, a boron transporter of Arabidopsis thaliana, regulated by boron availability. Proceedings of the National Academy of Sciences of the United States of America **102**(34): 12276–12281.

Takano J., Noguchi K., Yasumori M., Kobayashi M., Gajdos Z., Miwa K., Hayashi H., Yoneyama T., and Fujiwara T., 2002 Arabidopsis boron transporter for xylem loading. Nature **420**(6913): 337–340.

Thompson G.A. and Okuyama H., 2000 Lipid-linked proteins of plants. Progress in Lipid Research **39**: 19–39.

Tyerman S.D., Bohnert H.J., Maurel C., Steudle E., and Smith J.A.C., 1999 Plant aquaporins: their molecular biology, biophysics and significance for plant water relations. Journal of Experimental Botany **50**: 1055–1071.

Tyerman S.D., Niemietz C.M., and Bramley H., 2002 Plant aquaporins: multifunctional water and solute channels with expanding roles. Plant Cell and Environment **25**(2): 173-194.

van Duin M., Peters J.A., Kieboom A.P.G., and van Bekkum H., 1984 Studies on borate esters I Tetrahedron **40**(15): 2901–2911.

Verstraeten S.V., Lanoue L., Keen C.L., and Oteiza P.I., 2005 Relevance of lipid polar headgroups on boron-mediated changes in membrane physical properties. Archives of Biochemistry and Biophysics **438**(1): 103–110.

Wang Z.Y., Tang Y.L., Zhang F.S., and Wang H., 1999 Effect of boron and low temperature on membrane integrity of cucumber leaves. Journal of Plant Nutrition **22**(3): 543-550.

Watson P.A., 1991 Function follows form: generation of intracellular signals by cell deformation. FASEB J. **5**: 2013–2019.

Wei Y.Z., Bell R.W., Yang Y., Ye Z.Q., Wang K., and Huang L.B., 1998 Prognosis of boron deficiency in oilseed rape (Brassica napus) by plant analysis. Australian Journal of Agricultural Research **49**(5): 867–874.

Wimmer M., 2000 Untersuchungen zur Funktion von Bor im Apoplasten der Ackerbohne (Vicia faba L.). Bonn, Germany.

Wimmer M.A., Bassil E.S., Brown P.H., and Läuchli A., 2005 Boron response in wheat is genotype dependent and related to boron uptake, translocation, allocation, plant phenological development and growth rate. Functional Plant Biology **32**: 507–515.

Wimmer M.A., and Goldbach H.E., 1999 Influence of Ca^{2+} and pH on the stability of different boron fractions in intact roots of Vicia faba L Plant Biology **1**: 632–637.

Xu F.S, Wang Y.H., and Meng J., 2001 Mapping boron efficiency gene(s) in Brassica napus using RFLP and AFLP markers. Plant Breeding **120**(4): 319–324.

Ye Z.Q., Bell R.W., Dell B., and Huang L.B., 2000 Response of sunflower to boron supply at low root zone temperature. Communications in Soil Science and Plant Analysis **31**(11-14): 2379–2392.

Ye Z.Q., Huang L.B., Bell R.W., and Dell B., 2003 Low root zone temperature favours shoot B partitioning into young leaves of oilseed rape (Brassica napus). Physiologia Plantarum **118**(2): 213–220.

Yu Q., Baluska F., Jasper F., Menzel D., and Goldbach H.E., 2003 Short-term boron deprivation enhances levels of cytoskeletal proteins in maize, but not zucchini, root apices. Physiologia Plantarum **117**(2): 270–278.

Yu Q., Hlavacka A., Matoh T., Volkmann D., Menzel D., Goldbach H.E., and Baluska F., 2002 Short-term boron deprivation inhibits endocytosis of cell wall pectins in meristematic cells of maize and wheat root apices. Plant Physiology **130**(1): 415–421.

Yu Q., Wingender R., Schulz M., Baluska F., and Goldbach H.E., 2001 Short-term boron deprivation induces increased levels of cytoskeletal proteins in Arabidopsis roots. Plant Biology **3**(4): 335–340.

Part I

Boron in Plants

Physiology and Metabalism of Boron in Plants

Boron Modulation of Chilling and Freezing Tolerance in Leaf Cells of Warm Season Species

Longbin Huang[1], Richard W. Bell[1] and Bernard Dell[2]

[1]School of Land and Food Sciences, The University of Queensland, Brisbane Qld 4072, Australia

[2]Division of Science and Engineering, Murdoch University, WA 6150, Australia

Introduction

Historical accounts of possible B roles in the protection of tree and horticulture species against frost damage were reported as early as 1950s, with more field observations and experimental evidence since then. Although the initial reports (Anon 1958; Beltram 1958) were no more than anecdotal evidence due to the lack of proper comparative control experiments, later field and glasshouse studies have provided more reliable evidence about the involvement of B in the protection against frost damage – decreased frost-induced shoot-tip dieback or increased flowering and fruit yield, for example: in subtropical eucalypts *Eucalyptus grandis* (Cooling 1967; Cooling and Jones 1970) and *Eucalyptus grandis* x *Eucalyptus urophylla* (Lu and Huang 2003); apple, pear and blueberry (Blevins et al. 1996; Hanson and Breen 1985; Milovankic et al. 1990) and in birch, Scots pine, and Norway spruce (Braekke 1983). Frost-induced "white top" (bleached young leaves) has been observed in low temperature-sensitive *E. urophylla* and *E. grandis* in south China (Xu Daping, pers. comm.) where B deficient soils are common (Dell and Malajczuk 1994). Field observations have also suggested a link between low canopy temperature and enhanced leaf tissue damage (bleached patches) in oilseed rape grown in low B soil in south-east China (Ye et al. 1997).

In many cases, deficient B status already present in the trees grown in low B soils, resulted in exacerbated frost damage in young foliage and shoot tips. However, in prune trees with adequate B levels in midshoot leaves, foliar B spray (500 ppm Solubor) in the fall increased fruit set by 32% only in a cool spring when low temperature was present during flowering (Hanson and Breen 1985). The evidence has suggested a direct role of B deficiency in weakening frost tolerance in the shoot, particularly young growth (shoot tips) and flower buds, which often have weak B sink strength compared to youngest mature leaves and old leaves in the canopy profile. However whether "luxury" B supply has any protection against frost damage in plants already with adequate B supply remains to be established.

Boron deficiency is the most widespread of all the micronutrient deficiencies in many crop regions from tropical to temperate zones (Shorrocks 1997), where many crop, vegetable and tree species of tropical or subtropical origins are often grown during the warm seasons. Coincidently, these species are relatively high

F. Xu et al. (eds.), Advances in Plant and Animal Boron Nutrition, 31–46.

B-demanding, such as corn, sunflower, cucumber, tomato, and eucalypts, etc. (Bell 1997; Sakya et al. 2002). Chronic and marginal B deficiency has been reported in many crop species across large areas of field production (Shorrocks 1997). Field observations have suggested a link between chilling canopy temperature and enhanced leaf tissue damage (bleached patches) in oilseed rape grown in low B soil in Southeast, China (Ye 2005).

Low air temperature (ranging from chilling to freezing) is a common threat to the development of shoot (particularly, young leave and shoot tips), flowers and seeds/fruits in many plant species of tropical or subtropical origin during warm growing seasons (Lyons et al. 1979). On the basis of the threshold at which the initiation of ice crystallization occurs, low temperature is often categorised into non-freezing (or chilling, from -1.5 to 20°C) and freezing (below -1.5°C) (Graham and Patterson 1982). With looming climate change and increased unpredictability of sudden cold events in the atmosphere during the early growing season (such as the spring, late autumn), low-temperature induced damage to plant canopy may become increasingly common in these species, which would be further exacerbated by the presence of marginal or low B status due to inadequate B fertiliser input and/or low B availability induced by soil conditions (such as seasonal drought and high pH).

In nature, overwintering plant species (including annuals such as winter rape, and perrennials –such as evergreen woody species in temperate region have developed polygenic sensing mechanisms to adjust cellular metabolism and functions in response to low temperature and shorter photoperiod in the late autumn or early winter for inducing freezing tolerance, which is often termed as "cold acclimation" (Guy 1990; Xin and Browse 2000). However, chilling and frost (the mildest freezing) tend to impose a much greater risk to shoot tip/young leaves and buds of these warm season species during the growing seasons (e.g. early spring), as freezing tolerance in these plants is minimal or nonexistent (Guy 1990). Extended low temperature in the night followed by sunlight exposure leads to photo-oxidative damage in the leaves of chilling sensitive species, particularly non-acclimated plants (Kratsch and Wise 2000). In autumn or spring, mild frosts (about -5°C) often alternate with extended periods of warm night temperatures, which cause severe cell death and tissues damage to young leaves and shoot tips. Plants in the growing seasons are also prone to B deficiency due to inadequate soil B availability in dry topsoil and/or inadequate B uptake capacity of small root system. As a result, low night temperature or frost in the spring and fall poses a much greater threat to the growth and function of leaves.

On the basis of the above consideration of growing conditions for warm season species of tropical and subtropical origin, the present discussion will focus on how B nutrition modulates the tolerance of leaf cells to chilling and frost (the mildest form of freezing) stresses, through possible key physiological and biochemical mechanisms involving cellular membrane (plasma membrane and chloroplastic membrane) structure and functions, cytoplasmic anti-oxidation systems and photo-oxidation, cellular water relation and cryoprotectant (sucrose) (Figure 1). The interaction between continual root chill and B nutrition has been recently reviewed (Huang et al. 2005). Thorough reviews on plant responses to chilling and/or freezing temperature are available in the literature, in aspects of cellular structure, metabolism and genetic regulation in the leaves (Allen and Ort 2001; Hughes and Dunn 1996; Kratsch and Wise 2000; Xin and Browse 2000).

Cellular membranes

Boron deficiency can induce a great sensitivity to shoot chilling, in terms of gross structural changes (e.g. membrane leakage) and photosynthetic functions of leaf cells (Ye 2005), even in a chilling-tolerant species such as oilseed rape (Wise et al. 1983). In a chilling-sensitive cucumber cultivar, chilling temperature (7-8°C/5°C, day/night, under 300 μmol m^{-2} s^{-1} light) enhanced membrane leakage (K^+) induced by B-deficiency and chloroplast disruption and plasmolysis of mesophyll cells (Wang et al. 1999). Chilling-temperature around leaves induced a higher solute leakage in the youngest open leaf of low B status in oilseed rape (10-15 mg B kg^{-1} dry matter) than leaves containing adequate B (25 mg B kg^{-1} dry matter) (Ye 2005). Under high irradiance, bleached spots have been observed in leaves of sunflower plants subject to chilling or B deprivation treatments in our glasshouse experiments. Enhanced electrolyte leakage from leaf cells seems a common response to low temperature stress (Guy 1990; Kratsch and Wise 2000) and B deficiency (Cakmak and Romheld 1997; Goldbach et al. 2002). The evidence suggests cellular membranes (particularly plasma membrane and chloroplast membrane) in leaf cells are the primary interaction sites of B and low temperature.

Membrane functions

The functions and structural stability of plasma membrane and chloroplastic membrane show a high sensitivity to B deficiency (Cakmak and Romheld 1997; El-Shintinawy 1999; Goldbach et al. 2002) and low temperature (Allen and Ort 2001; Guy 1990; Kratsch and Wise 2000). The plasma membrane and chloroplast membrane are probably the primary sites of interactions between low temperature and B deficiency in leaf cells.

Rapid effects of B deprivation on plasmalemma enzyme functions in leaf and root cells are well known (refer to reviews by (Brown et al. 2002; Cakmak and Romheld 1997; Goldbach et al. 2002). These responses to B deprivation include depolarization of membrane potentials (Blaser-Grill et al. 1989; Goldbach et al. 1991), proton release (Goldbach et al. 1991; Goldbach et al. 1990; Roldán et al. 1998), H^+-ATPase (Ferrol and Donaire 1992; Obermeyer et al. 1996), and Fe-reductase (Barr et al. 1993; Ferrol and Donaire 1992; Wimmer and Goldbach 1999). Boron deficiency reduces the activity of proton-pumping ATPase and thus the

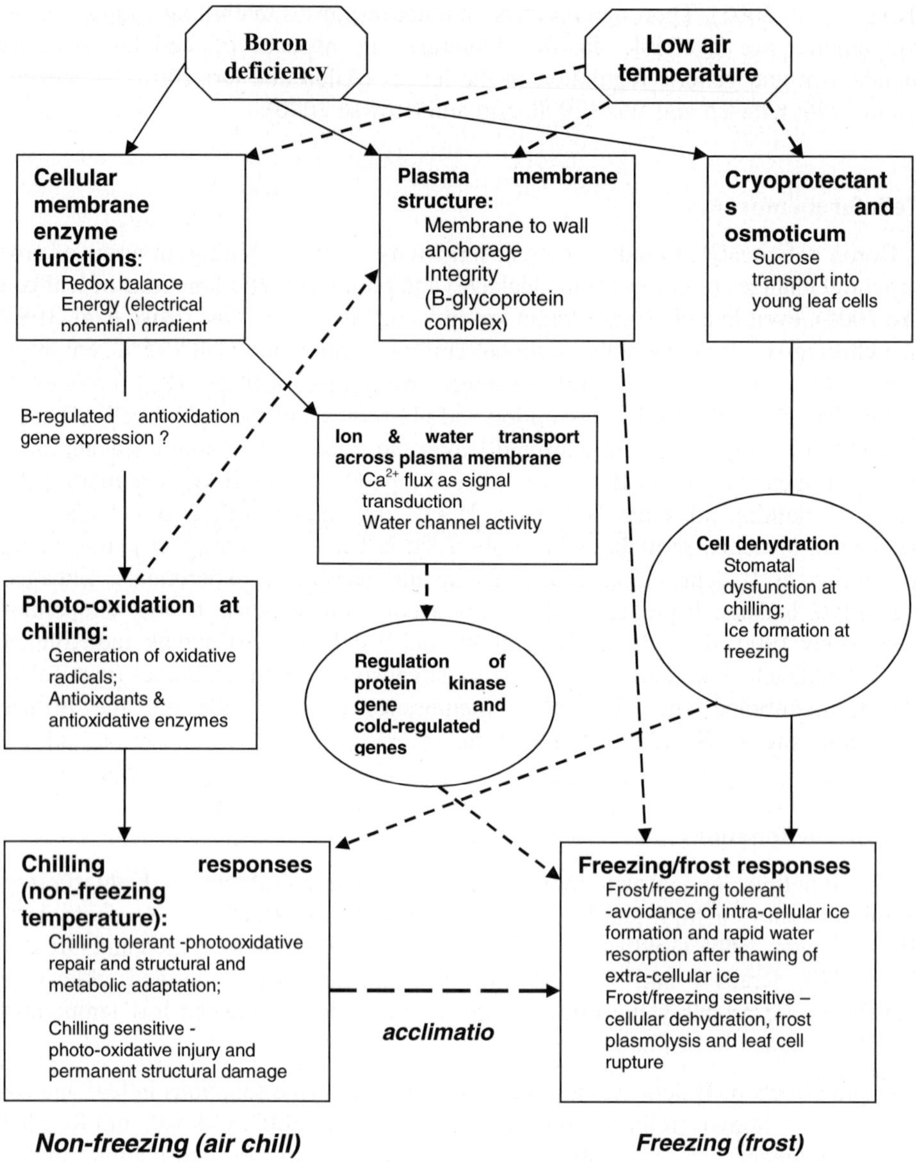

Figure 1. A conceptual diagram describing the possible mechanisms by which B modulates the responses of leaf cells to chilling and freezing temperatures in the canopy

proton gradient across the plasma membrane and inhibits oxidoreductase in the plasma membrane within minutes of B deprivation (Barr et al. 1993; Goldbach et al. 1991), which may disturb energy-dependent processes such as ion transport and energy transfer in redox reactions.

These early effects of B deficiency on membrane functons may disrupt the energy transfer in the thylakoid membrane (such as electron transfer through PS-II and PS-I) and substrate transport across chloroplast membrane (such as phosphate influx and glucose export (Marschner 1995), leading to the oxidative damage to the membrane, such as lipid peroxidation as observed in sunflower leaves (El-Shintinawy 1999). Both ATP and NADPH are products of photochemical electron transport in response to light reaction in chloroplasts (Melis 1999). Reduction in thylakoid electron transport is one of the earliest responses of leaf cells to low temperature, particularly under high irradiance, leading to the generation of oxidative free radicals, changed redox state of the stroma, inhibition of CO_2 assimilation and consequential photo-oxidation (Allen and Ort 2001). The reversibility of photoinhibition and the avoidance of permanent chloroplast damage will depend on the intrinsic anti-oxidation system to timely scavenge oxidative radicals (Allen and Ort 2001). As a result, B deficiency may exacerbate the sensitivity of leaf cells to low temperature through the roles of B in membrane functions, in addition to the pathway involving oxidative free radicals (Figure 1). Possible mechanism involved in anti-oxidation systems in response to low temperature and B deficiency will be discussed in later sections.

Ca transport and signal transduction

The up-regulation of many genes in plant cells at both transcriptional and post-transcriptional levels has been proposed as the genetic basis of cold-acclimation and cold tolerance (Guy 1990; Hughes and Dunn 1996; Thomashow 1999; Xin and Browse 2000). These genes encode water channel proteins, and enzymes involved in the biosynthesis of osmoprotectants (such as sugars) and in antioxidation systems (Shinozaki and Yamaguchi-Shinozaki 2000). One of the mechanisms triggering the regulation of chilling/freezing tolerance gene expression in response to low temperature is the Ca signal transduction pathway, in which cytoplasmic Ca^{2+} levels increase rapidly from the influx of apoplastic Ca^{2+} in response to low temperature (Thomashow 1999). The inhibition of Ca^{2+} influx by blocking the plasma membrane channels partially suppresses the expression of genes encoding protein kinase that induces the expression of cold-regulated genes, and the development of freezing tolerance in *Arabidopsis thaliana* (L.) Heynh (Tahtiharju et al. 1997) and alfalfa *cas15* (Thomashow 1999). This Ca^{2+}-dependent pathway may be disrupted by B deficiency-induced dysfunction of ion transport channels in the plasma membrane.

Boron deficiency seems to affect the distribution of free Ca^{2+} in the apoplastic pool and membrane-bound Ca, with an increase in apoplastic free Ca^{2+} after B deprivation, which was then reversed by B resupply (Muhling et al. 1998), though no cytosolic free Ca^{2+} levels were reported in the study. This B-induced regulation of cytoplasmic influx of Ca^{2+} may disrupt the Ca^{2+}-dependent regulation of cold-regulated gene expression, slowing or inhibiting cold acclimation and weakening cold tolerance of leaf cells (Figure 1).

Research on B-induced gene expression is still at its infancy. Recently, (Kobayashi et al. 2004) have demonstrated the up-regulation of some genes important in oxidative repair systems by short-term B deprivation, including genes for salicylate-inducible glucosyltransferase, glutamine synthetase, and glutathione *S*-transferase in cultured tobacco BY-2 cells. More research is required to test the above hypothesis that B-regulated cytoplasmic influx of free Ca^{2+} may disrupt the Ca-dependent gene regulation in cold acclimation and tolerance, by utilising transgenic mutants and target gene silencing.

Membrane structure and membrane-to-wall anchorage

Two key cellular responses involved in chilling and freezing tolerance are (1) the maintenance of membrane-to-wall adherence or anchorage to avoid cell collapse and plasmolysis; and (2) the reversible expansion of membrane structure in the freezing-thawing cycle to avoid cell rupture upon the water resorption from melted ice in the intercellular space. These may be closely related to the roles of B in membrane structure stablisation.

Unlike the direct evidence of B-rhamno-galacturonan II (RG-II) complex in cell walls, direct roles of B in membrane functions have been hypothesised on the basis of numerous evidence showing rapid responses of membrane integrity and membrane-bound enzyme functions of root and leaf cells (or protoplasts) to B deprivation with minutes or hours after commencing treatment, though some of these effects were attributed towards the secondary responses of B deficiency induced cell wall disruption (Cakmak and Romheld 1997; Goldbach et al. 2002). However, research evidence with yeasts (with pectin-free cell walls) (Bennett et al. 1999) and animal cells have further pointed to the direct functional role of B in the membrane (Eckhert, 1998; Eckhert and Rowe, 1999; Nielson, 2000).

Boron roles in membrane structure have been theoretically deduced from the biochemical constituents of membranes, such as glycoproteins and glycolipids, which have boric acid/borate complexing ligands (Goldbach, pers. comm.). Membrane constituents rich in hydroxyl groups have received particular attention, including glycosyl-phosphatidyl-inositol-anchor (GPI) to which membrane surface proteins are attached, arabinogalactanproteins (AGP, glycoproteins), and GPI-anchored proteins containing high density of B-binding sites (Goldbach, pers. comm.). Preliminary investigation by Goldbach's group in the University of Bonn, Germany has led to the first success in isolating B-binding membrane-bound proteins. Apart from the direct structural role of B in cellular membranes, recent evidence also suggests a possible function of B in transvacuolar cytoplasmic strands and cell membrane-to-wall anchorage (Bassil et al. 2004), though further evidence is necessary to define this postulated B function in plasma membrane and cell wall interaction.

These postulated B functions in cell membrane and membrane-to-wall anchorage would have important implications in chilling and freezing tolerance of leaf cells, in terms of the avoidance of cellular dehydration and cell plasmolysis.

Ice formation occurs initially in the intercellular spaces and cell walls (apoplastic spaces) in response to freezing temperature such as frost – the mildest form of freezing, as the freezing point in the intercellular space fluid is higher than the cytoplasmic solution (Guy 1990; Xin and Browse 2000). One consequence of this ice formation in the intercellular space is the dehydration of cells due to the efflux of water towards the ice triggered by the sudden drop of water potential (Guy 1990). As a result, there may be some common genes in plant responses to freezing stress, drought stress and salinity stress (Xin and Browse 2000). The extent of cell dehydration may be determined by the osmotic pressure in the cytoplasm produced by osmoticums such as sucrose (Guy 1990). Another important factor contributing the tolerance of freezing stress is the adhesion of cell membrane to cell wall to avoid cell plasmolysis (or called frost plasmolysis) (Guy 1990). In fact, cold treatment in *Brassica napus* specifically induced the expression of a gene encoding a putative cell wall-plasma membrane linker protein rich in proline (Goodwin et al. 1996). This cell membrane-to-cell wall adhesion may be disrupted by B deficiency because of the loss of transvacuolar cytoplasmic strands and cell membrane-to-wall anchorage (Bassil et al. 2004).

Freezing injury can also be caused by ice thawing process because of the effects of freezing stress on plasma membrane structure and function (Xin and Browse 2000). Freezing may cause the loss of plasma membrane surface area due to the invagination of the plasma membrane and budding off of endocytotic vesicles, leading to cell rupture due to the water influx to the cytoplasm upon thawing (Xin and Browse 2000). The stablization of membrane surface area in the freezing-thawing process is key to the avoidance of freezing injury in leaf cells, which may be greatly weakened by B deficiency due to the structural role of B in complexing glycoproteins and glycolipids (Figure 1).

Sugar metabolism and accumulation

The accumulation of low molecular weight compounds is a common response to non-freezing temperature or air chills, as one of the important process of cold acclimation and the induction of freezing tolerance (Guy 1990). Among the compounds (including proline, soluble sugars, and betaine), the accumulation of soluble sugars (particularly sucrose) is the most widely reported to be correlated with the development of freezing tolerance in leaf cells, though other compounds may also contribute to the full freezing tolerance (Guy 1990; Xin and Browse 2000). The protective roles of soluble sugars in cells may be related to its functions as cryoprotectants for specific enzymes, membrane-stablising molecules, and osmolyte for preventing cell dehydration (Guy 1990; Xin and Browse 2000). In leaves of spring and winter wheat, cold stress in the light (temperature decreased from 20 to 5°C) caused 5 to 10 fold increase in sucrose:starch ratio and increased the activity of neutral invertase activity by 2-3 fold (Savitch et al. 2000). Leaf photosynthesis in wheat leaves stressed by exposure to 5°C was only 45% of those at 20°C and day-time ^{14}C-export from the leaves at 5°C was only 35% of the 20°C leaves, without any response to increasing irradiance and CO_2 levels (Leonardos et al. 2003). The decrease in photochemical reaction and limited substrate export in leaf cells may lead to the lower efficiency of electron transport in the thylakoid membrane and reversible or irreversible oxidative damage to Photosystem II (PS-II), which is to be discussed in next section.

Therefore, the transport of sucrose into shoot tip and young leaf cells and its metabolism may influence their tolerance of both chilling and freezing temperature. One of the secondary effects of B deficiency is the inhibition of sucrose transport out of mature leaf cells, into the young leaf and shoot tips (Marschner 1995), which are often most vulnerable to chilling and frost damage. As a result, it is postulated that B-deficiency induced limitation of sucrose supply and metabolism in young leaf cells may weaken the capacity of cells to tolerate cell dehydration and the development of freezing tolerance (Figure 1). Although no ice is formed in leaf cells exposed to chilling temperature, chilling-induced stomatal dysfunction (locked open or delayed closure) can also cause leaf cell dehydration (Allen and Ort 2001). The accumulation and metabolism of sucrose in young leaf cells may be important to both chilling and freezing tolerance of plants.

Oxidative free radicals, antioxidation system and photoinhibition

Low temperature induced oxidative responses and associated photo-oxidative damage in leaf cells have received much greater attention in the literature, compared to other aspects of research on plant tolerance of low temperature (Allen and Ort 2001; Guy 1990; Kratsch and Wise 2000). In chilling or frost sensitive plants, photo-oxidation appears in the form of the light- and oxygen-dependent bleaching of photosynthetic pigments when being exposed to low temperature in the light (Wise, 1995). Increased generation of oxidative free radicals in leaf cells seem to occur within hours of exposure to B-deficiency and B deficiency may impair the energy transfer between PS-II and PS-I (Cakmak and Romheld 1997; Goldbach et al. 1991). As a result, rapid interactions between B deficiency and low temperature are most likely to occur in the process of oxidative radicals generation and anti-oxidation reactions in leaf cells, which ultimately lead to either reversible or irreversible photoinhibition.

Oxidative free radicals and anti-oxidation system

Low temperature tolerance in plant species is closely related to the capacity of anti-oxidation systems in leaf cells, which consist of oxidative free radical scavenging enzymes (e.g. SOD, ascorbate peroxidase (APX), and catalase (CAT)) and anti-oxidants (e.g. ascorbate, glutathione and carotene). Low temperature increases superoxide dismutase (SOD) activity, leading to increased production of H_2O_2, that with superoxide (O^{2-}), forms highly reactive OH^- radicals in leaf cells (Saruyama and Tanida 1995). In one of the chilling sensitive species, cucumber, leaves at 5°C in the light had increased H_2O_2 concentration, due to a substantial decline (by up to 80% compared to that at 25°C) in the activity of the thylakoid APX - a key enzyme in H_2O_2-scavenging (Terashima et al. 1998). The chilling tolerance of cucumber seedling radicles correlated positively with the level of APX and CAT activity, but negatively with activity levels of SOD, glutathione reductase and guaiacol peroxidase (Kang and Saltveit 2002). In inbred maize lines, tolerance to chilling stress increased with the levels of antioxidants in leaves, including ascorbate and carotene (Hodges et al. 1996). Photo-oxidative conditions induced by low temperature in the light can cause a range of ultrastructural distortion in chloroplasts, including thylakoid dilation and chloroplast swelling (Kratsch and Wise 2000). This enhanced oxidative free radical generation may be caused by the inhibited electron transfer between PS-II and PS-I and the decreased photochemcial quenching of light energy.

So far, there is no direct evidence about B roles in enzymes involved in plant anti-oxidation systems. However, B deficiency can decrease the levels of antioxidants in leaves, including ascorbic acid (reduced form), SH-compounds and glutathione reductase (Cakmak and Romheld 1997). The decrease in reduced-ascorbate level in leaves seems not to be a cause for B deficiency-induced membrane leakage which is rapidly reversed by B addition into the incubation solution of sunflower leaf discs (Pfeffer et al. 1998). In leaves of a subtropical Eucalyptus hybrid Eucalyptus grandis x E. urophylla hybrid, B deficiency treatments (0 and 5M B) for up to 96 hours at 5°C significantly increased the production rate of superoxide (O^{2-}) and polyphenol oxidase activity, but not in plants with adequate (15M) B supply at the same temperature (Lu and Huang 2003). (Lu and Huang 2003) also found that the activities of anti-oxidation enzymes (superoxide dismutase, peroxidase, catalase and ascorbate peroxidase) in leaves were decreased at 5°C with B deficiency (<10M B), but not at 15M B. Some initial information suggests that there may be molecular basis of oxidative reactions to B deficiency in leaf cells as B deficiency induced the up-regulation of some important anti-oxidation enzymes including genes for salicylate-inducible glucosyltransferase, glutamine synthetase, and glutathione S-transferase in cultured tobacco BY-2 cells (Kobayashi et al. 2004).

As a result, B deficiency may increase the sensitivity of leaf cells to low temperature, through enhanced generation of oxidative free radicals and weakened anti-oxidation capacity. However, published findings about B in photosynthesis were mostly made at organ or tissue level, which make the differentiation of causal events occurring at subcellular level (e.g chloroplastic membrane, thylakoid membrane) difficult, due to the inconsistency in leaf cell age, leaf cell water potential (transpiration) and B status within the same organ or tissue. We may ask whether B roles in photo-oxidation is the consequence of primary effects of B in chloroplast membrane integrity and functions or the result of metabolic process failure such as substrate feed back and suppressed anti-oxidation activity in the chloroplastic membrane. To achieve defined physiological and biochemical responses in leaf cells to B deficiency and/or chilling stress without the complications of the above inconsistency, protoplast or/and chloroplast culture may be used to explore the proposed mechanisms in relation to B roles in photosynthesis under chilling treatments over a time course.

Photosynthesis and photo-oxidation

Low temperature around leaves can disrupt key processes in photosynthesis, including thylakoid electron transport, carbon assimilation, and stomatal control (Allen and Ort 2001; Kratsch and Wise 2000). The severity of low temperature stress on photosynthesis can be particularly exacerbated by simultaneous or sequential exposure to light due to the enhanced risks of photo-oxidative damage (Allen and Ort 2001; Kratsch and Wise 2000). In *Phaseolus vulgaris* quantum yield was severely decreased by exposure to 6°C at 2000 μmol m^{-2} s^{-1} for 3 hours, but not at 12°C or under low photon flux density (Powles and Critchley 1980). In one of the most chilling-sensitive species *Gossypium hirsutum* (Kratsch and Wise 2000), leaf chlorosis was not observed after 144 h exposure to chilling (5°C) in the dark though severe leaf wilting appeared after 72 hours, but under 500 μmol m^{-2} s^{-1} light, the same temperature caused leaf chlorosis after 48 hours and permanent wilting of the fully expanded leaves by 24 hours (Wise et al. 1983). Ultrastructurally, chilling in the light for 144 hours caused the loss of starch granules and dilation of thylakoids in chloroplasts but not in the dark (Wise et al. 1983).

On the basis of very limited information, B deficiency induced a much greater sensitivity to shoot chilling than root chilling, in terms of gross structural changes (e.g. membrane leakage) and photosynthetic functions of leaf cells (Ye 2005), even in a chilling-tolerant species such as oilseed rape (Wise et al. 1983). In a chilling-sensitive cucumber cultivar, chilling temperature (7-8/5 °C, day/night, under 300 mol m^{-2} s^{-1} light) enhanced membrane leakage (K^+) induced by B-deficiency and chloroplast disruption and plasmolysis of mesophyll cells (Wang et al. 1999). Chilling-temperature around leaves induced a higher solute leakage in the youngest open leaf of low B status oilseed rape (10-15 mg B kg^{-1} dry matter) than leaves containing adequate B (25 mg B kg^{-1} dry matter) (Ye 2005). This gross structural damage may be the consequence of chronic photo-inhibition caused by chilling stress in leaf cells with low/deficient B status, and the further effects of high light flux density (Huang et al. 2002). The pre-exposure of leaf cells to suboptimal B nutrition may also lower the activity of anti-oxidation systems and decrease the possibility of recovery from photoinhibition in the photosystems after chilling stress.

Low temperature-induced reduction in photochemical reaction (CO_2 assimilation) and in inhibition of thylakoid electron transport may be the fundamental causes for photoinhibition at chilling air temperature (either in the dark or in the light), including dynamic (repairable) and chronic (photodamage caused by oxidative radicals) photoinhibition (Allen and Ort 2001). The primary event of photoinhibition is suggested to be the functional inactivation of the photosystem II (P680) reaction centre (Cleland 1988), where the steady state oxidation-reduction level of the primary quinone acceptor (Q_A) of PS-II must be maintained to avoid photo-oxidation of the light harvesting centre (Melis 1999). The extent of PS-II photoinhibition is closely related to the redox state of the electron acceptor Q_A in PS-II and the level of reduced Q_A increases with light flux density, leading to increased risks of photodamage in PS-II (Melis 1999; Sonoike 1999). When thylakoid electron transport is inhibited through the normal pathway of CO_2 assimilation, excitation energy stored in the reduced Q_A at PS-II is dissipated via a charge-recombination reaction with oxygen (rather than CO_2 fixation), which generates singlet oxygen (Melis 1999; Sonoike 1999). The highly reactive oxygen free radicals $O_2^-$and OH^-n the acceptor side of PS-II, which increase with photon flux density levels (Arato et al. 2004), may consequently cause photo-oxidation of chlorophyll and chloroplastic membranes (Melis 1999).

Plant chilling tolerance will depend on the intrinsic anti-oxidation systems in leaf cells, which may be weakened by B deficiency (Lu and Huang 2003). Due to the limitation of direct experimental evidence in the literature, the following discussion will focus on photo-oxidative processes in leaf cells, in which chilling and B nutrition are most likely to interact with each other, by reviewing update knowledge about their separate effects on photo-oxidative responses. Boron deficiency is likely to enhance chilling-induced photoinhibition in leaf cells, through its possible effects on photosynthesis. (Kastori et al. 1995) observed that relatively prolonged exposure (23 days) to B deficiency (1 mol B) significantly decreased photosynthetic oxygen generation in leaves, quantum yield and PS-II efficiency, which was attributed to reduced electron transport efficiency and the accumulation of soluble carbohydrates in leaves. In a follow-up study by the same group (Plesnicar et al. 1997), B deficiency was found to reduce the maximum rate of photosynthetic O_2 production, but had no effect on the efficiency of PS-II electron transport. The reduction in maximum quantum yield may be related to the reduction of chlorophyll content and photochemical quenching of electrons (Plesnicar et al. 1997). In sunflower at 0.02 □M B, the decline in oxygen generation rate and non-photochemical quenching in leaves accompanied a range of effects on other metabolism processes: including accumulation of sucrose, phenolic compounds and stimulated peroxidase activity and lipid peroxidation (El-Shintinawy 1999). The existing evidence of B deficiency-induced effects on leaf photosynthesis has been mostly obtained from long term (10 days or longer) treatments of plants with deficient B supply. These long-term B-deficiency effects on leaf photosynthesis make it hard to pinpoint the primary events induced by B-effects on thylakoid membrane, photochemistry or electron transport efficiency within the chloroplast.

In summary, on the basis of the above discussion on both factors in leaf photosynthesis, it is speculated that B deficiency may exacerbate chilling-induced photo-oxidative damage and weaken the anti-oxidative system for photo-inhibition recovery in leaf cells, through: (1) enhanced generation of oxidative free radicals, at least through B-deficiency induced disturbance to enzyme functions (such as H^+-ATPase and NADP-reductase) in thylakoid membranes and/or B-deficiency induced substrate feedback–inhibition of triose-phosphate export and starch accumulation in chloroplasts and, (2) weakening the recovery capacity from photo-oxidative damage by decreasing enzyme activities in oxidative free radical metabolism and the levels of antioxidants (e.g. ascorbate).

Concluding remarks

Low temperature effects on plant canopy in the light impairs photosynthesis processes in leaf cells, through a range of ultrastructural, biochemical and molecular events (such as oxygen free radical generation and anti-oxidation defence), which may eventually lead to irreversible photo-oxidative damage (Allen and Ort 2001; Kratsch and Wise 2000; Melis 1999). These changes may be significantly modulated by B nutrition in leaf cells as B plays an important role in cell membrane integrity through forming B-complexes with membrane constituents containing *cis-diol* groups and in antioxidation systems in leaf cells (see the review by Cakmak and Romheld 1997). As a result, poor B nutrition in the shoot may lower chilling tolerance of leaf cells, exhibited as photo-oxidative damage in the light, particularly in chilling-sensitive crop species. Carefully designed experiments are required to test these hypotheses. To test if low B status increases the sensitivity of leaf cells to photo-inhibition, chilling responses of photosynthetic characteristics should be monitored at the on-set of chilling treatment over a time course, including chlorophyll fluorescence and the quantum yield of PS-II photochemistry.

From a practical point of view, in chilling sensitive species, when the occurrence of cold air currents is expected, preventive measures may be taken to boost foliar B status by foliar B application to minimise the risk of B deficiency-induced sensitivity to photoinhibition in leaf cells. This may also help to alleviate chilling damage in the young canopy of many tropical/subtropical woody species (e.g. teak, eucalyptus) in soils of low B status.

References

Allen D.J. and Ort D.R., 2001 Impacts of chilling temperatures on photosynthesis in warm-climate plants. TRENDS in Plant Science **6**, 36–42.

Anon 1958 Borax as frost protection. Horticultural Advertiser September.

Arato A., Bondarava N., and Krieger-Liszkay A., 2004 Production of reactive oxygen species in chloride and calcium-depleted phtosystem II and their involvement in photoinhibition. Biochimica et Biophysica Acta **1608**, 171–180.

Barr R. and Bottger Crane F.L., 1993 The effect of boron on plasma membrane electron transport and associated proton secretion by cultured carrot cells. Biochemistry and Molecular Biology International **31**, 31–39.

Bassil E., Hu H., and Brown P.H., 2004 Use of Phenylboronic Acids to Investigate Boron Function in Plants. Possible Role of Boron in Transvacuolar Cytoplasmic Strands and Cell-to-Wall Adhesion. Plant Physiol. **136**, 3383–3395.

Bell R.W., 1997 Diagnosis and prediction of boron deficiency for plant production. Plant and Soil **193**, 149–168.

Beltram V., 1958 Bor als frostschutz (Boron as frost protector). Allgemeine Forstzeitschrift **10**, 147–148.

Blaser-Grill J., Knoppik D., Amberger A., and Goldbach H., 1989 Influence of boron on the membrane potential in Elodea densa and Helianthus annuus roots and H^+ extrusion of suspension cultured Daucus carota cells. Plant Physiology **90**, 280–284.

Blevins D., Scrivner C.L., Reinbott T.M., and Schon M.K., 1996 Foliar boron increases berry number and yield of two highbush blueberry cultivars in Missouri. J. Plant Nutr. **19**, 99–113.

Braekke F.H., 1979 Boron deficiency in forest plantations on peatland in Norway. Meddelesar fra Norsk Institutte for Skogforskning **35**, 213–236.

Braekke F.H., 1983 Micronutrients-prophylactic use and cure of forest growth disturbances. Proc. Int. Workshop on Growth Disturbances of Forest Trees. Finland, October 1982. Commun. Inst. Forest. Fenniae Helsinki **116**, 159–169.

Brown P.H., Bellaloui N., Wimmer M.A., Bassil E., Ruiz J., Hu H., Pfeffer H., Dannel F., and Romheld V., 2002 Boron in plant biology. Plant Biology **4**, 205–223.

Cakmak I. and Romheld V., 1997 Boron deficiency-induced impairments of cellular functions in plants. Plant and Soil **193**, 71–84.

Cleland R.E., 1988 Molecular events of photoinhibitory inactivation in the reaction centre of photosystem II. Australian Journal of Plant Physiology **15**, 135–150.

Cooling E.N., 1967 Frost resistance in Eucalyptus grandis following the application of fertiliser borate. Rhod. Zamb. Mal. J. Agric. Res. **5**, 97–100.

Cooling E.N. and Jones B.E., 1970 The importance of boron and NPK fertilisers to eucalypts in the southern province, Zambia. East Afr. Agric. & Forest. J. **36**, 185–194.

Dell B. and Malajczuk N., 1994 Boron deficiency in eucalypt plantations in China. Canadian Journal of Forest Research **24**, 2409–2416.

El-Shintinawy F., 1999 Structural and functional damage caused by boron deficiency in sunflower leaves. Photosynthetica **36**, 565–573.

Ferrol N. and Donaire J.P., 1992 Effect of boron on plasma membrane proton extrusion and redox activity in sunflower cells. Plant Science **86**, 41–47.

Goldbach H.E., Blaser-Grill J., Lindemann N., Porzelt M., Hörrmann C., Lupp B., and Gesser B., 1991 Influence of boron on net proton release and its relation to other metabolic processes. Current Topics in Plant Biochemistry and Physiology **10**, 195–220.

Goldbach H.E., Hartmann D., and Rötzer T., 1990 Boron is required for the stimulation of the ferricyanide-induced proton release by auxins in suspension-cultured cells of Daucus carota and Lycopersicon esculentum. Physiologia Plantarum **80**, 114–118.

Goldbach H.E., Wimmer M.A., Chaumont F., Matoh T., Volkmann D., Baluška F., Wingender R., and Schulz M.Q., 2002 Rapid responses of plants to boron deprivation. In Boron in Plant and Animal Nutrition, Bonn, Germany, 2002. Eds H.E. Goldbach, B. Rerkasem, M. Wimmer, P.H. Brown, M. Thellier and R.W. Bell. pp 167.

Goodwin W., Pallas J.A., and Jenkins G.I., 1996 Transcripts of a gene encoding a putative cell wall-plasma membrane linker protein are specifically cold-induced in Brassica napus. Plant Molecular Biology **31**, 771–781.

Graham D. and Patterson B.D., 1982 Responses of plants to low, nonfreezing temperatures: proteins, metabolism, and acclimation. Annual Review of Plant Physiology and Plant Molecular Biology **33**, 347–372.

Guy C.L., 1990 Cold acclimation and freezing stress tolerance: role of protein metabolism. Annual Review of Plant Physiology and Plant Molecular Biology **41**, 187–223.

Hanson E.J. and Breen P.J., 1985 Effects of all boron sprays and environmental factors on fruit set and boron accumulcation in 'Italian' prune flowers. Journal of American Society and Horticulture Science **110**, 389–392.

Hodges D.M., Andrews C.J., Johnson D.A., and Hamilton R.I., 1996 Antioxidant compound responses to chilling stress in differentially sensitive inbred maize lines. Physiologia Plantarum **98**, 685–692.

Huang L., Gherardi M., Bell R.W., and Dell B., 2002 High photon flux density increases external boron (B) requirements for leaf growth of sunflower (Helianthus annuus L. cv. Hysun 25) in B-buffered solution culture. In Boron in Plant and Animal Nutrition, Bonn, Germany, 2002. Eds H.E. Goldbach, B. Rerkasem, M. Wimmer, P.H. Brown, M. Thellier and R W Bell. pp 213.

Huang L., Ye Z., Bell R.W., and Dell B., 2005 Boron Nutrition and Chilling Tolerance of Warm Climate Crop Species. Ann Bot, **96**, 755–767.

Hughes M.A, and Dunn M.A., 1996 The molecular biology of plant acclimation to low temperature. Journal of Experimental Botany **47**, 291–305.

Kang H.-M. and Saltveit M.E., 2002 Effect of chilling on antioxidant enzymes and DPPH-radical scavenging activity of high and low vigour cucumber seedling radicles. Plant Cell & Environment 25.

Kastori R., Plesnicar M., Pankovic D. and Sakac Z., 1995 Photosynthesis, chlorophyll fluorescence and soluble carbohydrates in sunflower leaves as affected by boron deficiency. Journal of Plant Nutrition **18**, 1751–1763.

Kobayashi M., Mutoh T., and Matoh T., 2004 Boron nutrition of cultured tobacco BY-2 cells. IV. Genes induced under low boron supply. J. Exp. Bot. **55**, 1441–1443.

Kratsch H.A. and Wise R.R., 2000 The ultrastructure of chilling stress. Plant, Cell and Environment **23**, 337–350.

Leonardos E.D., Savitch L.V., Huner N.P.A., Oquist G., and Grodzinski B., 2003 Daily photosynthetic and C-export patterns in winter wheat leaves during cold stress and acclimation. Physiologia Plantarum **117**, 521–531.

Lu C.-q. and Huang B.-l., 2003 Effects of boron on membrane lipid peroxidation and endogenous protective mechanisms in leaves of Eucalyptus grandis x Eucalyptus urophylla under low temperature. Journal of Tropical and Subtropical Botany **11**, 217–222.

Lyons J.M., Graham D., and Raison J.K., 1979 Low Temperature Stress in Crop Plants: The Role of the Membrane. Academic Press, New York.

Marschner H., 1995 Mineral Nutrition of Higher Plants. Academic Press, London.

Melis A., 1999 Photosystem-II damage and repair cycle in chloroplasts: what modulates the rate of photodamage in vivo? TRENDS in Plant Science **4**, 130–135.

Milovankic M., Keserovic Z., and Vasiljevic Z., 1990 Studies on some chemicals used for prevention of frost damage to fertile buds and flowers of fruit crops. Yugoslovensko Vocarstvo **24**, 21–27.

Muhling K.H., Wimmer M., and Goldbach H., 1998 Apoplastic and membrane-associated Ca^{2+} in leaves and roots as affected by boron deficiency. Physiologia Plantarum **102**, 179–184.

Obermeyer G., Kriechbaumer R., Strasser D., Maschessnig A., and Bentrup F.W., 1996 Boric acid stimulates the plasma membrane H^{+}-ATPase of ungerminated lily pollen grains. Physiologia Plantarum **98**, 281–290.

Pfeffer H., Dannel F., and Römheld V., 1998 Are there connections between phenol metabolism, ascorbate metabolism and membrane integrity in leaves of boron-deficient sunflower plants? Physiologia Plantarum **104**, 479–485.

Plesnicar M., Kastori R., Sakac Z., Pankovic D., and Petrovic N., 1997 Boron as limiting factor in photosynthesis and growth of sunflower plants in relation to phosphate supply. Agrochimica **41**, 144–154.

Powles S.B. and Critchley C., 1980 The effect of photon flux density during growth on photo-inhibition of intact attached bean leaflets. Plant Physiology **65**, 1181–1187.

Roldán M., Belver A., Rodriguez-Rosales M.P., Ferrol N., and Donaire J.P., 1998 In vivo and in vitro effects of boron on the plasmamembrane proton pump of sunflower roots. Physiologia Plantarum **84**, 49–54.

Sakya A.T., Dell B., and Huang L., 2002 Boron requirements for Eucalyptus globulus seedlings. Plant and Soil **246**, 87–95.

Saruyama H. and Tanida M., 1995 Effect of chilling on activated oxygen-scavenging enzymes in low temperature-sensitive and tolerant cultivars of rice (Oryza sativa L.). Plant Science **109**, 105–113.

Savitch L.V., Harney T., and Huner N.P.A., 2000 Sucrose metabolism in spring and winter wheat in response to high irradiance, cold stress and cold acclimation. Physiologia Plantarum **108**, 270–278.

Shinozaki K. and Yamaguchi-Shinozaki K., 2000 Molecular responses to dehydration and low temperature: differences and cross-talk between two stress signaling pathways. Current Opinion in Plant Biology **3**, 217–223.

Shorrocks V.M., 1997 The occurrence and correction of boron deficiency. Plant and Soil **193**, 121–148.

Sonoike K., 1999 The different roles of chilling temperatures in the photoinhibition of photosystem I and photosystem II. Journal of Photochemistry and Photobiology B: Biology **48**, 136–141.

Tahtiharju S., Sangwan V., Monroy A.F., Dhindsa R.S., and Bor M., 1997 The induction of kin genes in cold-acclimating Arabidopsis thaliana. Evidence of a role for calcium. Planta **203**, 442–447.

Terashima I., Noguchi K., Itoh-Nemoto T., Park Y.-M., Kubo A., and Tanaka K., 1998 The cause of PSI photoinhibition at low temperatures in leaves of Cucumis sativus, a chilling-sensitive plant. Physiologia Plantarum **103**, 295–303.

Thomashow M., 1999 Plant cold acclimation: freezing tolerance genes and regulatory mechanisms. Annual Review of Plant Physiology and Plant Molecular Biology **50**, 571–599.

Wang Z.Y., Tang Y.L., Zhang F.S., and Wang H., 1999 Effect of boron and low temperature on membrane integrity of cucumber leaves. Journal of Plant Nutrition **22**, 543–550.

Wimmer M.A., and Goldbach H., 1999 Influence of Ca^{2+} and pH on the stability of different boron fractions in intact roots of Vicia faba L. Plant Biology **1** 632–637.

Wise R.R., McWilliam J., and Naylor A.W., 1983 A comparative study of low temperature-induced ultrastructure alterations of three species with different chilling sensitivities. Plant Cell & Environment **5**, 525–535.

Xin Z. and Browse J., 2000 Cold comfort farm: the acclimation of plants to freezing temperatures. Plant, Cell and Environment 2000, 893–902.

Ye Z., 2005 Effect of low temperature on plant boron nutrition. PhD thesis. Murdoch University, Perth, Australia.

Ye Z., Bell R., Huang L., Yang Y., and Dell B., 1997 Covering winter oilseed rape (Brassica napus cv. Eureka) increases yield on a low boron soil. In The International Symposium on Boron in Soils and Plants, Chiang Mai, Thailand, 7-11 September, 1997, 1997. Eds R. Bell and B. Rerkasem. pp 29–34.

Influence of Boron on Xylogenesis in *Pinus radiata* Organ Cultures

Tracy Putoczki, Juliet Gerrard[§] and Sandra Jackson

School of Biological Sciences, University of Canterbury, Priavte Bag 4800 Christchurch, New Zealand 8020. [§]E-mail: juliet.gerrard@canterbury.ac.nz

Introduction

Very little is known about the mechanisms that drive xylogenesis. In conifers, xylogenesis incorporates cambial cell division, subsequent xylem cell differentiation and culminates in cell death, marking the formation of a mature tracheid. Many researchers suggest that boron availability plays a critical role in the regulation of cell development. In particular, boron may be involved in maintaining the strength and integrity of the cell wall matrix, as well as the production and deposition of cell wall material (Hu 1994). We have a particular interest in the role of boron in xylogenesis and xylem cell development.

The phloem and xylem of wood tissue are formed through periclinal divisions of the meristimatic cambial cell layer. Through cell division the cambium maintains itself, produces outer layers of phloem cells, and inner radial files of xylem cells. The production of more xylem than phloem cells accounts for the formation of wood in trees. The signals that induce xylem formation are not known. The newly formed xylem cells are unable to divide, but undergo differentiation whereby they expand and elongate. During this process they produce secondary cell walls that thicken as cellulose microfibrils are deposited in organized layers and are impregnated with lignin (Higuchi 1997). These xylem cells ultimately undergo autolysis of symplastic cellular material which marks their final transition to a mature cell (Higuchi 1997). The process of differentiation into mature cells occurs as xylem cells are displaced from the cambium by newly divided cells, which allows for the identification of the "mother cell" by following the radial cell file. Since division and differentiation are presumed to be regulated by internal and external factors, this feature of xylem makes it ideal for the study of the influence of external factors, such as nutrition, on wood formation, as changes in more than one stage of differentiation can be observed simultaneously (Chaffey 1999).

We have studied the function of boron within the xylem cell by comparing growth responses of boron deficient cultures to those that were grown under optimal to supra-optimal boron conditions. We were particularly interested in the relationship between boron and the pectic polysaccharides, which are believed to be involved in maintaining the strength and stability of the cell walls (Hu 1994).

F. Xu et al. (eds.), Advances in Plant and Animal Boron Nutrition, 47–62.

Materials and methods

Plant material

Pinus radiata logs were harvested from Burnham and Rotorua, New Zealand between October 2003 and May 2005.

Preparation of the growth media

The organ culture media was prepared essentially as described by Savidge (1993). Appropriate volumes of stock solution aliquots were combined with NanoPure deionized water in autoclavable polypropylene containers. This concentrated organ culture media was then passed through Amberilite IRA-743 resin columns without boric acid, potassium iodide, sodium molybdate and NAA (naphthaleneacetic acid) since these are retained by the resin (Matoh 1992). Prior to autoclaving, the remaining organ culture media components were added, the volume was adjusted with boron free NanoPure water, the pH was adjusted, and the agar solution was added to give a final volume of 1 L. Sterilized organ culture media was poured into sterile Petri dishes in a horizontal laminar flow cabinet, sealed with plastic stips, and stored in a sterile and dark environment.

Preparation of organ cultures

The organ culture explant method was derived from Savidge (1993). *Pinus radiata* trees planted in 1994 were felled as required, and stem segments between whorls of branches were removed from a region of the tree approximately 1 – 3 metres above the ground. The stem segments were returned to the lab, and stored at 4°C. To prepare for culturing, a disc of approximately 10 cm was removed from the stem, washed and rinsed in cold water, and scrubbed with detergent, then rinsed with dilute bleach. The outer rhytidome was removed with razor knives, rinsed with ethanol and the disc was placed in a specially designed holder in a horizontal laminar flow cabinet. The surface of the disc was sprayed with ethanol and flame sterilized. Heat sterilized equipment was used to remove sections approximately 3 cm axial x 2 cm tangential x 0.5 cm radial. These explants contained surface sterilized phloem, cambium and xylem. The explant was placed on the growth media so that the exposed xylem surface was in direct contact with the media (Fig. 1). The Petri dish was resealed with plastic strips and placed in the growth room. An additional explant was kept as a control and stored in 4% paraformaldehyde in ¼ × phosphate-buffered saline.

Preparation for inductively coupled plasma mass spectometry (ICP-MS)

The phloem and the portion of the organ culture that was in direct contact with the growth medium were removed using equipment cleaned with boron free water. For analysis of the whole culture, samples from one tree were pooled and air dried after which they were ground to a homogenous powder using a Mixer Mill (Retsch GmbH and Co., KG) for 7 mins at 23 Hz. Approximately 1 g of each sample was sent to Hill Laboratories (Hamilton, New Zealand) for ICP-MS analysis to determine the amount of boron present in the wood.

Light microscopy

Explants were fixed in 4% paraformaldehyde in PBS. After dehydration in a series of 2-methylpropan-2-ol the samples were embedded in paraffin wax, and sections (10 μm thick) were cut with a Reichert-Jung 2040 rotary microtome.

Toludine blue

Sections mounted on glass slides were de-waxed in a xylol series, rehydrated in a graded ethanol series to water, stained for 1 min with 0.05% aq. toludine blue, dehydrated and mounted in DPX.

Safranin-fast green

Sections mounted on glass slides were de-waxed in a series of xylol and rehydrated with a graded series of ethanol, after which they were stained for 24 hours in a 1% aq. safranin solution. The sections were then rinsed in water, treated with a 95% ethanol and 0.5% picric acid solution, followed by a 100% ethanol and ammonium hydroxide solution. Following an additional 100% ethanol exposure, the sections were counter stained with 0.05% aq. fast green and dehydration was continued in a graded series of ethanol and xylol. The sections were mounted in DPX.

Observation of growth: cell counts

Toludine blue stained slides were observed using a light microscope and the number of cells present was counted in four radial cell files from four pictures of each sample (16 cell files in total). Three defined regions were used. The first region was defined as the 'cambial' region. This region was comprised of the cells actively undergoing division and newly divided cells, which were recognized by their small size and thin walls. The second area measured was defined as the region undergoing 'radial expansion'. This region was composed of cells undergoing enlargement in the radial dimension and included cells immediately following the cambial region and preceding the cells that were undergoing cell wall lignin deposition. The final region, termed the 'new secondary wall', consisted of the region of cells undergoing visible cell wall lignin deposition until the point of identification of a mature cell. Each of these regions was compared to control organ culture explants that were collected from the same disk and fixed on the day of culturing.

Observation of growth: lumen area and cell wall thickness

Image Pro Plus Software® (Media Cybernetics Inc., Silver spring U.S.A) was used to analyze images collected with a light microscope. These images were used to calculate the lumen area and the percent of the cell area occupied by the cell wall.

The results obtained were compared to control organ culture explants that were collected from the same disk and fixed on the day of culturing.

Transmission electron microscopy: uranyl acetate and lead citrate

Hand cut tissue blocks (3 mm tangential x 2mm radial x 2 mm longitudinal) were fixed in 4% glutaraldehyde in 100 mM cacodylate buffer and post-fixed in 1% osmium tetroxide. The samples were then dehydrated in an ethanol series, and transferred to 100% acetone before infiltrating and embedding in Spurr resin. Ultra thin transverse sections were cut using a diamond knife, placed on copper grids, and stained with uranyl acetate and lead citrate for observation using a Hitachi H-600 TEM at 75 kV.

Immunogold transmission electron microscopy: JIM5 and JIM7

Hand cut tissue blocks were fixed in 4% paraformaldehyde and ½% glutaraldehyde in 100 mM cacodylate buffer. The samples were dehydrated in an ethanol series after which they were infiltrated and embedded in L.R. White resin. Ultra-thin transverse sections were placed on formar coated nickel grids. Labelling was performed as per AURION immunogold reagent methods, using AURION reagents. Briefly, residual aldehyde groups were inactivated by incubation in a glycine/PBS solution. Grids were blocked using AURION blocking solution, washed in buffer and incubated in the appropriate dilution of primary antibody (JIM5 and JIM7; 1:20). Following incubation, the grids were washed in buffer and incubated in AURION Ultra Small Gold Conjugate Reagent, diluted in incubation buffer. The grids were then washed, post-fixed in glutaraldehyde and washed further with phosphate buffer. The grids were silver enhanced using an AURION R-GENT Silver Enhancing Kit, washed and stained with uranyl acetate and lead citrate, prior to viewing with a Phillips CM-100 TEM at 80 kV.

Statistical analysis

Statistix 8.0 (Analytical Software, Talbhase U.S.A) was use to generate randomized analysis of variance outputs from the data generated from each culture set. If significant results were obtained, a Tukey multi-variate comparison was perfermed.

Results

Organ culture viability and boron uptake

The cultures grown on medium containing varying concentrations of boron exhibited green callus formation along the perimeter of the culture throughout the growth period (Fig. 1). The presence of callus was determined to be an indication that the cultures were alive for the duration of the culture period. However, the amount of callus present varied greatly within each treatment and between each culture set. Moreover, no consistent differences in the colour of the callus were observed. While callus presence suggested a healthy culture, we determined that it was not a good indication of the rate of growth, or the effect the media treatments were incurring on the cultures. In order to determine if the uptake of boron in the organ cultures was successful whole cultures were collected and processed by ICP-MS. The results indicate that boron uptake with increasing boron concentration in the culture medium (Fig. 2).

Figure 1. Organ culture explant demonstrating green callus growth along the periphery of the culture. The presence of callus was used as an indicator of the viability of the culture. There were no distinguishable differences between the colour or amount of callus present between the different boron treatments. The image was collected using digital photography

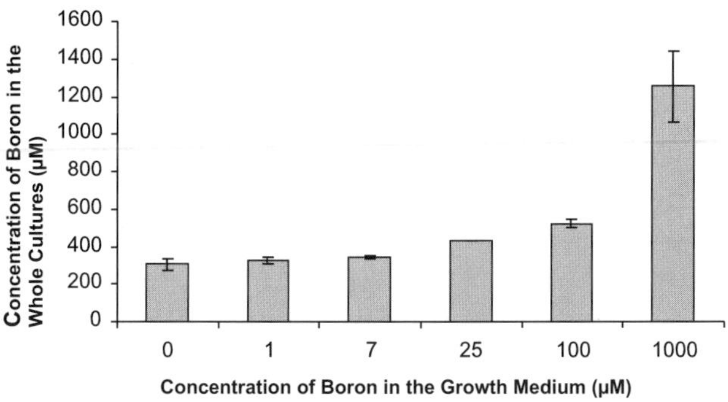

Figure 2. Boron uptake increases in whole cultures, as the concentration of boron in the growth medium increases. Values were obtained by ICP-MS and represent two trees. Bars indicate standard error

Boron can influence growth

Organ cultures from three trees grown on the high boron medium contained 27% more cambial cells on average, while those grown on the low boron medium only had a 5% average increase, when compared to the control explant collected at the time of culture ($P < 0.05$; Fig. 3). These results suggest that a surplus of boron may up-regulate cell division, while boron deficiency can limit it.

Cell size can be altered by boron

Cultures grown on the low boron medium generally displayed a decrease in lumen area in the radially expanding region, while those grown on the high boron medium had a slight increase (Fig. 3; data not shown). This suggests that altering the available boron can result in concentration dependent changes in the cell size. Changes in the area occupied by the cell wall were also possible. Cultures grown on the low boron medium had thicker cell walls, while those grown on the high boron medium had thinner cell walls ($P < 0.05$; Fig. 4).

Organization of the compound middle lamella can be altered by boron

The compound middle lamella of the low boron cultures appeared disorganized when compared to the cultures grown on the high boron medium, and control explants collected at the time of culture (Fig. 5). In particular, the region of the compound middle lamella / S_1 cell wall appeared to have darker striation deposits compared to the other cultures, suggesting potential changes in lignification. Furthermore, this region generally appeared larger, or possibly "swollen" when compared to the high concentration boron cultures, which appeared "tighter" (Fig. 5C).

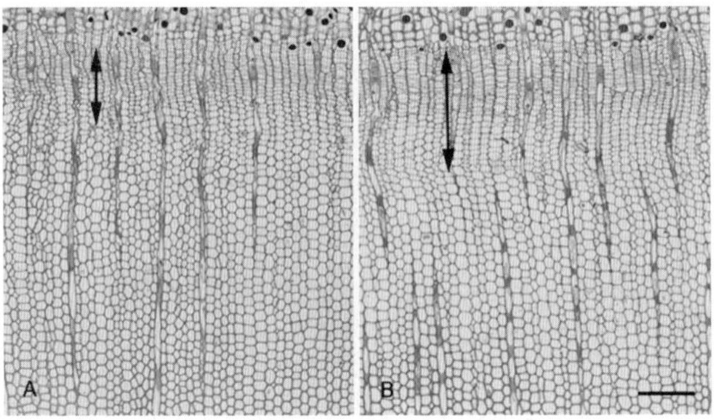

Figure 3. Cultures grown under high boron conditions experience an increase in growth. A) Organ cultures grown on low boron concentrations (0 μM) demonstrated a decrease in cell number and thus division compared to B) those grown under high boron concentrations (1000 μM) as indicated by vertical arrows. Transverse sections of P.radiata stained with toludine blue and observed by light microscopy. Images were collected on a zeiss axioshop camera. Scale bar = 100 μm

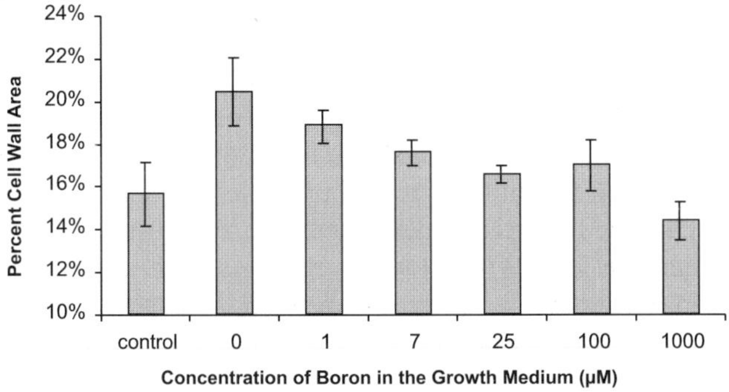

Figure 4. Culture cell walls become thicker when grown in boron deficient conditions. Cultures grown under low boron concentrations tend to have thicker cell walls. Values were obtained with Image Pro Plus software (Media Cybernetics Inc, Silver spring U.S.A.). Bars indicate standard error

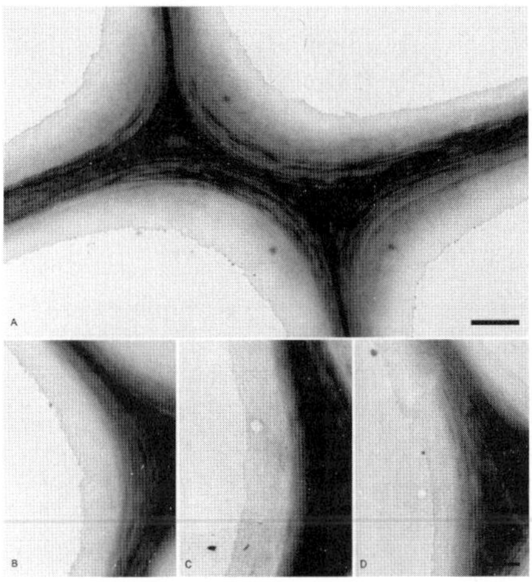

Figure 5. Alterations in the organization of the compound middle lamella occur under different boron conditions. Cultures grown under A), B) low boron concentrations (0 µM) present with an apparent increase in the size of the striated compound middle lamella / S_1 cell wall regions when compared to the cultures grown under C) higher boron concentrations (1000 µM). Furthermore, the cultures grown under low boron concentrations have darker striation patches compared to the D) control cultures. This suggests that the organization of the compound middle lamella is changing in response to the boron treatments. Transverse sections of P. radiata cultures, embedded in Spurr resin and stained with lead and uranyl acetate. Images were collected on a Hitachi TEM. Scale bars = 2 µm and 1 µm respectively

Changes in pectin esterification

In the newly divided and radially expanding cells, an increase in JIM 5 epitope expression was observed in the cultures grown on the low boron media (Fig. 6A), while it was drastically reduced in the cultures grown on the high boron media (Fig. 6C). Furthermore, under low boron conditions, the distribution of the JIM 5 epitope generally appeared clustered compared to the homogenous distribution seen in the control explants (Fig. 6E). It was determined that this was not an artefact of culture, as some control explants from other culture treatments also displayed clustering. JIM 7 expression appeared relatively homogenous in all of the samples, although there was a general reduction in the cultures grown under low boron conditions (Fig. 6B), while a moderate increase occurred under high boron conditions (Fig. 6D). This was expected, as an increase in JIM5 epitope suggests de-esterification has occurred, which should correlate with a decrease in esterified pectin, and thus JIM7 expression.

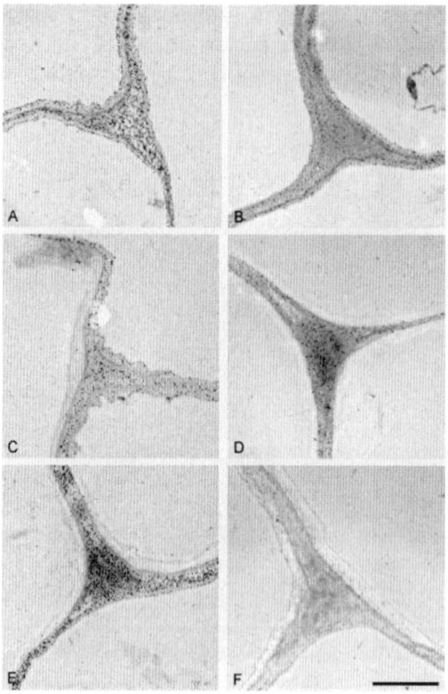

Figure 6. Cultures grown under low boron conditions experience an apparent increase in pectin de-esterification. Immunogold studies of transverse sections of P. radiata organ cultures using JIM5 and JIM7 monoclonal antibodies. JIM5 (A, C, E) is specific to unesterified pectin while JIM 7 (B, D, F) is specific to esterified pectins. Cultures grown under (A) low boron conditions (0 μM) present with an increase in the JIM5 epitope expression compared to cultures grown under (C) high boron conditions (1000 μM). Furthermore, the distribution of the JIM5 epitope in the (E) control explant appears much more homogenous than either of the boron treatments. The JIM7 epitope was most abundant in the (D) high boron cultures compared to the (B) low boron and (F) control explants. This suggests that the cultures grown under low boron conditions may experience an increase in pectin de-esterification and possibly deposition. Images were collected with a Megaview III Software Imaging System and a Phillips TEM. Scale bar = 2 μM

Lignification appears altered by boron

Slides stained with safranin and counter stained with fast green were used as an initial qualitative indication of lignin distribution. Safranin stains the lignin present in the cell wall red, while fast green stains the cellulose blue (Wardrop 1981). The decreased presence of a red colouration observed as the boron concentration was increased (Fig. 7) suggests that the deposition or polymerization of lignin is reduced in the organ cultures. Furthermore, under high boron concentrations, the presence of cellulose appears increased, suggesting that cellulose deposition may be up-regulated in response to lignin deficiency.

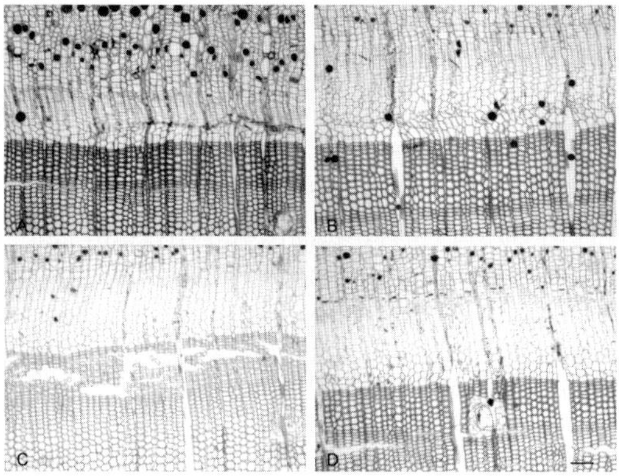

Figure 7. Cultures grown in low boron conditions appear to have more lignin. Transverse sections of P.radiata organ cultures stained with safranin/ fast green. Safranin stains lignin red. The decreased presence of red colouring in the sections suggests that the presence of lignin is reduced as the concentration of boron increases. Cambial organ cultures grown on media containing A) low boron concentrations (0 μM), B) adequate boron concentrations (7 μM), C) high boron concentrations (100μM) and D) the control explant collected at the time of culture. Images were collected on a light microscope with a Cool Snap Digital Camera. Scale bar = 100 μm

Discussion

Boron mobility

In the organ cultures boron is entering the wood (supplied in the form of boric acid) through passive diffusion. With no phloem in contact with the growth medium (Fig. 1) boron must enter the culture through the xylem. As such, a mechanism for boron transport into and between the cells is likely to exist within the network of xylem cells. We propose that boron is entering the organ cultures by following a concentration gradient within the xylem system (likely entering via the ray cells) that is determined by the amount of boron supplied in the medium. While this mode of boron absorption does not correspond to boron transport *in vivo* it does result in the successful uptake of boron into the organ cultures, which ultimately makes boron available to the newly dividing cells (Fig. 2). This has allowed us to determine how boron availability can impact the growth of new wood.

Influence on cell division and cell size

Boron deficiency hindered cell division and thus plant growth in our cultures, which is consistent with the boron deficiency responses observed in other plants. In squash roots for example, growth in low boron conditions allowed for the initiation of mitosis but cell division ceased shortly after (Cohen 1977). Studies on pumpkin demonstrated that growth was inhibited under boron deficient conditions (Ishii 2001). Moreover, studies on plums found that boron deficiency symptoms occur in the growing leaves but not in the fully mature boron depleted leaves, suggesting that boron directly influences new plant growth (Brown and Hu, 1997).

We also observed an increase in cell division in the cultures grown on the high boron medium. This was a surprise, as our high boron concentrations were presumed to be toxic to the cells. As far as we are aware, ours is one of the first studies to suggest that a surplus of boron can stimulate cell division and growth. Other studies, such as those on suspension cultured tobacoo cells, have found growth decreases at high boron levels (Ghanati 2001). However, the tobacco cells did have thinner walls than those grown at physiological boron concentrations, which is consistent with our findings (Ghanati 2001).

We have also observed a decrease in cell size under boron deficient conditions, suggesting that cell expansion is hindered. Some researchers believe that boron's primary role is in expansion rather than division, and that the absence of cell division under boron deficient conditions is the result of abnormalities in the formation of the cell wall, which prevent the cell from becoming organized for mitosis (Whittington 1959; Dell 1997; Fleischer 1999). Greater than 90% of cellular boron is localized to the cell walls of cultured tobacco BY2 cells (Kobayashi 1997). The localization of boron to plant cell walls implies that boron may have a structural role in the cell wall matrix, or may be required for the synthesis of new cell wall material (Hu 1994). As such, it is here that we should expect to see the first signs of alterations in cell development as a direct result of changes in boron availability. We suggest that our observed changes in division and growth are a secondary response to boron, stemming from changes in the ultra-structure of the cell wall.

Organization of the cell wall

Pectin is deposited in the cell wall during primary growth and as such may play a critical role in determining the physical characteristics of the primary cell wall (Northcote 1986). As the first wall layer deposited, changes in its composition may affect further cell development, including division and expansion. This idea is supported by the reduction in cross-linking of the pectic polysaccharide rhamnogalacturonan-II (RG-II) observed as a result of boron deficiencies (Matoh

2000). Reduced boron-RG-II cross-linking is accompanied by the formation of swollen walls in suspension cultured tobacco cells (Matoh 2000). In boron deficient pumpkins, a reduction in growth and increase in cell wall thickening is also accompanied by a decrease in boron cross links (Ishii 2001). However, growth can be restored by re-supplying borate to the boron deficient plants, thereby reducing wall thickening and increasing the number of RG-II cross links (Ishii 2001). Sunflower (Stark 1963; Lee 1966; Hirsh 1980), and tomato (Kouchi 1976) also demonstrate thickening of the cell wall accompanied by an increasing degree of disorder of the middle lamella under low boron conditions. We also observed at increase in cell wall thickening and disorganization of the compound middle lamella (Figs. 4 and 5) under boron deficient conditions, which could be the result of altered B-RG-II cross links. Moreover, the tightening of the compound middle lamella observed in high boron conditions may also be the result of altered B-RG-II cross links.

Pectin esterification

Esterification has been shown to change during cell wall maturation, with a decrease in esterified pectin being linked to the cessation of growth and induction of wall stiffening, as a result of the formation of calcium cross links (Catesson 1994; Willats 2001; Vincken 2003). In our low boron cultures, we have seen a decrease in esterified pectin (Fig. 6B). If a decrease in esterified pectin correlates to a cessation of growth, this follows with the decrease in cell division that we have observed in the low boron cultures. Moreover, our high boron cultures experience an apparent increase in esterification. This means that fewer growth inhibitory cross links can form, which may be related to the increase in cell growth that we have observed. Furthermore, the high boron cultures display a decrease in de-esterified pectin (Fig. 6C) further suggesting the hindering of calcium cross-links and wall stiffening thus allowing for cell growth. This implies that proper formation of the cell wall, ie. deposition and esterification of pectins, may be required for cell division and growth to occur.

Pectin methylesterases

Pectin deposition in the cell wall is a highly regulated process which strengthens the theory that pectin, and its various constituents may possess specific functions at the level of cell wall development (Vincken 2003). Pectin is in a methylesterified form when it is transported to the cell wall in Golgi vesicles to prevent premature ionic interactions with its carboxyl groups (Goldberg 1996). Once in the cell wall, the methyl esters are removed by pectin methylesterases (Goldberg 1996; Willats 2001) allowing for the formation of a covalently linked macromolecular pectin network (O'Neil 1996; O'Neil 2004). These methylesterases may themselves be developmentally regulated (Willats 2001).

In our cultures, the increase in de-esterified pectin in the newly formed wood may be the cell's way of encouraging ionic interactions, in particular the cross-linking of

RG-II by boron. It has been suggested that pectin methylesterases have the potential to modify matrix properties in specific regions of the cell wall, in response to functional requirements (Willats 2001), suggesting that perhaps in response to low boron these enzymes are upregulated. Furthermore, if boron is capable of encouraging the deposition of pectin, then under excess boron conditions, the cell may be responding by decreasing the availability of binding sites by down regulating pectin methylesterases. In other words, the cell may be responding to the decrease in boron availability by decreasing pectin. Each of these scenarios may be a survival response on the part of the plant.

Alterations in lignification

We have observed alterations in lignification of the compound middle lamella, as well as the S_1 secondary cell wall (data not shown). In general, we have found that under boron deficient conditions there is a trend towards increased lignification (Fig. 7). This contradicts previous suggestions that poor lignification occurs under boron deficient conditions (Wardrop 1981). However, this may be due to different definitions of what constitutes lignin, as the same report suggested that there was an accumulation of lignin precursors under boron deficient conditions and other reports have shown that these lignin precursors can be polymerized to lignin (Wardrop 1981; Lapierre 1988; Jouanin 2000). In addition, there are early reviews that suggest the accumulation of phenolic compounds under boron deficient conditions (Dugger 1983). We are currently analyzing our samples using the klason and acetyl bromide lignin assays, as well as with pyrolysis mass spectrometry, to determine if the observed changes in lignification are true, or the result of a cross reaction with the staining.

The process of lignification can be divided into two distinct parts: formation of monolignols, and the polymerization of monolignols into lignin, both of which can be influenced by boron (Wardrop 1981). Furthermore, boron can influence the metabolic pathways preceding the formation of monolignols. We propose that boron has the potential to impede the pathways involved in lignin formation, resulting in the indirect control of lignification through interference with metabolic branch points.

Interaction between boron, pectin and lignification

We have observed changes in both the pectin composition and lignification of cultures grown on different boron concentrations. This suggests that boron functions to regulate both of these cell wall elements, effectively altering the composition of the cell wall. We suggest that the interaction of boron with the cell wall is much more complex than just cross-linking the pectins and creating stronger walls (Wingender 2001). For example, the alteration of pectin composition alone may indirectly modify the cell wall by changing lignification. During lignification, the xylem contains acidic pectins, which can lower the pH of the cell wall favouring a more condensed lignin structure in the compound middle lamella than in the secondary wall (Terashima 1988).

We suggest that our observed changes in division, expansion and lignification are secondary growth responses resulting from the interaction of boron with RG-II. Under low boron culture conditions, nearly all of the available boron may be bound by RG-II. We believe that the cultures respond by encouraging ionic interactions, and boron uptake, through increased de-esterification. This would result in the increased formation of calcium cross-links creating stiffer primary cell walls that do not favour division or expansion. However, this does not account for the disorganization of the compound middle lamella or the increase in wall thickness that we have observed. We would expect an increase in calcium cross-links to 'tighten' the primary cell wall. It may be that calcium cross-links strengthen the primary wall, but due to the size of calcium and the nature of its interactions it does not pull the pectin network together. The interaction of boron with RG-II may be required for this to occur. As such, a reduction in available boron and thus B-RG-II cross links may account for the compound middle lamella swelling. Moreover, if boron can influence metabolic processes such as lignification, the reduction in free boron would result in these processes continuing un-regulated (Fig. 8A), accounting for in the increase in lignification that we have observed. Since we expect boron to interact at the level of the primary cell wall, the observed changes in overall cell wall thickness are likely due to the increase in lignification.

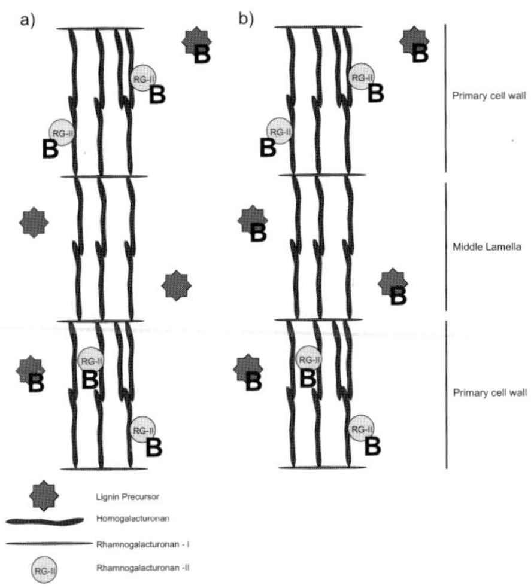

Figure 8. Simplified schematic of the potential relationship between boron, the pectin matrix and lignin. A) Under low boron conditions, nearly all of the available boron may bind to RG-II, resulting in a decrease in the boron available to regulate metabolic processes such as lignification. Likewise, under B) high boron conditions once all of the available RG-II is bound excess boron may be able to interact with lignin precursors resulting in a decrease in lignification

Likewise, under high boron culture conditions once all of the available RG-II is bound the excess boron would be able to interact with other components of the cell. We believe that the cultures respond by discouraging ionic interactions through increased esterification in an attempt to regulate boron uptake. This would result in a reduction in calcium cross-links, which would weaken the primary cell wall. Presumably, this would favour the increase in expansion that we have observed. Again, this does not account for the tightening of the compound middle lamella, or the reduction in overall wall thickness that we have observed. We would expect a decrease in calcium cross-links to result in a weaker pectin complex that would 'swell'. However, if boron is required to keep a compact pectin matrix, we would expect to observe a tighter compound middle lamella. Furthermore, excess boron may interact with lignin precursors resulting in decreased lignin production (Fig. 8B). This would account for the observed decrease in lignification we have observed under the high boron culture conditions, and is probably responsible for the decrease in wall thickness.

Acknowledgements

This project was supported by the Wood Quality Initiative Ltd., New Zealand. We would like to thank Dr. Paul Knox (University of Leeds) for providing the monoclonal antibodies JIM5 and JIM7, Bronwyn Smail (University of Otago Electron Microscopy Unit), Manfred Ingerfeld for technical assistance, and Dr. Juliet Gerrard and Dr. Brian Butterfield for assistance with preparation of the manuscript.

References

Catesson A.M., 1994 Cambial ultra-structure and biochemistry: changes in relation to vascular tissue differentiation and the seasonal cycle. International Journal of Plant Science. **155**, 251–261.

Chaffey N., 1999 Cambium: Old challanges - new opportunitues. Trees. **13**, 138–151.

Cohen M.S. and Lepper R., 1977 Effect of boron on elongation and division in squash roots. Plant Physiology. **59**(5), 884–887.

Dell B. and Huang L., 1997 Physiological response of plants to low boron. Plant and Soil. **193**, 103–120.

Dugger W.M., 1983 Boron in plant metabolism. Inorganic Plant Nutrition. A. Lauchi 627–650.

Fleischer A., O'Neil M.A., and Ehwald R., 1999 The pore size of non-graminaceous plant cell walls is rapidly decreased by borate ester cross-linking of the pectic polysaccharide rhamnogalacturonan-II. Plant Physiology. **121**, 829–38.

Ghanati F., Morita A., and Yokota H., 2001 Selection and characterization of a boron tolerant tobacco cell line. Soil Science and Plant Nutrition **47**, 405–410.

Goldberg R., Morvan C., Jauneau A., and Jarvis M.C., 1996 Methyl-esterfication, de-esterfication and gelation of pectins in the primary cell wall. Pectins and Pectinases. J. Visser, Voragen, A.G.J. Amsterdam, Elsevier Science, 151–172.

Higuchi T., 1997 Biochemisty and Molecular Biology of Wood Berlin.

Hirsh A.M. and a. T.J.G., 1980 Ultrastructural changes in sunflower root cell in relation to boron deficiency and added auxin. Canadian Journal of Botany **58**, 856–866.

Hu H. and Brown P.H., 1994 Localization of boron in cell walls of squash and tobacco and its association with pectin. Plant Physiology **105**, 681–89.

Ishii T., Matsunga T., and Hiyash N., 2001 Formation of rhamnogalacturonan II-borate dimer in pectin determines cell wall thickness in pumpkin tissue. Plant Physiology **126**, 1698–705.

Jouanin L., Goujon T., de Nadai V., Martin M.-T., Mila I., Vallet C., Pollet B., Yoshinaga A., Chabbert B., Petit-Conil M., and Lapierre C., 2000 Lignification in transgenic poplars with extremely reduced caffeic acid O-methyltransferase activity. Plant Physiology **123**, 1363–1373.

Kobayashi M., Ohno K., and Matoh T., 1997 Boron nutrition of cultured tobacco BY-2 cells. II. Characterization of the boron-polysaccharide complex. Plant Cell Physiology **38**(6), 676–683.

Kouchi H. and Kumazawa K., 1976 Anatomical responses of root tips to boron deficiency. III. Effect of boron deficiency on sub-cellular structure of root tips, particularly on morphology of cell wall and its related organelles. Soil Science and Plant Nutrition **41**, 53–71.

Lapierre C., Tollier M.T., and Monties B., 1988 A new type of constitutive unit in lignins from the corn bm3 mutant. Comptes Rendus de l'Academie des Sciences Serie III: Sciences de la Vie **307**, 723–728.

Lee S.G. and Aronoff S., 1966 Investigations on the role of boron in plants. III. Anatomical observations. Plant Physiology **41**, 1570–1577.

Matoh T., Ishigaki K., Mizutani M., Matsunga W., and Takabe K., 1992 Boron nutrition of cultured tobacco BY-2 cells. I. Requirement for and intracellular localization of boron and selection of cells that tolerate low levels of boron. Plant Cell Physiology **33**(8), 1135–1141.

Matoh T., Takasaki M., and Takabe K., 2000 Boron nutrition of cultured tobacoo BY-2 cells: III. Characterization of the boron rhamnogalacturonan Ii complex in cells acclimated to low levels of boron. Plant Cell Physiology **39**, 483–91.

Northcote D.H., 1986 Control of pectin synthesis and deposition during plant cell wall growth. Chemistry and function of pectin. M.L. Fishman, J.J. jen, Washington, American Chemical Society, 134–140.

O'Neil M.A., Ishii T., Albersheim P., and Darvill A.G., 2004 Rhamnogalacturonan-II: Structure and function of a borate cross-linked cell wall pectic polysaccharide. Annual Review of Plant Biology **55**, 109–39.

O'Neil M.A., Warrenfeltz D., Kates K., Pellerin P., Doco T., Darvill A.G., and Albersheim P., 1996 Rhamnogalacturonan-II, a pectic polysaccharide in the walls of growing plant cell, forms a dimer that is covalently linked cross-linked by a borate ester. Journal of Biological Chemsitry **271**(37), 22923–22930.

Savidge R.A., 1993 In vitro wood formation in 'chips' from merchantable stem regions of Larix larcina. International Association of Wood Anatomists Journal **14**(1), 3–11.

Stark J.R., 1963 Effect of boron on the cell wall structure in sunflower. Acta Societatis Botanicorum Poloniae **32**, 619–623.

Terashima N. and Fukashima K., 1988 Heterogeneity in formation of lignin - XI. An autoradiographic study on the heterogeneous formation and structure of pine lignin. Wood Science and Technology **22**, 259–270.

Vincken J.P., Schols H.A., Oomen R.J.F.J., McCann M.C., Ulvskov P., Voragen A.G.J., and Visser R.G.F., 2003 If homogalacturonan were a side chain of rhamnogalacturonan-I. Implications for cell wall architecture. Plant Physiology **132**, 1781–1789.

Wardrop A.B., 1981 Lignification and xylogenesis. Xylem cell development. Barnett. Kent J.R., Castle House Publications Ltd, 115–151.

Whittington W.J., 1959 The role of boron in plant growth. II. The effect of growth on the radicle. Journal of Experimental Botany **10**, 93–103.

Willats W.G.T., McCartney L., Mackie W., and Knox P., 2001 Pectin: cell biology and prospects for functional analysis. Plant Molecular Biology **47**, 9–27.

Willats W.G.T., Orfila C., Limberg G., Buchholt H.C., Van Alebeek G.W.M., Voragen G.J., Marcus S.E., Christensen T.M.I.E., Mikkelsen J.D., Murray B.S., and Knox J.P., 2001 Modulation of the degree and pattern of methyl-esterification pf pectic homogalacturonan in plant cell walls. Journal of Biological Chemsitry **276**(22), 19404–19413.

Wingender Q.Y., Schulz M., Baluska F., and Goldbach H.E., 2001 Short-term boron deprivation induces increased levels of cytoskeletal proteins in Arabidopsis roots. Plant Biology **3**, 335–340.

Influence of Boron on Al Absorption and Ca Release of Root Border Cells of Pea (*Pisum sativum*)

Min Yu[1] and Heiner E. Goldbach[2]

[1]Department of Horticulture, Foshan University, Foshan, 528000, Guangdong, China, E-mail: yumin@fosu.edu.cn;

[2]INRES-Plant Nutrition, University of Bonn, Karlrobert–Kreiten–Str. 13, D-53115 Bonn, Germany, E-mail: h.goldbach@uni-bonn.de

Introduction

Root border cells possess special physiological properties and biological significance, being distinct from the root cap cells. This cell population detaches from root tips and rapidly suspends in water once the root tips are immersed. They can survive in a wide range of osmotic pressure including distilled water for a relatively long time without bursting or any visible damage (Brigham et al. 1995). These unusual cells form a biological interface between root tip surface and soil. Most tested plant species produce root border cells, and usually over 90% of them remain viable when detached (Brigham et al. 1995).

Both the early responses of B deficiency and the most prominent responses of higher plants to Al toxicity are expressed as an inhibition of root elongation. Since an early response to Al toxicity and B deprivation is the disorganization and destabilization of the cytoskeleton (Blancaflor et al. 1998; Sivaguru et al. 1999; Schwarzerova et al. 2002; Yu et al. 2002, 2003), the cytoskeleton-plasma membrane-extracellular matrix continuum is assumed to be a primary target of Al^{3+} toxicity and B deficiency (Horst et al. 1999; Goldbach et al. 2001; Schwarzerova et al. 2002; Bassil et al. 2004). It has been observed that addition of boron beyond the metabolic need of the plant prevented Al toxicity in some species (Lenoble et al. 1991, 1996 a, b). It is thus hypothesized that B may alleviate Al toxicity in root border cells. These cells might have crucial roles in protecting root tips from Al toxicity (Hawes et al. 2000; Miyasaka and Hawes, 2001). We reported about the alleviation of Al-induced border cell death by elevated B levels (Liu et al. this volume), and below, we will focus on the effects of B and Al on Ca release and Al absorption of border cell suspensions." Border cells were isolated in large quantities by a newly designed simple device (Yu et al. 2006, in print).

Materials and methods

Germination of root tips of pea

Pea (*Pisum sativum*) seeds were germinated according to Brigham et al. (1995), modified in the manner described subsequently. The seeds were immersed in 95% ethanol for 10min, thereafter in 5.25% sodium hypochlorite for 30min, and then rinsed 6

F. Xu et al. (eds.), Advances in Plant and Animal Boron Nutrition, 63–68.

times in deionised water. Only unaltered seeds were kept and immersed in 0.5mM $CaSO_4$ for 6 h and then evenly spread on the mesh screen of the mist culture device and germinated in the semi-closed plastic tank with 30 sec' fog exposition every 5 min. for 48h (Yu et al. unpublished data). 0.5mM $CaSO_4$ was added as well for fog producing ultra-pure water to avoid membrane and cell wall damage.

Influence of B on Al absorbed and Ca released in border cells of Pea

B level was varied by pre-soaking seeds in 0 (control) and 1 mM boric acid (+B) and then germinated in 0 (control) and 0.1mM boric acid (+B), respectively. Border cells were harvested by immersing the root tips in B free ultra pure water and gently shaking the root tips. The detached border cells present in the solution were concentrated by centrifugation at 300 g for 5 min. The pellets were rinsed by ultra-pure water for 3 times. The number of purified border cells was counted by a light microscope. Purified border cells (approximately 3×10^5cells.replication^{-1}) were incubated in 4 ml 0, 50 μM 500 $AlCl_3$ solution (pH 4.5, 4 replication) for 90min (4 replicates). Thereafter, border cells were rinsed once in 50 volumes of millipore water and pelleted by centrifugation at 300 g for 5 min. The pelleted cells were wet digested in Teflon pressure bombs at 90°C for 12h with 1ml ultra-pure concentrated nitric acid. Al and Ca in the ash solutions and in the supernatant were determined by flameless AAS in a graphite furnace (Perkin-Elmer Zeeman/3030 AAS) and AAS (Eppendorf ELEX 6361). Boron contents were determined colorimetrically by the miniaturized curcumin method (Wimmer and Goldbach, 1999a).

Results

Boron content in germinated pea seeds

Boron content was significantly ($P<0.01$) increased by pre-soaking seeds in 1 mM boric acid solution and germinating them in 0.1 mM boric acid fog (Table 1).

Table 1. Boron content (mg·kg^{-1}) in germinated pea seeds with and without B addition

	Control	+B
Boron content (mg/Kg)	6.05 ± 0.35	10.22 ± 0.38

Al absorbed by root border cells at different B levels

Aluminium absorbed by border cells was calculated by the differences of Al concentration before and after incubating these cells in 0, 50 and 500 μM $AlCl_3$. Significantly less Al ($P<0.01$) was absorbed by border cells in +B compared to the control (Fig. 1). Al absorbed by border cells was also significantly higher at 500 than at 50 μM $AlCl_3$.

Al content in root border cells at different B levels

Aluminium content in border cells was determined after rinsing once in 50 volumes of ultrapure water, which releases Al that is only loosely absorbed on the cell surface (see Materials and Methods). The addition of B resulted in a significantly ($P<0.01$) lower Al content in border cells (Fig. 2). This indicates that less Al was adsorbed by border cells at the higher B level.

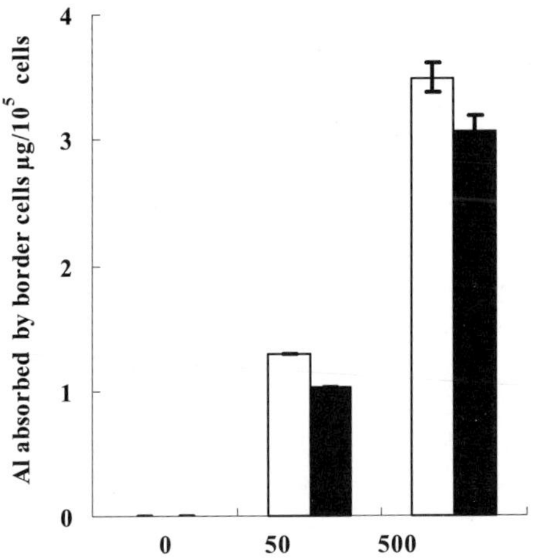

Figure 1. Al (µg) absorbed per 10^5 border cells as calculated by differences of Al concentration before and after 90 min. incubation of border cells at a ratio of 3×10^5 cells per 4 mL incubation solution at 0, 50, and 500 µM $AlCl_3$ (X-axis) (pH 4.5) Open columns: pre-treatment without, closed columns: soaking in 1mM B and germination in 0.01 mM B

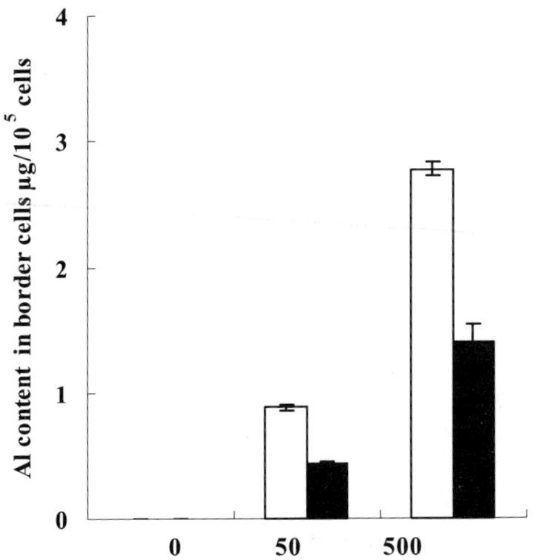

Figure 2. Content of Al (µg) in 10^5 cells of pea after 90 min. incubation of border cells at a ratio of 3×10^5 cells per 4 mL incubation solution at 0, 50 , and 500 µM $AlCl_3$ (X-axis) (pH 4.5). Open columns: pre-treatment without B, closed columns, soaking in 1 mM B and germinating in 0.01 mM B

Ca released from root border cells at different B levels

When incubating border cells with B (+B), significantly ($p<0.01$) less Ca was released into the incubation solution containing 50 and 500μM $AlCl_3$ (Fig. 3). Thus, addition of B apparently reduced the exchange of Ca by Al. The reduction of Ca release was more pronounced at 50 μM Al than at 500 μM ($p<0.01$), and the proportion of Ca released at 500 μM Al was much lower. This might be attributed to the very dense and thick mucilage formed around the pea border cells at high Al levels, which also reduced the penetration of larger dye molecules (Yu et al. 2005).

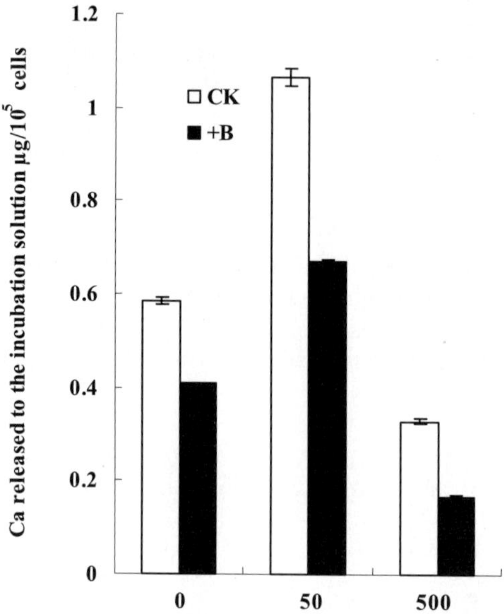

Figure 3. Ca (μg) released from 10^5 border cells of pea after 90 min. incubation of border cells at a ratio of 3×10^5 cells per 4 mL incubation solution at 0, 50 , and 500 μM $AlCl_3$ (X-axis) (pH 4.5). Open columns: pre-treatment without B, closed columns, soaking in 1mM B and germinating in 0.01 mM B

Discussion

Displacement of Ca^{2+} ions may be one (or even the only?) major factor of Al toxicity (Hanson, 1984; Kinraide and Parker, 1987; Kinraide et al. 1994; Horst, 1995; Ryan et al. 1997). It is likely that this affects the cytoskeleton-plasma membrane-extracellular matrix continuum (Horst et al. 1999; Goldbach et al. 2001; Schwarzerova et al. 2002; Bassil et al. 2004). In border cells, too, we observed a displacement of Ca, similar to the effect observed for the remaining root system (Hanson, 1984; Kinraide and Parker, 1987; Kinraide et al. 1994; Horst, 1995; Ryan et al. 1997). Higher B levels reduced the Ca displacement at both 50 and 500 μM $AlCl_3$ to some extend and increased cell viability (Liu et al. this volume). This indicates that higher B levels are able to stabilize the bonding of apoplastic Ca, likely by a higher degree of cross-linking of cell wall polymers (RGII?) and thus the integrity of border cell (wall) structure (see also Wimmer and

Goldbach, 1999b; Mühling et al. 1998). This may provide an explanation for the alleviation of Al toxicity by enhanced B supply for entire root systems (LeNoble et al. 1991, 1996 a,b), especially in a soil system where border cells remain mostly attached to the root surface (in contrast to hydroponic systems). It still needs to be tested, whether there are further ligands, eventually other than RGII, to complex with B in the cell wall of border cells.

The proportion of Ca released into the incubation solution was found to be significantly lower at 500μM $AlCl_3$. This might be attributed to the significantly thicker mucilage which possibly provides a physical barrier for solute exchange *in situ* (Liu et al. this volume; Yu et al. 2005).

Border cells are likely involved in the alleviation of Al toxicity. The underlying mechanism(s) as well as their apoplastic B binding ligands still need further research.

Acknowledgement

Financial support by National Natural Science Foundation of China (NSFC, No 30200167 and 30471035) and Institute of Plant Nutrition, University of Bonn and the scholarship from DAAD are gratefully acknowledged.

References

Bassil E., Hu H., and Brown P.H., 2004 Use of phenylboronic acids to investigate boron function in plants. Possible role of boron in transvacuolar cytoplasmic strands and cell-to-wall adhesion. Plant Physiol. **136**, 3383–3395.

Blancaflor E.B., Jones D.L., and Gilroy S., 1998 Alterations in the cytoskeleton accompany aluminium induced growth inhibition and morphological changes in primary roots of maize. Plant Physiol. **118**, 159–172.

Brigham L.A., Woo H.H., and Hawes M.C., 1995 Root border cells as tools in plant cell studies. Methods Cell Biol. **49**, 377–387.

Goldbach H.E., Yu Q., Wingender R., Schulz M., Wimmer M.A., Findeklee P., and Baluška F., 2001 Rapid response reactions of roots to boron deprivation. Plant Nutr Soil Sci **161**, 173–178.

Hanson J.B., 1984 The functions of calcium in plant nutrition. In: P.B. Tinker, and A. Läuchli: Advances in Plant Nutrition, Vol 1.Praeger Scientific, New York, pp. 149–208.

Hawes M.C., Gunawardenaa U., Miyasaka S., and Zhao X., 2000 The role of root border cells in plant defense. Trends in Plant Sci. **5**, 128–133.

Horst W.J., 1995 The role of the apoplast in aluminium toxicity and resistance of higher plants: a review. Z. Pflanzenernaehr Bodenkd **158**, 419–428.

Horst W.J., Schmohl N., Kollmeier M., Baluška F., and Sivaguru M., 1999 Does aluminium affect root growth of maize through interaction with the cell wall-plasma membrane-cytoskeleton continuum? Plant Soil **215**, 163–174.

Kinraide T.B. and Parker D.R., 1987 Cation amelioration of aluminum toxicity in wheat. Plant Physiol. **83**, 546–551.

Kinraide T.B., Ryan P.R., and Kochian L.V., 1994 Al^{3+}-Ca^{2+} interactions in aluminum toxicity. II. Evaluating the Ca^{2+} displacement hypothesis. Planta **192**, 104–109.

LeNoble N., Blevins D.G., and Miles J.R., 1991 Mineral composition of alfalfa roots in high aluminum soils with additional boron. In D.D. Randall et al. (eds) Current topics in Plant Biochemistry and Physiology **10**: 312.

LeNoble N., Blevins D.G., and Miles J.R., 1996 a Prevention of aluminium toxicity with supplemental boron. 1. Maintenance of root elongation and cellular structure. Plant Cell Environ. **19**: 1132–1142.

LeNoble N., Blevins D.G., and Miles J.R., 1996 b Prevention of aluminium toxicity with supplemental boron. 2. Stimulation of root growth in acidic, high-aluminium subsoil. Plant Cell Environ. **19**: 1143–1148.

Miyasaka S.C. and Hawes M.C., 2001 Possible role of root border cells in detection and avoidance of aluminum toxicity. Plant Physiol. **125**, 1978–1987.

Mühling K.H., Wimmer M., and Goldbach H.E., 1998, Apoplastic and membrane-associated Ca^{2+} in leaves and roots as affected by boron deficiency. Physiol. Plant. **102**, 179–184.

Ryan P.R., Reid R.J., and Smith F.A., 1997 Direct evaluation of the Ca^{2+}-displacement hypothesis for Al toxicity. Plant Physiol. **113**, 1351–1357.

Schwarzerova K., Zelenkova S., Nick P., and Opatrny Z., 2002 Aluminum-induced rapid changes in the microtubular cytoskeleton of tobacco cell lines. Plant Cell Physiol **43**, 207–216.

Sivaguru M., Baluška F., Volkmann D., Felle H.H., and Horst W.J., 1999 Impacts of aluminum on the cytoskeleton of the maize root apex: short-term effects on the distal part of the transition zone. Plant Physiol **119**, 1073–1082.

Wimmer M.A. and Goldbach H.E., 1999 a A miniaturized curcumin method for the determination of boron in solutions and biological samples. J. Plant Nutrition Soil Sci. **162**: 15–18.

Wimmer, M.A. and Goldbach H.E., 1999 b Influence of Ca^{2+} and pH on the stability of different boron fractions in intact roots of Vicia faba L. Plant Biol. **1**: 632–637.

Yu M., Feng Y.M., and Goldbach H.E., 2006 Mist culture for mass harvesting of root border cells: Al effects. J. Plant Nutrition Soil Sci. (In print).

Yu M., Feng Y.M., Wen H.X., Xiao H.D., Zhang Y.H., and He L.L., 2005 Influence of Al on cell viability and mucilage secretion of root border cells of pea (Pisum sativum). In FS Zhang et al. (eds). Plant Nutrition for Food Security, Human Health and Environmental Protection. The proceeding of 15th IPNC, Beijing, China, pp. 726–727.

Yu Q., Hlavaćka A., Matoh T., Menzel D., Volkmann D., Goldbach H.E., and Baluška F., 2002 Short-term boron deprivation inhibits endocytosis of cell wall pectins in meristematic cells of maize and wheat root apices. Plant Physiol. **130**, 415–421.

Yu Q., Baluška F., Jasper F., Menzel D., and Goldbach H.E., 2003 Short-term boron deprivation enhances levels of cytoskeletal proteins in maize, but not zucchini, root apices. Physiol Plant. **117**, 270–278.

Influence of Boron and Aluminum on Production and Viability of Root Border Cells of Pea (*Pisum sativum*)

Jiayou Liu[1, 2], Min Yu[*2], Changquan Wang[1] and Yingming Feng[2]

[1]College of Resources and Environment, Sichuan Agricultural University, Ya'an, China

[2]Department of Horticulture, Foshan University, Foshan, Guangdong, China;
*Correspondence: yumin@fosu.edu.cn

Introduction

Root border cells are released from the root apices of most plant species, and function crucially in the rhizosphere through their ability to modify its chemical and physical properties. After detachment from the root cap, root border cells have already served several functions. They can dramatically alter the behavior of populations of rhizosphere microflora (Hawes and Brigham 1992; Hawes et al. 1998; Hawes et al. 2000). These detached cells can reduce the mechanical friction of the growing root under some conditions (Hawes et al. 1998; Iijima et al. 2004) and protect the tip by repelling bacteria.

It is believed Al toxicity is the main factor limiting plant productivity on soils with pH ($CaCl_2$) below 4.5 (Foy 1988), which constitute > 40% of the world's arable land (Kochian, 1995). The initial and most dramatic symptom of Al toxicity is inhibition of root growth, which results in a stunted and damaged root system and can lead to mineral deficiencies and water stress (Ryan et al. 1993; Delhaize and Ryan, 1995; Kochian, 1995). The root apex is the primary target of Al toxicity, and the reduction in root growth is detectable within minutes after Al addition (Ryan et al. 1993; Jones and Kochian, 1995). Horst et al. (2000) showed that Al-induced injury was closely correlated to the loss of cell viability.

Boron is an essential micronutrient for growth and development of higher plants (Marschner 1995). Lenoble et al. (1996) found that soil B application promoted root growth of alfalfa in an acidic, high-Al subsoil and supra-optimal B concentrations in nutrient media prevented inhibition of root and shoot growth of Al-stressed squash. On the other hand, Taylor and MacFie (1994) showed that B did not alleviate Al toxicity symptoms in wheat. Blevins and Lukaszewskj (1998) claim that inhibition of plant Al toxicity as a result of B depends on B level in nutrient solution and plant species. Recently, the responses of higher plants to Al toxicity were related to organic acids and large molecules released by roots, but thousands of metabolically active border cells that are a major constituent of the root tip capsule were ignored in these studies. Border cells are released from healthy young roots and their production is tightly regulated by the root in response to environmental and endogenous signals (Miyasaka and Hawes, 2001). Since the root tip is the target of A1 toxicity and B deficiency, it was postulated that border cells might function crucially in the

F. Xu et al. (eds.), Advances in Plant and Animal Boron Nutrition, 69–74.

response to A1 toxicity and B supply. In the current study, production and viability of root border cells of pea (*Pisum sativum)* pre-treated by mist containing different levels of Al for 24 h at different levels of B (0, 0.1, 1, 2.5 mM) were examined to test the hypothesis that B could alleviate Al induced cell death.

Materials and methods

Preparation of the seeds and culture of root border cell

Pea (*Pisum sativum*) seeds were germinated according to Brigham et al. (1995), modified in the manner described subsequently. The seeds were immersed in 95% ethanol for 10 min, thereafter in 5.25% sodium hypochlorite for 30 min, and then rinsed 6 times in deionised water. Only unaltered seeds were retained. Surface sterilized seeds were soaked in 0, 0.1, 1, and 2.5 mM B solution for 8 h, then germinated by mist culture (Yu et al. in print) with / without identical boric acid solution containing 0.5 mM $CaSO_4$ in a 24°C growth chamber for 24 h, finally different Al levels (0, 50, 500 μM, pH 4.5) were added to the mist solution for another 24 h.

Sample analysis

Root border cells from three tips of the intact root were collected in 1.5 mL sterile distilled water containing 0.5 mM $CaSO_4$. The production and viability of root border cells of pea was detected under microscopic examination (100 X magnification) by the trypan blue exclusion test immediately after staining. Boron contents were determined colorimetrically by the miniaturized curcumin method (Wimmer and Goldbach, 1999). Analysis of variance was conducted using Statistical Analysis Systems (6.12) programs to determine treatment and interaction effects.

Results

Effect of B supply on content of B in pea (Pisum sativum) seeds

Boron concentrations in germinated seeds was significantly increased with the increase of B levels in seed soaking solution for 8 h with or without 48 h mist with B solutions of the same concentrations ($P<0.05$), and it displayed a dose dependent increase. Boron concentrations in seed after adding B in soaking solution and mist solution was significantly ($P<0.05$) higher than that only treated with seed soaking (Fig. 1).

Influence of B and Al on viability of root border cells of pea

The higher levels of B in germinated seeds resulted in significantly ($P<0.05$) higher viability of border cells in mist culture in a dosage dependent manner and it was highest at 0.1 mM B. The viability of border cells was significantly ($P<0.05$) decreased by 50 and 500 μM Al in mist, with a lowest viability at 50 μM Al. Interactions between B and Al overall, and at any B and Al level was significant (at least P<0.05). These results indicated that cell viability was enhanced by B at proper levels and B alleviated Al induced cell death.

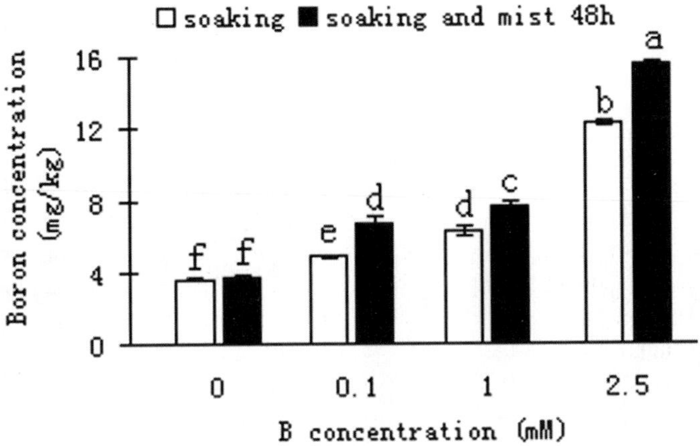

Figure 1. Concentration of B in pea seedlings treated with different B levels (0, 0.1, 1, 2.5 mM) by seed soaking and / or mist culture. Means with different lowercase letters are significantly different (P<0.05, F test). Vertical bars are standard errors of five independent replicates

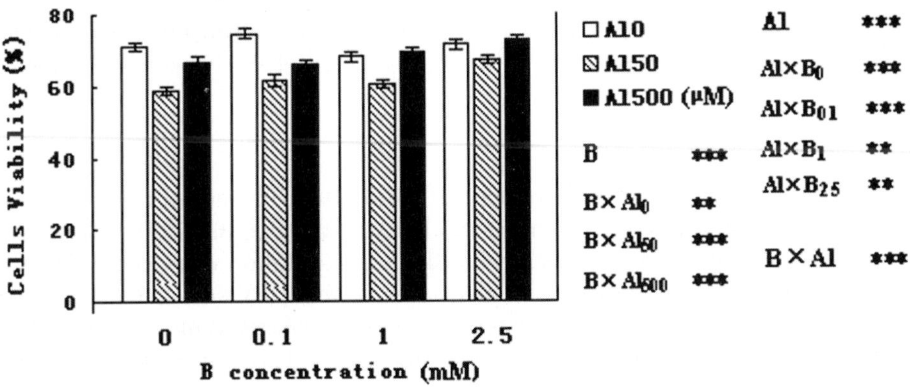

Figure 2. Viability of root border cells as affected by B and Al supply in mist solution containing 0.5 mM $CaSO_4$, pH 4.5. Vertical bars are standard errors of five independent replicates. Significant differences between mean values were found (F test), n=5. *, **, and***, significant at the P<0.05, 0.01, and 0.001 according to the F test. ns, Not significant

Influence of B and Al on production of root border cells of pea

The number of border cells significantly (P<0.05) decreased with increasing solution B, suggesting that higher B levels inhibited border cell production. Cell number dropped with the addition of Al in mist solution at 0 B and it was significantly lower at 50 μM Al. This indicated that Al inhibited border cell production with the benefit of added solution B. No interactions between B and Al were found.

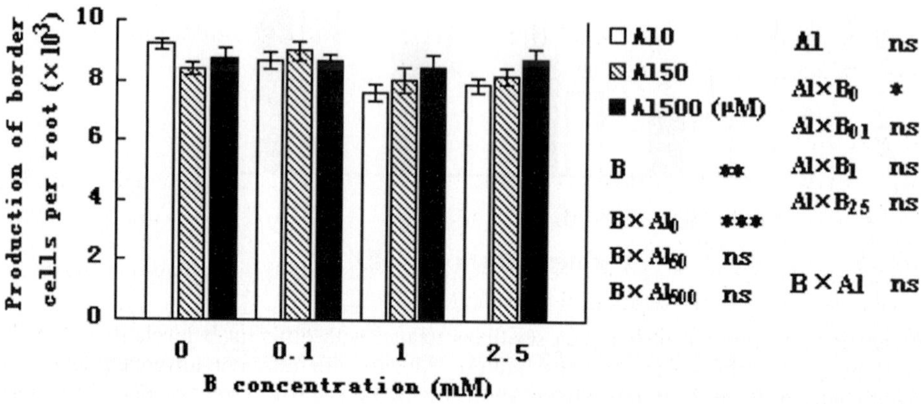

Figure 3. The production of root border cells as affected by B and Al concentrations in mist solution containing 0.5 mM $CaSO_4$, pH 4.5. Vertical bars are standard errors of five independent replicates. Significant differences between mean values were found (F test), n=5. *, **, and***, significant at the P<0.05, 0.01, and 0.001 according to the F test. ns, Not significant

Discussion

Evidence suggesting an involvement of B in the avoidance of Al toxicity has gradually been strengthening (Lenoble et al. 1996; Yang et al. 2004). Our research provided evidence that B pretreatment with in mist solution enhanced B levels in seedling. An important advantage of mist culture is that effects of Al induced border cell death and the decline in production in a dosage dependent manner (more at 50 than at 500 μM) was more precise than previous reports (Miyasaka and Hawes, 2001; Zhu et al. 2003). Interestingly, it was found that B enhanced border cell viability but inhibited its production in a dosage dependent manner (more at 0.1 mM than at 1 and 2.5mM) and there was significant interaction between B and Al in border cell viability but not in their production. Thus it can be concluded that B alleviated Al induced border cell death which might be related to their effects on function of the cell wall—the target of B deficiency and Al toxicity (Horst et al. 1999; Goldbach et al. 2001; Schwarzerova et al. 2002; Bassil et al. 2004). It was speculated that

RG-II-B formed in the cell wall was required to maintain the normal pore structure of the wall matrix and to mechanically stabilize the cell wall (Fleisher et al. 1998; Fleisher et al. 1999; Bastias et al. 2004). More KDO were found with B addition which might preclude Al from entering into the cell (Matoh et al. 1996; Liu et al. unpublished data). The involving of B and Al in cell wall function of border cells, however, needs further investigation.

Acknowledgement

Financial support by National Natural Science Foundation of China (NSFC, No 30471035 and 30200167) are gratefully acknowledged.

References

Blevins D.G. and Lukaszewski K.M., 1998 B in plant structure and function. Annu. Rev. Plant. Physiol. Plant. Mol. Biol. **49**: 481–500.

Bassil E., Hu H., and Brown P.H., 2004 Use of phenylboronic acids to investigate boron function in plants. Possible role of boron in transvacuolar cytoplasmic strands and cell-to-wall adhesion. Plant Physiol. **136**: 3383–3395.

Bastías E.I. et al., 2004 Zea mays L. amylacea from the Lluta Valley (Arica-Chile) tolerates salinity stress when high levels of boron are available. Plant and Soil. **267**: 73–84.

Brigham L.A., Woo H.H., and Hawes M.C., 1995 Root border cells as tools in plant cell studies. Methods In Cell Biology. **49**: 377–387.

Brown P.H. and Hu H., 1994 B uptake by sunflower, squash and cultured tobacco cells. Physiologia Plantarum. **91**: 435–441.

Delhaize E. and Ryan P.R., 1995 Aluminium toxicity and tolerance in plants. Plant. Physiol. **107**: 315–321.

Fleischer A., Titel C., and Ehwald R., 1998 The B requirement and cell wall properties of growing and stationary suspension-cultured Chenopodium album L cells. Plant Physiol. **117**: 1401–1410.

Fleischer A., O'Neil M.A., and Edwald R., 1999 The pore size of non-graminaceous plant cell walls is rapidly decreased by borate easter cross-linking of the pectic polysaccharide rhamnogalacturonan II. Plant Physiol. **121**: 829–838.

Foy C.D., 1988 Plant adaption to acid, aluminum-toxic soils. Commun. Soil. Sci. Plant. Ann. **19**: 959–987.

Goldbach H.E., Yu Q., Wingender R., Schulz M., Wimmer M.A., Findeklee P., and Baluška F., 2001 Rapid response reactions of roots to boron deprivation. Plant Nutr Soil Sci. **161**: 173–178.

Hawes M.C. and Brigham L.A., 1992 Impact of root border cells on microbial populations in rhizosphere. Adv. Plant. Pathol. **8**: 119–148.

Hawes M.C. et al., 1998 Function of root border cells in plant health: pioneers in the rhizosphere. Annu Rev Phytopathol. **36**: 311–327.

Hawes M.C, Gunawardena, and Miyasaka S., 2000 The role of root border cells in plant defense. Trends in Plant Science. **5**: 128–133.

Horst W.J. and Sdhmohl N., 2000 Cell wall pectin content modulates aluminum sensitivity of Zea mays (L) cells grown in suspension culture. Plant Cell and Environment. **23**: 735–742.

Horst W.J., Schmohl N., Kollmeier M., Baluška F., and Sivaguru M., 1999 Does aluminium affect root growth of maize through interaction with the cell wall-plasma membrane-cytoskeleton continuum? Plant Soil. **215**: 163–174.

Iijima M., Higuchi T., and Barlow P.W., 2004 Contribution of root cap mucilage and presence of an intact root cap in maize (Zea mays) to the reduction of soil mechanical impedance. Ann. Bot (Lond). **4**: 473–487.

Jones D.L. and Kochian L.V., 1995 Aluminum inhibition of the 1,4,5-trisphosphate signal transduction pathway in wheat roots: a role in aluminum toxicity. Plant Cell. **7**: 1913–1922.

Kochian L.V., 1995 Cellular mechanisms of aluminum toxicity and resistance in plants. Annu. Rev. Plant. Physiol. Mol. Biol. **46**: 237–260.

Lenoble M.E., Blevins D.G., and Miles R.J., 1996 Prevention of aluminum toxicity with supplemental B. II. Stimulation of root growth in an acid, high-aluminum subsoil. Plant Cell Environ. **19**: 1143–1148.

Marschner H., 1995 Mineral Nutrition of Higher Plants. 2nd Ed Academic Press, London, pp 889.

Matoh T., Kawaguchi S., and Kobayashi M., 1996 Ubiquity of a borate-rhamno-galaacuronan-Il complex in the cell walls of higher plants. Plant Ce11 Physiol. **37**: 636–640.

Miyasaka S.C. and Hawes M.C., 2001 Possible role of root border cells in detection and avoidance of aluminum toxicity. Plant. Physiol. **125**: 1978–1987.

Ryan P.R., DiTomaso J.M., and Kochian L.V., 1993 Aluminum toxicity in roots: an investigation of spatial sensitivity and the role of the root cap. J. Exp. Bot. **44**: 437–446.

Schwarzerova K., Zelenkova S., Nick P., Opatrny Z., 2002 Aluminum-induced rapid changes in the microtubular cytoskeleton of tobacco cell lines. Plant Cell Physiol. **43**: 207–216.

Seresinhe P.S.J.W. and Oertli J.J., 1991 Effects of B on growth of tomato cell suspensions. Physiologia Plantarum. **81**: 31–36.

Taylor W.K. and MacFie S.M., 1994 Modelling the potential for B amelioration of aluminum toxicity using the Weibull function. Can. J. Bot. **72**: 1187–1196.

Wimmer M.A. and Goldbach H.E., 1999 A miniaturized curcumin method for the determination of boron in solutions and biological samples. Plant Nutrition Soil Sci. **162**: 15–18.

Yong-Hua Yang et al., 2004 Effects of B on aluminum toxicity on seedlings of two soybean cultivars. Water, Air & Soil Pollution. **154**: 239–248.

Yu M., Feng Y.M., and Goldbach H.E., 2006 Mist culture for mass harvesting of root border cells: Al effects. Plant Nutrition Soil Sci. (In print).

Zhu M.Y., Ahn S.J., and Matsumoto H., 2003 Inhibition of growth and development of root border cells in wheat by Al. Physiologia Plantarum. **117**: 359–367.

Differences in Membrane Lipid Peroxidation, Activities of Protective Enzymes and Polyamines Contents in Leaves between Two Cotton Cultivars with Different Boron Efficiency

Mingjian Geng, Lishu Wu, Xiangyun Cao and Wuding Liu

Microelements Research Center, Huazhong Agricultural University, Wuhan, China

Introduction

Boron has been proved an essential element for higher plants growth since 1923 (Warington 1923), but its primary physiological function in plants is yet unclear. Increasing number of experiments supported the idea that boron play an important role in plasma membrane integrity by binding membrane compounds containing *cis*-diol groups, or affecting activities of some enzymes related to membrane function or structure, or participating phenolics metabolism (Blevins and Lukaszewski 1998, Brown et al. 2002, Goldbach et al. 2002). Cakmak et al. (1997) reported that the major defense system of cells against toxic active oxygen was reduced in response to boron deficiency. It is possible that supplied boron might protect plasma membranes against peroxidative damage by the toxic active oxygen.

Boron deficiency caused accumulation of polyamines in tobacco plants (Camacho-Cristobal et al. 2004, 2005). It was reported that polyamines might stabilize cell membrane by promoting the activities of protective enzymes or directly scavenging the toxic active oxygen generated in cell metabolism processes (Jiang et al. 1993; Drolet et al. 1986). Nevertheless, it was not clear whether the polyamines play a role in protecting plant cell membrane from peroxidative damage under boron deficiency stress.

Experiments were conducted to determine membrane permeability, membrane peroxidation, activities of protective enzymes and contents of polyamines in two cotton cultivars with different boron efficiencies under boron deficient and sufficient conditions.

Material and methods

Plant materials and culture

The cotton seedlings were grown on soil. The boron-efficient cotton cultivar '9702' and boron-inefficient one '9706' were selected from 103 upland cotton cultivars (*Gossypium hirsutum* L.) during the period of 1994 to 1996 (Cao et al. 1996). The soil used in the pot experiments was purple calcareous arenosol sampled from Xinzhou County, Hubei Province, China. The basic agrochemical property of the soil was: pH 8.0, organic matter 4.9 g/kg, total N 0.27 g/kg, alkali-hydrolyzable

F. Xu et al. (eds.), Advances in Plant and Animal Boron Nutrition, 75–81.

N 2.0 mg/kg, available P 7.0 mg/kg, available K 69 mg/kg, hot water soluble B 0.13 mg/kg.

The germinated cotton seeds were sowed in plastic boxes with 7.0 kg soil. Every kilogram soil was applied 0.70 g $(NH_4)_2SO_4$, 0.29 g KH_2PO_4, 0.16 g KCl, 0.25 g $MgSO_4.7H_2O$, 1.81 mg $MnCl_2.4H_2O$, 0.22 mg $ZnSO_4.7H_2O$, 0.08 mg $CuSO_4.5H_2O$ and 0.09 mg $Na_2MoO_4.2H_2O$, all fertilizers were analytical reagents. The cotton plants were treated with 2 boron levels, 0.13 mg B kg^{-1} (-B) and 1.0 mg B kg^{-1} (+B), each treatment was replicated 4 times. The cotton plants were grown in a greenhouse, irrigated with distilled water. When the plant was at the 5-leaves-stage, the leaves were sampled for determining the physiological index as following.

Determination of membrane lipid peroxidation

According to the method reported by Li et al. (2000), the extent of membrane lipid peroxidation was denoted with the content of malondialdehyde (MDA), an end product of lipid peroxidation. 0.5 g leaves sample was homogenized in 5 mL 5% trichloroacetic acid (TCA), then centrifuged at 4000 g for 10 min. 2 mL of the supernatants were mixed with 2 mL 5% TCA containing 0.6% thiobarbituric acid and boiled for 15 min, followed by quick cooling and centrifuged at 4000 g for 10 min again. The supernatant was measured at 532 nm and 450 nm with a spectrophotometer, the content of MDA was calculated with its extinction coefficient.

Determination of membrane permeability

Referring to the method of Li et al. (2000), circle slices of 1 cm in diameter were cut from the leaves of cotton and incubated in 20 mL distiller water for 3h at room temperature. After the incubation, the electrical conductivity of the solution was determined. The samples were then boiled for 20 min, the electrical conductivity of the solution was determined again. The membrane permeability of cotton leaves was expressed as percentage of the electrical conductivity of the solute leaked from the leaves before and after boiling.

Determination of composition of membrane fatty acid

The membrane fatty acid was extracted by the method reported by Su et al. (1980). A half gram of scissoring leaves were kept in oven at 100 °C for 5 min to inactivate enzymes and homogenized in the mixture of chloroform and methanol (1:2). The homogenate was repeatedly rinsed with petroleum ether until neutral fat was removed. The polar fat was methylated and then concentrated under the condition of decompression. The composition of membrane fatty acid was determined by gas chromatography.

Determination of protective enzyme activities

The activities of protective enzymes were measured according to the methods described by Li et al. (2000). One gram of leaf samples was homogenized in 5 mL of cold phosphate buffer solution (50 mM, pH7.8) in an ice bath, the homogenate was centrifuged at 10,500 g for 20 min, the supernatant was used as the enzyme extract. The activity of SOD was expressed by its ability to inhibit the photochemical reduction of nitro blue tetrazolium (NBT), one unit of SOD activity was defined as the amount of enzyme per a gram of fresh sample that caused 50% inhibition of the rate of NBT reduction. The CAT activity was measured by the method of sodium hyposulfite titration. One unit of CAT activity was defined as the amount of H_2O_2 decompounded by the enzyme per gram of fresh sample per min. The POD activity was determined by the method of guaiacol oxidation, one unit of POD activity was defined as the amount of H_2O_2 oxidated by the enzyme per a gram of fresh sample per min.

Determination of contents of polyamines

According to the method of Flores et al. (1982), 0.5 g of fresh leaves was homogenized with cold perchloric acid in an ice bath for 1 h, and then centrifuged at 26,000 g for 20 min at 4 ℃. The supernatant was derivatized with dansyl chloride, and the dansyl polyamines were extracted in 0.5 mL benzene. The composition of the dansylated polyamines in the extract was separated by thin-layer chromatography, then every band corresponding to polyamines was scraped and dissolved in acetoacetic ester, they were measured with fluorescence spectroscopy.

Results

Effects of boron deficiency on content of MDA and membrane permeability

MDA was an end product in the process of membrane lipid peroxidation, its concentration was in positive correlation with the extent of membrane lipid peroxidation. The concentration of MDA in leaves of two cotton cultivars at seedling stage were significantly increased under boron deficiency (Table 1), but the extent of increase of the boron-efficient cultivar ‘9702’ (57.2%) was less than the boron-inefficient one ‘9706’ (92.0%). Moreover, the relative electrical conductivity, which might express the impairment extent of membrane integrity, in leaves of the inefficient cultivar was markedly enhanced at the stress of boron deficiency, but it was not significantly different at $p < 0.05$ in the efficient one.

Table 1. Effects of boron on MDA content and membrane permeability in leaves of two cotton cultivars at seedling stage. Data in table were averages of three replications, different letters represent significant difference to each other at 0.05 level

Cultivars	9702		9706	
Treatments	−B	+B	−B	+B
MDA content（nmol/g,FW）	72.8 ± 8.5a	46.3 ± 3.6b	81.0 ± 10.8a	42.2 ± 7.3b
Relative electrical conductivity (%)	22.3 ± 1.4ab	20.7 ± 1.0b	25.0 ± 1.8a	20.4 ± 0.7b

Effects of boron deficiency on composition of membrane fatty acid

The fatty acid in cotton leaf cell membrane mainly is composed of octadecatrienoic acid (18:3), octadecadienoic acid (18:2), octadecenoic acid (18:1), octadecanoic acid (18:0) and hexadecanoic acid (16:0). The content of octadecatrienoic acid in the fatty acid composition was highest, hexadecanoic acid was second, and octadecanoic acid was lowest (Table 2). The double bond index (DBI) and the ratio of unsaturated fatty acid to saturated fatty acid (UFA/SFA) in leaf membrane fatty acid of both cotton cultivars were decreased by boron deficiency, but an extent of decrease was distinctly different between the cultivars. For example, the DBI and UFA/SFA at a low boron supply were 97.4% and 87.8%, respectively, of those at a sufficient boron supply in the boron-efficient cultivar, but they were only 89.0% and 67.7% in the boron-inefficient one, respectively. The differences in the extents of decrease of DBI and UFA/SFA between both cultivars were mainly brought on by the different change of the content of octadecatrienoic acid contained 3 unsaturated bonds, the content of octadecatrienoic acid in the inefficient cultivar was reduced 15.4% by boron deficiency, only 3.0% in the efficient one.

Table 2. Effects of boron on membrane fatty acid composition in leaves of two cotton cultivars at seedling stage. Data in table were averages of three replications

Cult -ivars	Treat -ments	Fatty acid composition (%)					DBI*	UFA/SFA**
		16: 0	18: 0	18: 1	18: 2	18: 3		
9702	−B	20.9 ± 1.7	1.4 ± 0.2	4.4 ± 0.5	9.4 ± 1.8	64.0 ± 5.2	215.2	9.7
	+B	19.1± 1.2	1.0 ± 0.1	5.1 ± 0.3	8.9 ± 2.5	66.0 ± 2.7	220.9	11.0
9706	−B	20.9 ± 0.8	3.3 ± 0.6	7.8 ± 1.3	10.2 ± 0.7	57.8 ± 1.9	201.6	8.3
	+B	17.8 ± 2.4	0.6 ± 0.2	5.0 ± 1.1	8.3 ± 2.7	68.3 ± 3.5	226.5	12.3

*DBI: Double bond index = 18:1 + 18:2 × 2 + 18:3 × 3
**UFA/SFA: Ratio of unsaturated fatty acid to saturated fatty acid = (18:1 + 18:2 × 2 + 18:3 × 3) / (16:0 + 18:0)

Effects of boron deficiency on activities of protective enzymes

When boron supply was deficient, the activities of CAT and POD in the leaves of the boron-inefficient cotton cultivar were markedly decreased, not significantly different at $p<0.05$ in the boron-efficient one at the same time (Table 3). The activity of SOD in the leaves of both cotton cultivars was decreased by boron deficiency, but the extent of decrease in the boron-inefficient cultivar (33.2%) was obviously more severe than in the boron-efficient one (14.8%).

Effects of boron deficiency on content of polyamines

When boron supply was deficient, the putrescine (Put) content in both cultivars was reduced, especially in the boron-inefficient one (Table 4). On the other hand, boron deficiency caused a significant increase in the content of spermidine (Spd), spermine (Spm) and the total content of polyamines (Spd+Spm+Put) in the boron-efficient cultivar, but only an upward trend in the boron-inefficient cultivar. For example, the contents of Spd, Spm and Spd+Spm+Put at low boron supply were 1.89, 0.61 and 0.45 times greater than at sufficient boron supply in the efficient cultivar, only 0.41, 0.13 and 0.02 times in the inefficient cultivar, respectively.

Table 3. Effects of boron on activity of CAT, POD and SOD in leaves of two cotton cultivars at seedling stage. Data in table were averages of three replications, different letters represent significant difference to each other at 0.05 level

Cultivars	9702		9706	
Treatments	−B	+B	−B	+B
CAT (μmol·min^{-1}·g^{-1},FW)	3002.8 ± 181.9b	3177.8 ± 155.6b	3123.8 ± 110.2b	3573.0 ± 224.5a
POD (μmol·min^{-1}·g^{-1},FW)	201.0 ± 15.6b	183.2 ± 44.0b	223.8 ± 41.6b	356.0 ± 53.3a
SOD (U·g^{-1},FW)	172.0 ± 9.5b	201.8 ± 17.2a	103.1 ± 24.7c	154.4 ± 13.3b

Table 4. Effects of boron on polyamines content in leaves of two cotton cultivars at seedling stage (nmol.g^{-1}, FW). Data in table were averages of three replications, different letters represent significant difference to each other at 0.05 level

Cultivars	9702		9706	
Treatments	−B	+B	−B	+B
Put	543.3 ± 31.5b	622.7 ± 27.6a	467.1 ± 38.7c	608.2 ± 18.3a
Spd	617.9 ± 56.1a	213.8 ± 34.2c	468.0 ± 72.8b	331.5 ± 78.1b
Spm	536.9 ± 81.6a	332.8 ± 50.6b	276.1 ± 37.4bc	244.2 ± 14.0c
Put+Spd+Spm	1698.1 ± 119.2a	1169.3 ± 144.1b	1211.2 ± 98.9b	1183.9 ± 78.6b

Discussion

More and more experiments supported the idea that boron play an important role in plasma membrane integrity, but the mechanisms that boron participate the structure or function of plasma membranes were not still clear. Some reported boron seemed to play a critical structural role in biomembranes by binding membrane compounds containing *cis*-diol groups, some pointed out boron influence membrane function by changing the activities of enzymes related to membrane, the others indicated boron might affect membrane integrity by participating phenolics metabolism (Blevins and Lukaszewski, 1998, Brown et al. 2002, Goldbach et al. 2002). However, these hypotheses were short of the direct evidences.

When plant was suffered from the abiotic stress or consenescence, the toxic active oxygen was accumulated in its cell as a result of impairments in antioxidative defence systems, and thus destroyed the biomembranes structure or function by the peroxidation of membrane lipid. Cakmak et al. (1997) reported the stress of boron deficiency might also result in membrane damage by a similar mechanism in sunflower. The hypothesis was further supported by our experiments, a decrease of the activities of protective enzymes such as SOD, CAT and POD in cotton leaves (Table 3) under the stress of boron deficiency might result in the accumulation of the toxic active oxygen in cell, which would further result in peroxidation of the membrane lipid. For example, the UFA/SFA and the DBI were decreased (Table 2), the concentration of MDA and the relative electrical conductivity were increased (Table 1). Compared with the boron-efficient cotton cultivar, the activities of protective enzymes in the leaves of the boron-inefficient one was decreased more obviously by boron deficiency, this might be one of the reasons why membrane impairment extent in the inefficient cultivar was greater than the efficient one at the same stress of boron deficiency. But it was yet unclear how boron affected the activities of the protective enzymes and why the influences were different between the boron-inefficient and boron-efficient cotton cultivar.

Polyamines appeared to be ubiquitous in living cells and had been implicated in a variety of regulatory processes ranging from promotion of growth and cell division to inhibition of ethylene production and senescence (Smith, 1985), in membrane stability (Roberts et al. 1986). It had been reported exogenous polyamines might maintain the stability of cell membrane structure and function by enhancing the activities of protective enzymes (Jiang et al. 1993) or directly scavenging the toxic active oxygen generated in cell metabolism process (Drolet et al. 1986). A significant increase of the polyamines content was proved in the boron-efficient cotton cultivar when boron supply was deficient, but no obvious change in the boron-inefficient one (Table 4). We speculated on more polyamines promote protective enzyme activity and further maintain membrane stability in the boron-efficient cultivar in comparison with the boron-inefficient one. But more researches were needed on how to interact between boron and polyamines and why the polyamines metabolism was different between the boron-inefficient and boron-efficient cotton cultivar at same stress of boron deficiency.

Acknowledgments

We are grateful to the National Natural Science Foundation of China (Project number 39870520) and the Doctoral Scientific Foundation of the National Education Committee of China (Project number 970404).

References

Blevins D.G. and Lukaszewski K.M., 1998 Boron in plant structure and function. Annu. Rev. Plant. Physiol. Plant. Mol. Biol. **49**: 481–500.

Brown P.H., Bellaloui N., and Wimmer M.A. et al., 2002 Boron in plant biology. Plant Biology. **4**(2): 205–233.

Cakmak I. and Römheld V., 1997 Boron deficiency-induced impairments of cellar functions in plants. Plant and Soil. **193**: 71–73.

Camacho-Cristobal J.J., Lunar L., and Lafont F. et al., 2004 Boron deficiency causes accumulation of chlorogenic acid and caffeoyl polyamine conjugates in tobacco leaves. Journal of Plant Physiology. **161**: 879–881.

Camacho-Cristobal J.J., Maldonado J.M., and Gonzalez-Fontes A., 2005 Boron deficiency increases putrescine levels in tobacco plants. Journal of Plant Physiology. **162**: 921–928.

Cao X.Y., Song S.W., and Liu W.D., 1996 A study on the utilizing efficiency and B-deficiency susceptibility of different cotton cultivars. Journal of Huazhong Agricultural University. **15**(6): 559–562.

Drolet G., Dumbroff E.B., and Legge R.L. et al., 1986 Radical scavenging properties of polyamines. Phytochemistry. **25**(2): 367–371.

Flores H.E. and Galston A.W., 1982 Analysis of polyamines in higher plants by high performance liquid chromatography. Plant Physiol. **69**: 701–706.

Goldbach H.E., Wimmer M.A., and Chaumont F. et al., 2002 Rapid responses of plants to boron deprivation: Where are the links between boron's primary role and secondary reactions? In: Goldbach H.E., Rerkasem B., and Wimmer M.A. et al. (eds.). Boron in Plant and Animal Nutrition. New York, Kluwer Academic/Plenum Publishers. 167–180.

Jiang L., Shen Z.Y., Zhang Z.L., and Yan J.Q., 1993 The effects of polyamines on metabolism of active oxygen in detached leaves of Hordeum. Vulgare var. nudum Hook. f. Acta Phytophysiolotica Sinica. **19**(4): 367–371.

Li H.S., 2000 Principle and Technique of Plant Physiological and Biochemical Experiments. Beijing, Higher Education Publishers. 167–169; 260–263.

Roberts D.R., Dumbroff E.B., Thompson J.E., 1986 Exogenous polyamines alter membrane fluid-basis for potential misiuter-potention of their physiological role. Planta. **167**: 395–400.

Smith T.A., 1985 Polyamines. Ann. Rev. Plant. Physiol. **36**: 117–143.

Su W.A., Wang W.Y., and Li J.S., 1980 Analysis method of plant lipoid and fatty acid——TLC-GLC. Plant Physiol. Comm. (3): 54–60.

Warington K., 1923 The effect of boric acid and borax on the broad bean andcertain other plants. Ann. Bot. **37**: 629–671.

Update on Boron Toxicity and Tolerance in Plants

Robert Reid

School of Earth and Environmental Sciences, University of Adelaide, Adelaide 5005, Australia.

Introduction

Boron presents a challenge to agronomists. Management of boron in soil is made difficult by its high mobility, being easily leached under high rainfall conditions, leading to deficiencies in plants that grow there. Under low rainfall conditions, the opposite is often true, that it is not sufficiently leached and therefore may accumulate to levels that become toxic to plant growth. This is often exacerbated by irrigation to compensate for the low rainfall, because of the high boron concentrations that characterise many irrigation waters. As will be discussed in more detail in this paper, it is also the nutrient for which the plant has the least control over uptake. All essential plant nutrients except boron are acquired as ionised solutes (except perhaps N supplied as urea), which limits their membrane permeability, and allows a high degree of control by the activation, induction or repression of membrane transporters. To use the vernacular, boron is a 'slippery customer' which has so far largely resisted scientific attempts to fully understand its physiology, agronomic attempts to improve plant performance on B-deficient and B-toxic soils, and molecular attempts to engineer plants that are able to tame boron.

The purpose of this paper is to provide an update on recent progress in the understanding of why boron is toxic to plants, and to highlight advances in defining the mechanisms by which some plants are able to resist the toxicity imposed by high concentrations of soil boron.

Boron toxicity

Appearance and consequences

The physiological effects of boron toxicity include reduced root cell division (Liu et al. 2000), decreased shoot and root growth (Lovatt and Bates 1984: Nable et al. 1990), decrease in leaf chlorophyll, inhibition of photosynthesis, lower stomatal conductance (Lovatt and Bates 1984), deposition of lignin and suberin (Ghanati et al. 2002), reduced proton extrusion from roots (Roldan 1992), increased membrane leakiness, peroxidation of lipids and altered activities of antioxidation pathways (Karabal et al. 2003). These toxicity symptoms are slow to develop, or are only observed with extreme B treatments. Toxicity effects appear to be loosely correlated with the accumulation of high concentrations of B in the shoot, which is a function both of the concentration of B in soil and the time of exposure. In leaves, necrosis develops in older leaves, and in the older parts of those leaves. In barley, this occurs

F. Xu et al. (eds.), Advances in Plant and Animal Boron Nutrition, 83–90.

when the local tissue concentration exceeds about 20 mM (Stangoulis et al. 2001). A complicating factor is that B can be leached from leaves by rainfall (Nable and Moody 1992), thereby reducing the B concentration in leaves that have previously been subjected to levels of B sufficient to cause symptoms to develop. It was noted by Nable and Moody (1992) that leaf symptom expression of toxicity is often not a good predictor of yield in cereal crops. Instead, the effect on fruit and grain is likely to be indirect, being mediated by the longer-term accumulation of B. For example, inhibition of root growth by B will limit the ability of the plant to take up essential mineral nutrients and water, while necrosis in leaves will limit the ability to supply photosynthate to developing fruits, seeds and storage organs.

Targets for B disruption of growth

One of the problems in searching for sites of B toxicity is that we know very little about what roles B plays in plant cells. In comparison to most other nutrients, B is relatively unreactive. Its biological chemistry is based on the formation of complexes with compounds having two hydroxyl groups in the cis-conformation (cis-diols). The strongest complexes are formed with cis-diols attached to a furanoid ring (Hunt 2002). In plants, the only widely accepted role for B in nutrition is as a structural component of primary cell walls where it cross-links pectic polysaccharides by binding to the furanoid sugar, apiose, in the rhamnogalacturonan II (RGII) complex. The plant requirement for B is closely related to the RGII content of the cell walls of different plant types (Matoh 1997). The only other compound of biological significance with cis-diols on a furanoid ring is ribose, which is a component of several important metabolites. Strong complexes are also formed with sugar alcohols such as mannitol and sorbitol, but high concentrations of sugar alcohols are only observed in a relatively small number of species (Bieleski 1982). Weaker binding to other metabolites with single and double hydroxyl groups has also been reported (Hu et al. 1997; Pfeffer et al. 1999) but it seems unlikely that at B concentrations found under optimum B nutrition, the formation of complexes with any of these compounds would significantly interfere with metabolism. However, elevated internal B concentrations associated with toxicity conditions could conceivably draw enough of these compounds from the metabolic pool to inhibit normal biosynthetic activities or energy transduction, leading to impaired growth and development.

Based on what is known of the chemistry of B in relation to molecules of biological importance, the most likely mechanisms of toxicity revolve around the possible disruption of cell wall development due to excess binding to apiose, or metabolic disruption due to binding to ribose, either as the free sugar or as a component of key compounds such as RNA, ATP, NADH or NADPH. Reid et al.

(2004) also considered the possibility that the accumulation of high concentrations of B in leaves might cause osmotic problems, but concluded from measurements of total leaf ion concentrations that this was unlikely to be a factor.

There is no strong evidence that high B exerts a toxic effect apoplastically. Dannel et al. (1998) found that the B content of cell walls of sunflower at high B was the same as that at adequate B levels, consistent with there being a limited number of binding sites for B, in the RGII complex, and once saturated no further binding of B can occur. On the other hand, Ghanati et al. (2002) found that levels of suberin and lignin in cultured tobacco cells exposed to 10 mM B increased, which could potentially cause a stiffening of the cell wall matrix and limit extensibility in meristematic tissues.

Metabolic disruption due to binding to ATP, NADH or NADPH is supported by NMR data that demonstrate complexation of each of these compounds by B, as well as to several related metabolites (Hunt 2002). *In vitro* experiments with enzymes for which two of these compounds are substrates, malate dehydrogenase (NADH) and isocitrate dehyrogenase (NADPH), are generally consistent with the NMR data, insofar as they demonstrate an increase in K_m with increasing B at B concentrations greater than about 5 mM (Reid et al. 2004), which could be explained by a reduction in the available substrate concentration due to complexation. V_{max} for the reactions was much less sensitive which suggests that B targets the substrate rather than the enzyme itself. This view is supported by a study with acid phosphatase, an enzyme whose substrates have low affinity for B, in which neither K_m nor V_{max} was significantly affected by up to 50 mM B (Reid et al. 2004).

The consequences of this binding have been more difficult to establish *in vivo*. Reid et al. (2004) demonstrated that neither photosynthesis, respiration nor protein synthesis was particularly sensitive to B. In leaf slices from barley, photosynthesis was unaffected by 50 mM B, and inhibited by only 23% at 100 mM B. Respiration was slightly more sensitive, being reduced by 37% at 50 mM B and by 60% at 100 mM B. Experiments with giant algal cells also do not support toxicity being directly due to a reduction in the supply of energy. Reid et al. (2004) showed that mature cells were able to withstand 60 mM B for at least 7 days if kept in the dark, but the mortality increased with increasing light, which would presumably have stimulated, rather than decreased, photosynthetic output, and indirectly, energy supply. The role of light in this instance is not clear, but high B might reduce the capacity of plant cells to resist photooxidative damage.

These limited effects of B on the metabolism of mature cells contrast with the severe inhibition of growth of roots as well as a range of cultured cells at concentrations of 1-5 mM (Reid et al. 2004). There appear to be significant differences in the responses of meristematic and mature tissues with respect to their sensitivity to high B, which might give clues to the primary targets for B in reducing growth. Apart from a higher overall metabolic activity, the feature that distinguishes mature and meristematic tissues is that in the latter, cells are dividing. There is therefore a much greater dependence on such processes as DNA replication and translation. Deoxyribose in DNA does not have the necessary cis-diol groups to bind boron, so disruption of DNA replication does not seem feasible. Similarly, one of the hydroxyl groups of ribose in RNA is used to link the nucleotide bases together and so is also not available for cis-diol bonding to B. However, both of the hydroxyl groups of ribose are exposed at the 3' end of RNA molecules, a region which may or may not be important in either transcription or translation. What is potentially more significant in terms of B binding at the 3' end is that in plants and animals RNA undergoes extensive splicing, during which ribose is presumably transiently exposed to B. Shomron and Ast (2003) have demonstrated that B does in fact inhibit one of the steps in splicing of mRNA.

Boron tolerance

Genotypic variation in B tolerance – can it be explained by a single mechanism?

There exists considerable variation in the ability of plants to grow in B-toxic soils. Screening of genotypes has identified important crop varieties that can cope with relatively high soil B concentrations, and also crop varieties that are inappropriate for use on these soils. While many agronomists seem happy to accept the benefits that arise by these simple screening trials, researchers have tried to understand these differences in tolerance in the hope that it would enable a more targeted approach to the development of tolerance. The early work of Nable, Paull and their co-workers established that in cereals, tolerance is associated with the ability to restrict B uptake into the plant (Nable 1988; Paull et al. 1988). Subsequently, this was shown to be a common feature in other plant species (e.g. Paull et al. 1992), and may well be the only feature that generates significant control over B toxicity.

Since these early observations of lower B accumulation in tolerant genotypes, progress in discovering how this is achieved at the physiological level has been slow. However, work over the past few years has shed new light on the phenomenon. The key to understanding how some varieties are able to restrict B accumulation was to understand how B enters plants. Predictions of membrane permeability of B based on ether:water partition coefficients (Raven 1980) suggested it should have a relatively high permeability. Such predictions were supported by studies with artificial lipid bilayers (Dordas and Brown 2000), but it was not until the work of Stangoulis et al. (2001) with giant algal cells that a definitive measurement of B permeability across the plasma membrane of an intact plant cell was made available. These measurements indicated that B entry into cells was very rapid indeed, and that equilibration of B between the internal and external phases could occur in these large cells within a few hours. In smaller cells such as those of roots with a much higher surface area, equilibration would be expected to be even more rapid. Such high permeability meant that it would be unnecessary, in fact futile, for B uptake to be mediated by a transport protein.

So the question was posed – how could tolerant varieties accumulate less B? Various mechanisms have been investigated, including differences in lipid composition of membranes that might reduce the permeability to B, and genotypic differences in transport from root to shoot, possibly mediated by differences in the efficiency of transpiration. While alteration of lipid composition did result in altered membrane permeability to B (Dordas and Brown 2000), much larger changes would be needed to prevent equilibration of B with the B concentration in the external medium. Similarly, differences in B:H_2O delivered in the transpiration stream were found to be insufficient to explain the large differences in shoot concentration between tolerant and sensitive genotypes (Nable 1988; Nable et al. 1997).

Hayes and Reid (2004) confirmed that the equilibration observed by Stangoulis et al. (2001) with giant algal cells, also occurred in the roots of B-sensitive barley cultivars. Perhaps more importantly, they established that when B was added to the solution around the roots, the initial uptake, and therefore permeability to B, were similar in both sensitive and tolerant genotypes. The difference however, was that although a steady state internal B concentration was achieved in both cultivars with a half-time of around 8 minutes, the steady state concentration in the tolerant cultivar was maintained at approximately half that of the sensitive variety. In other words, at equilibrium, the B concentration in the sensitive cultivar was similar to that in the solution, while in the tolerant cultivar, the internal concentration was much lower. This difference in concentration clearly must require an input of energy in order to maintain the concentration gradient. The obvious conclusion from these studies is that B is actively pumped from the cells in the tolerant varieties. The identity of the transporter that generates the efflux of B has not been established, but given the rapid influx of B, efflux through this transporter must occur at a similar rate to the influx.

Genetic targeting of B tolerance genes

Known B transporters

The only published description of a transporter involved in the transport of B across plant cell membranes is for BOR1 which catalyses the loading of B from xylem parenchyma cells into the xylem of Arabidopsis (Takano et al. 2002). Bor1 is therefore a B effluxer and may use a similar mechanism to that required for B efflux from other root cells involved in tolerance to B. However, Takano et al. (2005) have reported that although *Bor1* is expressed under B toxic conditions, the Bor1 protein is degraded at high B. Therefore it seems an unlikely participant in a B tolerance mechanism. There appear to be a number of *Bor1*-like genes in other plant species with different tissue expression patterns that have yet to be characterised fully. Potentially, one or more of these genes could encode a B efflux protein that remains active at high B concentrations.

Bor1 has a high degree of homology to anion exchangers in animal cells. On thermodynamic grounds, efflux of B could feasibly be achieved by exchange of internal borate anions for either OH^- or HCO_3^- (Hayes and Reid 2004).

Chromosome regions conferring tolerance to B

Molecular marker techniques have identified loci in wheat and barley that are correlated with B tolerance (Jefferies et al. 1999, 2000). In barley, loci on chromosomes

4 and 6 are related to reduced leaf B accumulation. On the basis of the studies described above, the gene(s) at these loci may well encode B efflux transporters. Hybridisation of wheat- barley addition lines with *Bor1*-like sequences uncovered only a single homologue located on chromosome 5 (JE Hayes and RJ Reid unpublished). These results however do not preclude the possibility that the barley genome contains other *Bor* genes on chromosomes 4 and 6 that were not detected by the primers used in the initial study. Alternatively, the loci on chromosomes 4 and 6 may contain regulatory genes that control the activity or expression of transporter genes located on other chromosomes.

Screening of plant genes in yeast

cDNA from wheat and lupin was cloned into yeast and the transformants examined for their tolerance to B (JE Hayes, unpublished). Six B-tolerant clones were obtained from lupin, but none of the cloned genes encoded a membrane protein, nor did any of the tolerant yeast clones show reduced accumulation of B. This result seems to suggest that there may be mechanisms for conferring tolerance to B other than the ability to lower the internal B concentration. It is perhaps significant that of the 6 lupin genes that showed tolerance in yeast, 5 were either transcription factors or ribosomal proteins. If, as discussed above, B toxicity is primarily due to disruption of RNA splicing, then these proteins could conceivably act by protecting the splicing sites from attack by B. Could this mechanism work in plants? Possibly yes, but it is important to understand that the yeast genome in much simpler than that of higher plants, most notably in that only a small proportion of the yeast genes contain introns, and usually only one. This contrasts with plants in which genes typically have large numbers of introns that need to be spliced out to generate mature RNA.

Where to now?

The demonstration that many B-tolerant species must possess a mechanism for efflux of B has provided a rationale for pursuing genes related to boron efflux. The search for such genes will undoubtedly proceed from two directions – through the discovery and characterisation of homologues of *Bor1*, and through the finer mapping of chromosome regions known to confer B tolerance. The alternative to tolerance based on reduced B accumulation is some form of internal tolerance, such as that demonstrated for yeast. Such a mechanism has not so far been clearly described in plants but it is possible that some species do possess such tolerance but that it occurs simultaneously with reduced B accumulation and is therefore difficult to detect.

References

Bieleski R.L., 1982 Sugar alcohols. In Encylopedia of Plant Physiology. New Series (eds F.A. Loewus & W. Tanner), Vol. 13A, pp 158–192. Springer Verlag, New York.

Dannel F., Pfeffer H., and Römheld V., 1998 Compartmentation of boron in roots and leaves of sunflower as affected by boron supply. Journal of Plant Physiology **153**: 615–622.

Ghanati F., Morita A., and Yokota H., 2002 Induction of suberin and increase of lignin content by excess boron in tobacco cells. Soil Science and Plant Nutrition **48**: 357–364.

Hayes J.E. and Reid R.J., 2004 Boron tolerance in barley is mediated by efflux of B from the roots. Plant Physiology **136**: 3376–3382.

Hu H., Penn S.G., Lebrilla C.B., and Brown P.H., 1997 Isolation and characterization of soluble boron complexes in higher plants. Plant Physiology **113,** 649–655.

Hunt, C.D., 2002 Boron-binding-biomolecules: a key to understanding the beneficial physiologic effects of dietary born from prokaryotes to humans. In Boron in Plant and Animal Nutrition (eds H.E. Goldbach et al.), pp. 21–36. Kluwer Academic, New York, USA.

Jefferies S.P., Barr A.R., Karakousis A., Kretschmer J.M., Manning S., Chalmers K.J., Nelson J.C., Islam, AKMR and Langridge P., 1999 Mapping of chromosome regions conferring boron tolerance in barley (Hordeum vulgare L.). Theoretical and Applied Genetics **98**: 1293–1303.

Jefferies S., Pallotta M., Paull J.G., Karakousis A., Kretschmer J., Manning S., Islam A., Langridge P., and Chalmers K., 2000 Mapping and validation of chromosome regions conferring boron toxicity tolerance in wheat (Triticum aestivum). Theoretical Applied Genetics **101**: 767–777.

Karabal E., Yücel M., and Hüseyin A.Ö., 2003 Antioxidant responses of tolerant and sensitive barley cultivars to boron toxicity. Plant Science **164**: 925–933.

Liu D., Jiang W., Zhang L., and Li L., 2000 Effects of boron ions on root growth and cell division of broadbean (Vicia faba L.). Israel Journal of Plant Sciences **48**: 47–51.

Lovatt C.J. and Bates L.M., 1984 Early effects of excess boron on photosynthesis and growth of Cucurbita pepo. Journal of Experimental Botany **35**(152): 297–305.

Matoh T., 1997 Boron in plant cell walls. Plant and Soil **193**, 59–70.

Nable R.O., 1988 Resistance to boron toxicity amongst several barley and wheat cultivars: a preliminary examination of the resistance mechanism. Plant and Soil **112**: 45–52.

Nable R., Bañuelos G., and Paull J., 1997 Boron toxicity. Plant and Soil **193**: 191–8

Nable R.O., Cartwright B., and Lance R.C., 1990 Genotypic differences in boron accumulation in barley: Relative susceptibilities to boron deficiency and toxicity. Genetic Aspects of Plant Mineral Nutrition Bassam. N. El., Dambroth M., and Loughman. Dordrecht B., Kluwer Academic Publishers**:** 243–251.

Nable R.O. and Moody D.B., 1992 Effects of rainfall on the use of foliar analysis for diagnosing boron toxicity in field-grown wheat. Plant and Soil **140**: 311–314.

Paull J.G., Cartwright B., and Rathjen A., 1988 Responses of wheat and barley genotypes to toxic concentrations of soil boron. Euphytica **39**: 137–144.

Paull J.G., Nable R.O., Lake A.W.H., Materne M.A., and Rathjen A.J., 1992 Response of annual medics (Medicago spp.) and field peas (Pisum sativum) to high concentration of boron: genetic variation and the mechanism of tolerance. Australian Journal of Agricultural Research **43**: 203–213.

Pfeffer H., Dannel F., and Römheld V., 1999 Isolation of soluble boron complexes and their determination together with free boric acid in higher plants. Journal of Plant Physiology **154**: 283–288.

Raven J.A., 1980 Short- and long-distance transport of boric acid in plants. New Phytologist **84**: 231–249.

Reid R.J., Hayes J.E., Post A., Stangoulis J.C.R., and Graham R.D., 2004 A critical analysis of the causes of boron toxicity in plants. Plant Cell and Environment **27**: 1405–1414.

Reid R.J. and Walker N.A., 1983 Adenlyate concentrations in Chara: Variability, effects of inhibitors and relationship to protoplasmic streaming. Australian Journal of Plant Physiology **10**: 373–383.

Roldán M., Belver A., Rodríguez-Rosales P., Ferrol N., and Donaire J.P., 1992 In vivo and in vitro effects of boron on the plasma membrane proton pump of sunflower roots. Physiologia Plantarum **84**: 49–54.

Shomron N. and Ast G., 2003 Boric acid reversibly inhibits the second step of pre-mRNA splicing. FEBS Letters **552**: 219–224.

Stangoulis J.C., Reid R.J., Brown P.H., and Graham R.D., 2001 Kinetic analysis of boron transport in Chara. Planta **213**: 142–146.

Takano J., Miwa K., Yuan L.X., von Wiren N., and Fujiwara T., 2005 Endocytosis and degradation of BOR1, a boron transporter of Arabidopsis thaliana, regulated by boron availability. Proceedings of the National Academy of Sciences of the United States of America **102:** 12276–81.

Takano J., Noguchi K., Yasumori M., Kobayashi M., Gajdos Z., Miwa K., Hayashi H., Yoneyama T., and Fujiwara T., 2002 Arabidopsis boron transporter for xylem loading. Nature **420**: 337–339.

Boron Nutrition and Boron Application in Crops

Plant Boron Nutrition and Boron Fertilization in China

Yunhua Wang[1], Lei Shi, Xiangyun Cao and Fangsen Xu

Microelement Research Center, College of Resources and Environment, Huazhong Agricultural University, Wuhan 430070, China; [1]Corresponding author

Introduction

In the past 30 years, research on plant boron (B) nutrition has progressed significantly and the application of B fertilizer has become a standard measure in many B deficient regions. In China, this promoted the development of agriculture, especially the production of cotton and oilseed rape. In this paper, we will review the respective progress and report about the Chinese studies on the physiological function of B, diagnosis of B status of plants and techniques of B fertilizer application.

Discovery of plant B deficiency in China

In 1970s, the 'disorders' of "flowering without seed setting" in oilseed rape (*Brassica napus*) and "budding without flowering" in cotton were discovered successively in a large area of Hubei province in China. These disorders were proven to be caused by B deficiency by Prof. Husheng Ren and Prof. Changzhi Liu from Oil Crop Research Institute, Agricultural Academy of China, and Prof. Yunhua Wang from Huazhong Agricultural University of China, respectively (Wang et al. 1976). The discovery of crops suffering from B deficiency initiated a new research area for plant B nutrition and B fertilization in China. B was not only found to cure these disorders efficiently and increase the yields of rapeseed and cotton, but also to improve their quality (Wang et al. 1978). Thereafter, the application of B fertilizer according to available B concentration in soils and plant demand became one of the important methods to improve the production and quality of rapeseed and cotton in China.

In November, 1981, the National Economical Committee, the Ministry of Agriculture, the Ministry of Chemical Industry and the Ministry of Metallurgical Industry in China held together the first national conference on 'Microelement Fertilizers in China' in Zhangzhou city, Fujian Province. In this meeting, participants exchanged their experiences on research and application of microelement fertilizers, and the 'National Collaboration Group on the Research and Application of Microelement Fertilizers in China' was founded at that occasion. Since then, the interrelated Ministries and Committees had paid much attention to the research on microelement fertilization. They organized the members of the Collaboration Group to discuss research proposals many times and published related books, journals and information on the principles of microelement fertilization. At the same time, they

F. Xu et al. (eds.), Advances in Plant and Animal Boron Nutrition, 93–101.

commissioned the Microelement Laboratory, Huazhong Agricultural University to hold the training classes on microelement research and fertilization more than ten times and promoted the popularization of the microelement fertilizers all over the country.

The result of field trials from all of the country showed that the yield of cotton and rape could be increased by 8-12% when B was added. In seriously B deficient soils, the yield increases of crops could reach or even surpass 20% and additionally improve the quality (Wang et al. 1982). Thus, the farmers readily adopted the use of B fertilizer because of its low cost, high economic benefit and simple application.

In the past, B fertilizer was applied mainly to intermediate or low yielding cropland, on B deficient soils and sensitive crops. Nowadays, a lot of crops, besides rape and cotton, such as sugar beet, sunflower, pepper, cauliflower, soybean, lettuce, citrus, apple, mulberry, wheat etc. receive B fertilizer regularly, and the area of B fertilization was enlarged annually along to the understanding of the importance of B for agricultural production.

Distribution of soil available B in China and diagnosis standard of B nutrition

Distribution of soil available B in China

The progress of application of B fertilizer in agriculture was directly related to the ongoing research on B fertilizer use. The first important work of the basic research was to determine the microelement's available concentration in soil and their distribution in the cultivated land in China. In 1980, a soil B map of China (1:1, 000, 000) was published by Liu Zheng. The map divided cultivated land into two regions: one is western inland which is rich in hot water soluble B(HWSB) (more than 0.5 mg kg^{-1}), the other is the Eastern area, in which B supply is very low in HWSB. Generally, ultisol and oxisol regions, in the south of China were very low in B; and also the aridisol in Northern China and the inceptisol of the Yellow River Valley lacked B. Soil B concentration of the former was slightly higher than in the latter. The seriously B deficient soils in China are mainly distributed in Guangdong, Hainan, Fujian provinces, the central and west regions of Zhejiang province, the east and south of Jiangxi province, the central and south of Hunan province and the east of Hubei Province (Liu et al. 1982, 1983).

The microelement investigation of cultivated land was carried out by each of provinces, autonomous regions and municipalities in the second soil survey from 1979-1988 in China, which provided more exact details of soil microelement levels, resulting in soil microelement maps of China (scale 1:1, 000, 000). Based on that survey, the B deficient cultivated area in 20 provinces amounted to more than 50%; the area in 8 provinces exceeded 80%; and a total of more than 33.3 million hectares of cultivated land suffered from B deficiency.

Wang et al. (1987) published the soil available B map of the cotton-cultivated land to show the regions where B application for cotton production was required or not. Based on the two maps, the areas for cotton production were divided into four classes according to the likeliness of B application to increase yields. The HWSB of the class with the highest likeliness to improve yields by B application ranged less than 0.2 mg kg^{-1}. Here, distinct B-deficiency related disorders appeared as pointed out above with clearly visible symptoms without B application. These regions were mainly distributed in east of Hubei province and the centre of Zhejiang province, where B application could increase the cotton yield by more than 20%. The soil available B ranged from 0.2-0.5 mg kg^{-1} in relatively B responsive regions, which were mainly distributed in the Henan and Hebei provinces, Shandong peninsula, the southwest of Shandong province and provinces in Yangtze River drainage area. The cotton plants there exhibit the typical B-deficiency symptoms of darker green ring-stripes at the petiole and B supply could raise yields by over 10%. The soil available B of intermediately deficient regions ranged from 0.5 mg kg^{-1} to 0.8 mg kg^{-1} and B could increase the yield here by more than 5%. These areas are located basically in Henan, Hebei, Shanxi, and southwest of Shandong Provinces. At HWSB levels above 0.80 mg kg^{-1}, responses of cotton to B application are erratic, such as in Yugao and Haian in Jiangsu province, and Changde and Shimen in Hunan province. These regions were referred to as the probably responsive regions, where additional factors, such as soil moisture, drought, chill etc. may influence responsiveness to B supply (Wang et al. 1989; Zhou et al. 1993).

Diagnosis of B nutrition

To select and adopt the methods and parameters diagnosing the plant's B status became very important for a proper B fertilizer application. These methods included plant shape diagnosis, soil nutrient diagnosis, and plant nutrient diagnosis. Among them, diagnosis of morphological characters is very simple. Meanwhile, the B deficiency symptoms of oilseed rape, cotton, sugar beet, sunflower, sesame, ramie, wheat, citrus, apple and *Arabidopsis* are clearly described, but most of these symptoms can be observed only when the plants grow in serious B deficiency. Under slightly or moderately B deficient conditions, however, visible symptoms are hardly found. Wang et al. (1985) found the ratio of ring-stripes on the petioles of cotton was closely related to potential B deficiency. If the percentage of plants showing ring-stripes was above 8% in the Yangtze River valley, or more than 3% in the Yellow River valley, the soil B was deficient and B application was essential for cotton to grow normally. The rate was confirmed to be an efficient index to estimate soil B status and was introduced into all cotton-production regions in China as a criterion for the application of B fertilizer.

Soil nutrient diagnosis is recommended before sowing or transplanting. HWSB has been introduced as a standard for the diagnosis of B deficiency (Table 1). The plant B nutrition diagnosis was summarized in Table 2. Moreover, Huang (1996) found that the growing young leaves, such as the young open leave(YOL) is the reliable part for B deficient diagnosis and the B concentration, 10-14 mg kg^{-1} dry matter should be kept for normally flower and set seed.

Table 1. Soil available B (HWSB) diagnosis standards for various crops

HWSB (mg B kg^{-1} soil)	Serious B deficiency	B deficiency	Potential B deficiency	B toxicity
Cotton	< 0.20	< 0.50	< 0.80	> 5
Oilseed rape (*Brassica napus*)	< 0.25	< 0.70	–	–
Sunflower	< 0.25	< 0.80	–	> 8
Sesame	< 0.20	< 0.50	< 0.50	> 2

Table 2. Plant B nutrition diagnosis standards for various crops in China

Leaf B Concentration (mg kg^{-1} DW)	B deficiency	B toxicity
Oilseed rape (*Brassica napus*)	< 20	-
Cotton(*Gossypium hirsutum*)	< 20 (the fourth leaf from the top)	> 100 (the fourth leaf from the top)
Sunflower (*Helianthus annnus*)	< 50 (the fourth leaf from the top)	> 200 (the fourth leaf from the top)
Orange tree (*Citrus L.*)	< 15 (the fourth leaf from the top of spring branch)	-
Apple tree (*Malus pumila*)	< 15-20 mg kg^{-1} (the fourth leaf from the top of spring branch)	-
Sesame (*Sesamum indicum*)	< 10 mg kg^{-1} (top stem)	> 200 (top stem)

Application technique of B fertilizers

Application of boron fertilizer is the mean to increase yield and quality of crops sensitive to B deficiency. Nowadays, borax is the most popular B fertilizer in China, which is usually used as foliar spray with a B concentration of 0.2%. Sometimes boric acid is also used at a concentration of 0.1% (w/v). 'The Application Standards of B and Zinc Fertilizers in Several Main Crops' were established and recommended all over China by the Ministry of Agriculture. According to the standard, for oilseed rape (*Brassica napus* L) seriously B deficient, 7.5 kg of borax per ha should be applied as basic fertilizer dressing, and for seriously deficient cotton, the recommended rate is 6.0 kg of borax per ha. Under mild B deficiency, foliar application with 0.2% borax at seedling and at budding is recommended for rapeseed, and for cotton, foliar application at budding early flowering and flowering-boll period is recommended. Combination of B with other fertilizers, such as NKBZn in cotton, NPB in oilseed rape, sesame and sunflower, and NPKBZn in citrus could significantly promote the uptake of nutrients and increase crop yield (Pi et al. 1989).

Song et al. (1992) studied the movement of B in the profile of Inceptisol in Hubei Province and found that losses of B were mainly caused by leaching. At the same time, some B was adsorbed and accumulated in the plow layer. The results from four years field trials showed that the absorption of B could reach up to 22-37% of the total B when 4500-6000 g of borax per hectare was applied annually. In these experiments, a B balance was calculated under without B fertilization. The input of B was 136.0 g ha^{-1}, and the output 366.8 g ha^{-1}. B losses amounted to 230.8 g ha^{-1} in a cotton and rape rotation. For a cotton– wheat rotation, the output was 277.0 g ha^{-1} and the loss amounted to 141.0 g ha^{-1}. When B was applied at 218.0 g ha^{-1}, the output increased to 812.5 g ha^{-1} and the loss to 594.5 g ha^{-1} in the cotton – rape rotation, and for the cotton – wheat rotation, the output was 446.9 g ha^{-1} and the losses 228.0 g ha^{-1}. Losses and output have to be compensated by B fertilizers in order to maintain a balanced B nutrition. Application of B on B impoverished soils will improve the B utilization efficiency by reducing the amount of B fixed by the soil. The ratio of B utilization was over 50% in the field trials in the cotton-wheat rotation or cotton–rape rotation with foliar spray of 3.0 kg and 4.5 kg borax ha^{-1} annually, respectively. After three years, the application of B was not necessary for the 4th year crop growth because B in soil was sufficient for high yields.

The function of B in plants

Responses of crops to B deficiency

Generally, the B concentration in leaves could be used as a parameter to diagnose whether the crop suffered B deficiency or not, but it was not suitable for sesame. The leaves of cotton, sunflower, ramie, and rape became thickened under B deficiency, whereas leaves of sesame became thinner. B concentration of the sesame‘s stem was often used to diagnose the B nutrition.

Budding stage of cotton was found to be the most sensitive to B deficiency. When plants with six leaves were transferred from a B free solution to sufficient concentration (0.5 mg L^{-1}), they could bud, flower and bear bolls normally. If transferred from sufficiency to B free solution at the same stage, they could bud, but did not flower and bear bolls. Thus, B should be applied as basal dressing on seriously B-deficient soils, whereas, foliar spray of B increases yield under B deficiency or slightly B deficient soils.

When plant suffered from B deficiency, the leaves became deformed, and the stomata were closed, the number of veins and vascular bundles decreased. At serious B deficiency, the mesophyll cells lessen and deformed chloroplasts. The grana developed rapidly, their amount in the chloroplasts decreased and the grana lamellae piled up tightly. The mitoplast of mesophyll cells was scarce, the amount of cristae increased significantly and the space in the cristae enlarged (Wei et al. 1989; 1992; 1993). Zhou et al. (1993) found that the formation of the cotton petiole ring-strips involved in six stages from the vascular cambium thickened abnormally to deformed pith cells. In the petiole ring-strips, some substance caused by B deficiency blocked up the holes of the sieve plates, which not only blocked up the translocation of B, but also decreased the content of Ca in the sieve tubes, vessels and pitches, and largely increased the content of S. Under B deficiency, anatomic structure of the cotton flower stalks, calyxes, petals, pistils, stamens were abnormal, the tapetum swollen and its disappearance delayed, the pollen became deformed, and the development of thorns in the pollen was inhibited (Xie et al. 1991). Development of ramie pollen mother cells from dyad to quadrant were inhibited by B deficiency (Zheng et al. 1989). Wei et al. (1993) found that the shape of thorns in the pollen of sunflower changed with the B levels.

Boron uptake and utilization

Knowledge about B uptake, translocation and distribution in cotton has progressed significantly. When cotton plants with two leaves were incubated in a solution containing ^{10}B for 60 minutes, ^{10}B could be detected in the top and all other leaves, which indicated that B uptake and translocation by the seedlings was very fast. When ^{10}B was spread on the surface of the cotyledon, it apparently could not be utilized and translocated as seen by the typical symptoms: the cotyledon lamina still was thickened, the cotyledon veins became protuberant and split longitudinally, the apical point became necrotic and axillary buds grew thickly. However, when on the surface of true leaves, some B was absorbed by the first and second true leaves and was translocated to new leaves; from the third leaf to the top leaf, the B translocation to the new leaf increased along with the distance to the top and at last reached a constant value. There were two directions in B translocation, one was towards the apex, and the other to boll-branches and buds. Although the true leaves of cotton took up ^{10}B, their B content only reached 60-70% of that grown at normal B supply, and the plants still suffered B deficiency. Thus, foliar B application to cotton could not compensate fully for extreme B deficiency. However, it could delay the bud falling of cotton grown in the serious B deficient medium. These results showed however that foliar B application could both increase leaf B content and improve the transport of B to the growing center. Foliar spray for two or three times was necessary to improve the B nutrition of the whole cotton plant (Xie et al. 1992).

Plant B deficiency not only affects the uptake of B, but also the acquirement of other nutrients. Generally, B deficiency results in an increase of plant N, P, K content, but a decrease in total uptake. At the optimum B level, the N, P and K levels decrease due to increased growth and the so-called 'dilution effect' (Pi et al. 1989; Wang et al. 1994).

Boron nutrition and C and N metabolism

B deficiency decreased the amount of chloroplasts, inhibited their development, and reduced the photosynthesis. Zheng et al. (1989) found that B affected the carbohydrate metabolism significantly, including a positive effect on chlorophyll contents, shorter time of midday depression, higher efficiency and rate of photosynthesis, and an enhanced translocation of photosynthesis products. Total carbohydrate levels in B deficient cotton leaves were lower, but the content of soluble sugar was higher than that in normal plants. The reason might be that the sugar translocation was hampered. Nitrogen metabolism was also affected by B deficiency. When B was deficient, the content of protein-N decreased, but the content of non-protein nitrogen increased.

Boron nutrition and enzyme activity

The activities of plant enzymes were also affected by B deficiency. When cotton grew under B deficient conditions, the activities of nitrate reductase in leaves decreased, the activities of proteinase and polypeptidase in leaves increased, the activities of glucose-6-phosphate dehydrogenase and peroxidase in the petiole increased, the activities of the cellulase, pectinase and peroxidase in the abscission-layers of the flower bud stalks increased, and the content of total phenols increased. The activity of cellulose in the abscission-layers of young B-deficient citrus fruits was 20 times higher compared to fully supplied citrus.

Boron nutrition and phytohormone metabolism

More and more research was focused on the relationship between B and plant growth regulators. The endogenous phytohormones in B deficient plants changed significantly, such as, the contents of ethylene in the blades, petioles, buds of cotton increased from two to ten times when it suffered B deficiency (Wang et al. 1994). Zhao et al. (1998b) reported that in cotton and cucumber, B deficiency inhibited the transport of IAA from young leaves and apex to lower parts, promoted the accumulation of phenols, and decreased the activity of IAA in young leaves. However, the phenomenon of IAA accumulation in young leaves and apex was not found in *Brassica napus*.

Boron nutrition and plant nitrogen fixation

Xiong et al. (1995; 1996) reported that roots of horse bean were very weak at seedling stage under B deficiency, and the dry weight was reduced by about 40%. B deficiency reduced growth and development of nitrogen-fixing root nodules, and thus the shoot exhibited both B and N deficiency symptoms. This suggested that Inadequate B nutrition can hamper the development of rhizobia and decrease the fixation of nitrogen.

Genetics and molecular biology of plant B nutrition

B efficiency may be attributed to a higher B uptake efficiency, B utilization efficiency or re-translocation efficiency, which is genetically controlled. Eight highly B-efficient and two low B-efficient cultivars were screened from 210 cultivars of oil seed rape (*Brassica napus*) by a two-step-method (Wang and Lan, 1995; Chu et al. 1999). Detailes about studies on B efficiency mechanisms of *Brassca napus* were reviewed by Xu and Wang in this book.

Acknowlwdgements

Financial support for this work was provided, amongst other sources, by the Ministry of Science and Technology, China (the former State Science & Technology Commission), the former State Economic Commission, China; the National Nature Science Foundation of China; the Ministry of Agriculture, China, the former Ministry of Chemistry & Industry, China, General Administration of Civil Aviation of China. The authors highly acknowledge Prof. Henier Goldbach from University of Bonn, Germany for carefully review the manuscript and give good suggestions and Dr. Hu Hong Qing from Huazhong Agricultural University for soil classification comparison between China and FAO.

References

Liu Z., Zhu Q.Q., and Tong L.H., 1980 Boron deficient soils and their distribution in China. Acta Ped. Sin. **17**, 228–239.

Liu Z., Zhu Q.Q., and Tong L.H., 1983 Microelements in the main soils in China. Soil Sci. **135**, 40-46.

Liu Z., Zhu Q.Q., Tong L.H., Xu J.X., and Yen C.L., 1982 Geographical distribution of trace elements deficient soils in China. Acta Ped. Sin. **19**, 209–223.

Pi M.M., Liu W.D., Wu L.S., and Wang Y.H., 1989 Effect of combined application of boron and nitrogen, phosphorous, or potassium on yield and nutrient uptake of cotton. In Proceedings of the International Symposium on Balanced Fertilization, pp. 361–369. Press of Agricultural, Beijing, China.

Song S.W. and Wang Y.H., 1992 The balance and adjustment of boron in cotton ecosystem. In The Characteristics and Utilization of Soil Resource, pp. 330–358. Press of Beijing Agricultural University, Beijing, China.

Wang Y.H. and Zhou X.F., 1992 The Effects of boron on the antomical structure of cotton petioles. In Proceeding of the International Symposium on the Role of Sulphur Magnesium and Micronutrients in Balanced Plant Nutrition, pp. 81–84. Hongkong.

Wang Y.H. and Zhou X.F., 1994 Effects of boron on some elements, phenol, enzyme activity and hormone in cotton petiole. J. Plant Nutrition and Fertilizer Science **1**: 61–66.

Wang Y.H., Liu W.D., and Pi M.M., 1985 Study on the latent boron dificiency in cotton and the effective application of boron to cotton. Scientia Agricultura Sinica **2**, 62–70.

Wang Y.H., Liu W.D., and Pi M.M., 1987 The available boron content in soils on the cotton production region in the Changjiang Valley and the technique of boron application. China Cotton, 5, 29–32.

Wang Y.H., Liu W.D., and Pi M.M., 1982 Study on the boron dificiency in cotton and the effective application of boron to cotton. Industry of Inorganic Salts, 24–32.

Wang Y.H., Liu W.D., Pi M.M., and Wang Z.R., 1988 Boron deficiency in cotton and division of B-application in important producing cotton area of China. J Huazhong Agricultural University **6**, 153–156.

Wang Y.H., Tao W.Y., Liu W.D., Pi M.M., and Tang G.H., 1976 The first study of shrink of cotton leaves and failure of cotton acquire to flower in Shigushan Region. Cotton, 35–36.

Wang Y.H., Tao W.Y., Liu W.D., Pi M.M., and Tang G.H., 1978 The relation between failure of cotton square of flower and boron nutrition. Scientia Agricultura Sinica **3**, 61–64.

Wei W.X. and Wang Y.H., 1992 Effects of boron on the anatomical structure of Sesame. J Huazhong Agricultural University **11**(2), 164–165.

Wei W.X., Wang Y.H., and Sun X.Z., 1993 Effect of boron deficiency on leaf and pollen structure of sunflower. J Huazhong Agricultural University **12**(5), 464–467.

Wei W.X., Wang Y.H., Sun X.Z., and Qu. B., 1989 Effects of boron on chloroplasts and mitochondria of leaf mesophyll cells of sunflower. J Huazhong Agricultural University, **8**(4), 361–363.

Xie Q., Wei W.X., and Wang Y.H., 1991 Effects of boron on the anatomical structure of reproductive organs in cotton.J Huazhong Agricultural University **10**(2): 171–179.

Xie Q., Wei W.X., and Wang Y.H., 1992 Studies on absorption, translocation and distribution of boron in cotton. Acta Agronomica Sinica **13**(1), 31–37.

Xiong S.L., Wu L.S., and Wang Y.H., 2001 Study on relationship btween B-deficiency symptoms of cucumber and changes of endogenous hormone. J Plant Nutrition and Fertilizer Science **7**(2) 194–198.

Xiong S.L., Wu L.S., Wang Y.H., and Lan L.F., 1995 Absorption and distribution of boron in different varieties of Rape (Brassica napus L). J Huazhong Agricultural University **21**: 85–91.

Xiong S.L., Wu L.S., Zhao Z.Q., and Wang Y.H., 1996 Effects of boron on the growth and nodule nitrogenase activity of faba bean. J Huazhong Agricultural University **15**(5): 447–449.

Zhao Z.Q., Wang Y.H., and Wu L.S., 1998a Effect of boron deficiency on ethylene release of cotton, cucumber and rape, **15**(2), 63–66.

Zhao Z.Q., Wang Y.H., and Wu L.S., 1998b Effect of boron deficiency on auxin metabolism in cucumber, J Huazhong Agricultural University, **17**(3): 232–236.

Zheng W., Pi M.M., and Liu W.D., 1989 A study on the effects of boron on the carbon metabolism of Ramie. J Huazhong Agricultural University **8** (4), 354–360.

Zhou X.F. and Wang Y.H., 1993 Studies on effects of boron on cotton petiole. In Proceedings of International Cotton Symposium pp. 448–453.

An Alarming Boron Deficiency in Calcareous Rice Soils of Pakistan: Boron Use Improves Yield and Cooking Quality

A. Rashid[1]*, M. Yasin[1], M.A. Ali[2], Z. Ahmad[3] and R. Ullah[1]

[1]National Agricultural Research Center, Islamabad-45500, Pakistan Correspondence:

E-mail: abdul.rashid@comsats.net.pk;

[2]Adaptive Research, Agriculture Department, Lahore, Pakistan;

[3]Engro Chemical Pakistan Ltd., Thatta, Pakistan

Introduction

Rice (*Oryza sativa* L.), the world's leading staple food, is grown on 2.3 M ha alluvial, calcareous, low organic matter soils of Pakistan, giving an average paddy yield of 2.000 t ha^{-1} (GOP 2004). As fertilizer use in rice predominantly pertains to nitrogen (N), and to a lesser extent to phosphorus (P) and zinc (Zn) (Rashid et al. 2000), one major cause of low farm-level productivity compared with much higher potential yields (i.e. 4.00–4.50 t ha^{-1}, M. Akram, personal communication) is inadequate and imbalanced nutrient management

Despite being categorized tolerant to boron (B) deficiency (Shorrocks 1997; Savithri et al. 1999), yield increases with B use have been observed in Pakistan and elsewhere. In Pakistan, the first-ever positive responses of rice to B application were observed about three decades ago (Chaudhry et al. 1977). Though the observed yield increases with B use, in 16 field experiments on two predominant rice cultivars (i.e., *Basmati-370* and IR-6) were appreciable, it is only recently that this micronutrient disorder has received adequate research attention. While reviewing micronutrient constraints in rice systems, Savithri et al. (1999) have stated that B deficiency is not a severe problem in rice. However they postulated that B deficiency may be a problem in calcareous, sodic, and excessively permeable soils in reverine flood plains. In the mean time, Dunn et al. (2005) have reported rice yield increases with B use in Missouri, USA. This paper reports the impact of B nutrition on rice productivity, milling return and cooking quality, observed during 2002–2004 in our extensive field experiments carried out in major growing areas of Pakistan.

Materials and methods

Field experiments were carried out during 2002–2004 in the traditional rice-growing areas of the Pakistan's Punjab and Sindh provinces, using elite fine-grain basmati-type aromatic rice cultivars as well as medium-long grain coarse grain cultivars. Boron treatments in two sets of field experiments were:

F. Xu et al. (eds.), Advances in Plant and Animal Boron Nutrition, 103–116.

Experiment 1. The control (no B applied) and 1.0 kg B ha^{-1} use experiments carried out in Punjab province

Year	Cultivar	Field sites
2002	*Basmati*-385	3
	Super Basmati	3
2003	*Basmati*-385	5
	Super Basmati	5
	KS-282	5

Experiment 2. Graded B levels experiments carried out in Punjab and Sindh provinces

Year	Province	Cultivar	B applied (kg ha^{-1})	Field sites
2003	Punjab	*Super Basmati*	0, 0.5, 1.0, 2.0	8
2004	Punjab	*Super Basmati*	0, 0.5, 1.0, 1.5	8
	Sindh	IR-6	0, 0.5, 1.0, 1.5	3

Experiment 1 was laid out in a completely randomized design and was non-replicated; locations were treated as replications. Experiment 2 was randomized complete block, with four replications. Soil properties of field experimental sites are presented in Table 1. Soil B was determined by 0.1 M HCl extraction (Rashid et al. 1994) and colorimetry using Azomethine–H (Bingham 1982).

In all field experiments, B was broadcast applied as borax (10.5% B), prior to transplanting, along with basal fertilizers. Blanket fertilization consisted of 120 kg N ha^{-1} as urea, 44 kg P ha^{-1} as DAP, and 10 kg Zn ha^{-1} as zinc sulfate. All nutrients, except for N, were applied prior to transplanting. Nitrogen was applied in three splits: one-third each prior to transplanting, 21 days after transplanting and 45 days after transplanting. Average plot size was about 1000 m2 and crop management was the same in all treatments.

Experimental data included plant height, productive tillers, panicle sterility, and yield of paddy and straw at maturity. Flag leaves sampled at heading and mature paddy were analyzed for B by dry ashing (Gaines and Mitchell 1979) and colorimetry using Azomethine–H (Bingham 1982). Also, rice grain were analyzed for physico-chemical characteristics, organoleptic parameters, and eating quality.

Table 1. Soil properties of field experimental sites in Punjab and Sindh, Pakistan

Parameter	2002		2003		2004			
	Punjab, 5 sites		Punjab, 6 sites		Sindh, 3 sites		Punjab, 8 sites	
	Range	Mean	Range	Mean	Range	Mean	Range	Mean
pH (1:1)	7.9–8.4	8.1	7.9 – 8.8	8.2	8.1 – 8.8	8.5	7.9–8.3	8.1
Organic matter (%)	0.4–1.1	0.6	0.8 – 1.8	1.1	0.7 – 0.9	0.8	0.9–1.5	1.0
$CaCO_3$ (%)	1.9–4.5	2.9	1.5 – 5.7	2.5	11.5–13.5	12.5		
Electrical conductivity (1:1) (dS m^{-1})	0.4–0.6	0.5	0.3 – 1.5	0.8	2.4 – 2.5	2.4	0.3–1.2	0.6
AB-DTPA extractable (mg kg^{-1})								
P	1.8–10.4	6.3	5.3– 14.0	10.2	3.4 – 7.3	5.4	3.4–6.8	5.4
K	86–186	131	88 – 200	146	58 – 180	130	58–180	109
Zn	0.8–2.4	1.8	0.7 – 2.5	1.5	0.9 – 3.1	1.8	0.9–3.1	1.0
Dilute HCl extr. B (mg kg^{-1})	0.26–0.44	0.35	0.15–0.42	0.34	0.43–0.51	0.45	0.21–0.44	0.29

Results and discussion

Paddy yield increases with boron application

In Experiment no.1, having 0 and 1.0 kg ha^{-1} B rates, paddy yield increases with B use were 17–20% in *cv*. Super Basmati, 20–23% in cv. Basmati-385, and 14% in *cv.* K-282 ($P \leq 0.05$; Table 2). Yield increases were consistent in all rice cultivars at all field sites during both experimental years. In Experiment No. 2, carried out during 2003 and 2004 at 14 locations employing graded levels of fertilizer B, paddy yield increases with B application were substantial in both rice varieties (*cv*. Super Basmati and IR-6) at all field sites ($P \leq 0.05$; Figs. 1-3). In all cases, maximum paddy yield was mostly obtained with 1.0 kg B ha-1, and no further yield increase was observed with greater B rates (Figs. 2-3). Paddy yield increases with B were 11–31% during 2003 and 7–30% during 2004 in *cv*. Super Basmati, and 18–34% in cv. IR-6. Mean paddy yield increases with B use were 26% in cv. Super Basmati and 27% in *cv.* IR-6. However, near-maximum (95% of maximum) paddy yield was associated with 0.75 kg B ha^{-1} for *cv* Super Basmati and 0.85 kg B ha^{-1} for cv. IR-6 (Figs. 2-3).

Table 2. Paddy yield, agronomic traits and plant B content as affected by B application in calcareous soils of Pakistan

Cultivar	B applied (kg ha^{-1})	Yield (t ha^{-1})		Panicle sterility (%)	Plant height (cm)	Productive tillers hill^{-1}	1000-grain weight (g)	B concentration (mg kg^{-1})		B uptake (g ha^{-1})	B use efficiency (%)
		Paddy	Straw					Leaves	Paddy		
					2002 (5 Sites)						
Super Basmati	0	3.23	5.15a	23a	116a	18.4	19.0a	5.5a	1.73a	17.2a	
	1	3.89	5.88b	14b	122b	20.1	20.2b	8.5a	2.51b	34.0b	1.68
	LSD (0.05)	0.68	0.41	4	2	NS	0.8	0.6	0.29	1.4	
Basmati-385	0	3.77a	5.43a	28a	134a	14.3	19.4a	5.3a	1.33a	17.4a	
	1	4.72b	6.63b	16b	140b	16.1	20.1b	8.2b	2.49b	35.2b	1.78
	LSD (0.05)	0.45	0.29	8	5	NS	0.6	0.5	0.18	1.5	
					2003 (6 Sites)						
Super Basmati	0	3.78b	9.74b	18a	131b	19b	21.2	7.2b	1.60b	31.1b	3.1
	1	4.39a	11.30a	12b	135a	22a	21.9	9.3a	2.74a	61.7a	
	LSD (0.05)	0.60	0.36	6	1	3	NS	1.3	0.24	12.9	
Basmati-385	0	3.69b	9.46b	24a	158b	16b	19.4	7.2b	1.63b	29.2b	-
	1	4.54a	10.86a	19b	160a	19a	20.3	8.8a	2.60a	57.1a	2.8
	LSD (0.05)	0.79	0.87	5	2	2	NS	1.0	0.20	13.7	
KS-282	0	4.82b	12.17b	15a	109b	12b	24.5	6.9b	1.69b	35.4b	
	1	5.48a	13.30a	12b	113a	15a	25.2	8.3a	2.58a	69.6a	3.4
	LSD (0.05)	0.64	0.94	5	3	2	NS	1.2	0.08	8.9	

Figure 1. Substantial improvement in paddy (*cv.* Super basmati) yield with B use in a calcareous rice soil of Pakistan

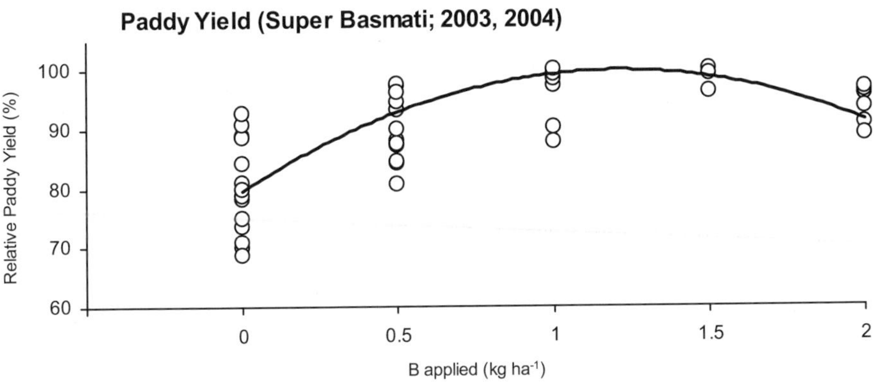

Figure 2. Relationship between fertilizer B rate and paddy yield of *cv. Super Basmati* (maximum yield: 2003, 3.51 t ha^{-1}; 2004, 3.61 t ha^{-1})

In Experiment 1, straw yield increases with B application ($P \leq 0.05$; Table 2) were slightly lesser than paddy yield increases. For example, mean increases in straw yield were 14–16% in *cv.* Super Basmati, 15–22% in cv. Basmati-385, and 9% in *cv.* KS-282. In Experiment No. 2, straw yield increases with B use were 15–36% during 2003 and 6–23% during 2004 in *cv.* Super Basmati (Fig. 4) and 18–34% in *cv.* IR-6 during 2004 (Fig. 5). Mean increase in straw yield with B use was 19% in *cv.* Super Basmati and 23% in cv. IR-6. Maximum straw yield increase of *cv.* Super Basmati was 24% during 2003 and 19% during 2004. The cv. IR-6 straw yield increase with B use was 23%. In short, overall straw yield increases were not much different than paddy yield increases (Table 2; Figs. 4, 5). Thus, contrary to the general perception that B deficiency in cereals hampers grain set more than vegetative growth (Rerkasem and Jamod 1997), our results affirm that B deficiency affects vegetative growth and paddy yield almost equally. This is also in conformity to our earlier findings with rice (Rashid et al. 2002c), wheat (Rashid et al. 2002a), sorghum (Rashid et al. 1997) and rapeseed (Rashid et al. 2002d).

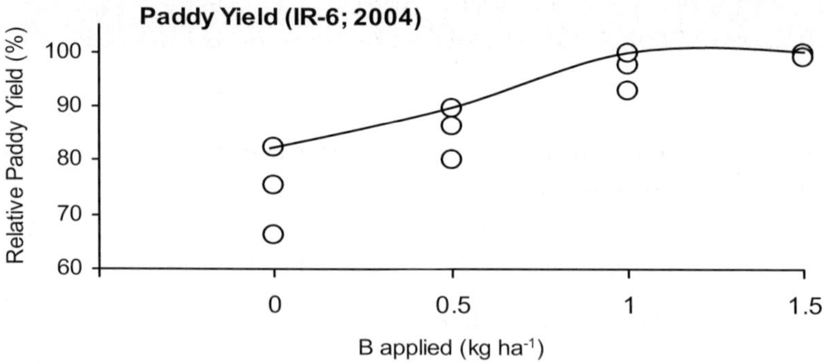

Figure 3. Relationship between fertilizer B rate and paddy yield of *cv.* IR-6 (maximum yield: 5.81 t ha^{-1})

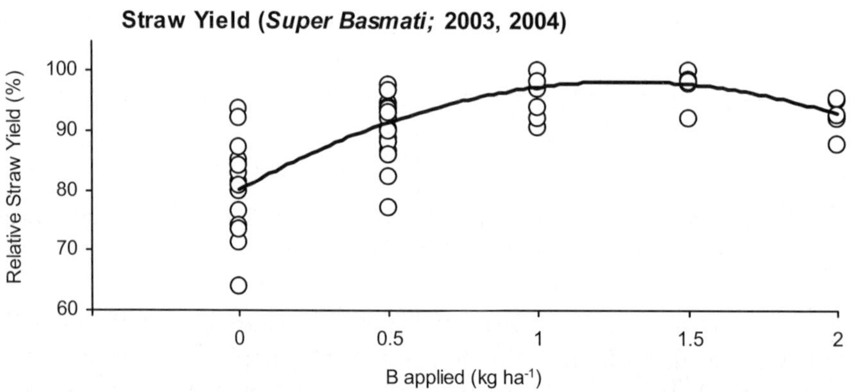

Figure 4. Relationship between fertilizer B rate and straw yield of rice (maximum yield: 2003, 8.97 t ha^{-1}; 2004, 5.80 t ha^{-1})

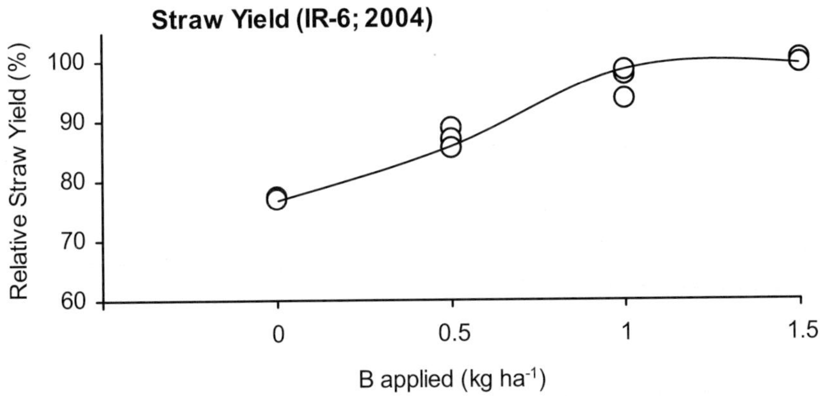

Figure 5. Relationship between fertilizer B rate and straw yield of *cv.* IR-6 (maximum yield: 10.12 t ha^{-1})

The generally suggested critical level of B in youngest mature leaves of rice is 6 mg B kg^{-1} (Jones et al. 1991; Reuter et al. 1997).Whereas B concentration in flag leaves of Basmati cultivars was lesser than 6 mg kg^{-1} during 2002, it was 7.2 mg kg^{-1} in leaves of both cultivars and 6.9 in leaves of cv. KS-282. Thus, internal B requirement in rice leaves as affected by environmental conditions warrants investigation. Boron concentration in mature paddy is hardly reported in the literature. Moreover, nutrient concentration in mature grain was believed to remain unaffected by nutrient status of the soil. However, in this investigation B concentration in rice paddy increased appreciably with B application ($P \leq 0.05$; Table 2). This is in conformity to our earlier finding with Zn nutrition of a number of field crops (Rashid and Fox 1992; Rafique et al. 2006).

Thus, this multi-location, multi-year field research, conducted through the major rice growing areas of Pakistan, revealed that B deficiency is a field-scale problem in fine-grain Basmati rice (*cv*. Super Basmati and Basmati-385) as well as in coarse-grain rices (*cv.* KS-282 and cv. IR-6).

Though Dunn et al. (2005) observed maximum rice yields in acid soils of Missouri having hot water extractable (HWE) B levels as low as 0.25–0.35 mg B kg^{-1} soil, the generally suggested threshold level of HWE soil B for adequate B nutrition of most field crops is >0.5 mg B ha^{-1} (Rashid et al. 1994). And this has been true in this extensive field investigation as well, because appreciable yield increases with B use were observed in calcareous soils having B levels as high as 0.35–0.51 mg kg^{-1} soil (Tables 1, 2; Figs. 1-5). Thus, all the field soils included in this study were deficient in B to varying degrees. However, the literature generally suggests a possibility of B toxicity in salt-prone calcareous soils (Sillanpaa 1982; Yau 1997). Contrarily, in our experience, rice crop grown in high pH sodic soils (pH up to 8.8; Table 1) suffered with B deficiency. In fact, incidence B deficiency has also been observed in rice grown in salt-affected (i.e., saline as well as saline-sodic) calcareous soils of Pakistan (Aslam et al. 2002). Previously, we had observed that cotton grown in saline calcareous soils (EC >4 dS m-1) also suffered with B deficiency rather than toxicity (Yasin et al. 2002). Therefore, crops grown in alluvial, calcareous soils of Pakistan are prone to B deficiency problem rather than toxicity.

Boron-bearing soil mineral, tourmaline, is believed to be the ultimate source of B in soil; however, soil organic matter (SOM) is the immediate source of B supply to plant roots (Berger and Truog 1944). Because of peculiar agro-climatic conditions in rice-wheat belt of Pakistan, the soils are inherently low in organic matter (Table 1). Moreover, SOM has got reduced badly, post Green Revolution, from an average of 1.02% during 1967–1974 to 0.59% during 1985–1994 (Byerlee at el., 2003). This sharp decline in SOM appears to be a major cause of soil B stock decline and, hence, increased incidence of B deficiency in rice crop. Additionally, we postulate that enhanced mining of soil B with more intensive cropping and greater biomass production per unit field area, during the post-green revolution era, coupled with the suspected leaching of B beyond the root zone, are responsible for more widespread and severe B deficiency in rice.

Boron nutrition and panicle sterility in rice

Though B application improved plant height and rice grain rice ($P \leq 0.05$; Table 2), paddy yield increases accrued primarily because of appreciable reduction in panicle sterility (Table 2; Figs. 6, 7) and increases in productive tillers per hill (Fig. 8). Boron deficiency in cereals is already known to induce panicle sterility in cereal crops including wheat (Rerkasem and Jamod et al. 2004). However, there is hardly any report of B-induced panicle sterility in rice. Though sterility in rice may be caused by many biotic and abiotic stresses, this research adequate establishes that B deficiency induced panicle sterility is a major cause of low rice productivity in calcareous soils.

Figure 6. Alleviation of rice panicle sterility and post-harvest shedding by using B in a calcareous soil of Pakistan

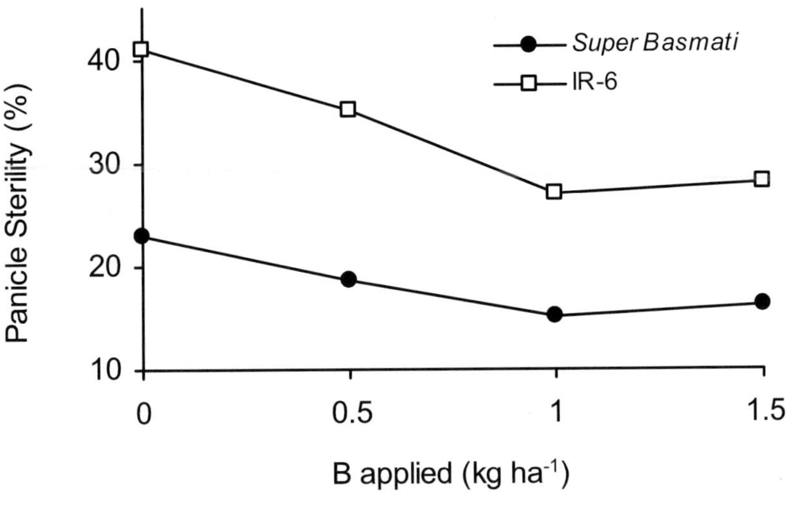

Figure 7. Panicle sterility in rice as affected by boron fertilization

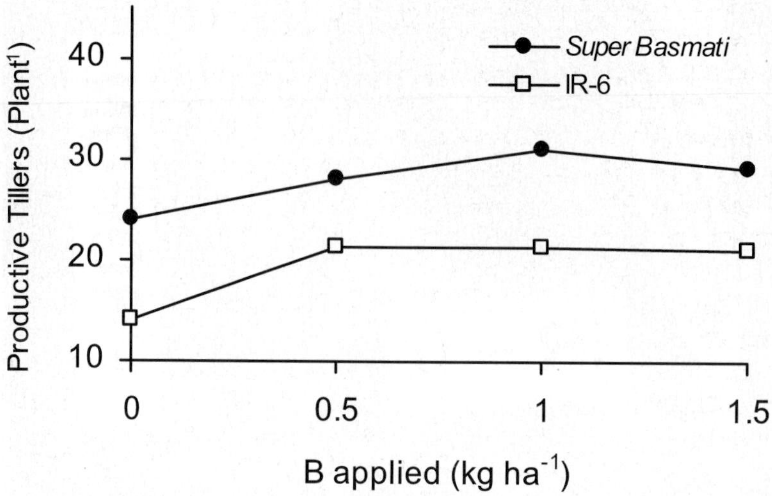

Figure 8. Productive tillers in rice as affected by boron fertilization

Milling return and grain quality improvements with boron use

The present research revealed that B deficiency in calcareous soils not only hampers paddy yield but also impairs grain quality. In our experience, fertilizer B application to rice improved milling return as well as head rice recovery ($P < 0.05$; Table 3). An improvement in B nutrition of rice plants also improved desirable cooking quality traits like quality index, kernel elongation ratio upon cooking, brusting-upon-cooking, and alkaline spreading value ($P \leq 0.05$; Table 3). Improvement in grain quality traits is attributed to better grain filling, facilitated with adequate B nutrition of rice plants. Rice is not only a staple cereal food for the local population and a delicacy at local festivals, it is also a major foreign exchange earner for the country. Its improved milling and cooking quality with adequate B nutrition — particularly of fine grain Basmati type rices — is of paramount significance both for local as well as international markets. Thus, adequate B fertility of calcareous soils, or B fertilizer use in deficient situations, is a prerequisite for harvesting optimum yields of good quality rice.

Boron use efficiency and residual effect of fertilizer boron

As a consequence of substantial increase in B content in plant tissues (Table 2) as well as enhanced biomass production with B application (Table 2; Figs. 1-5), total B uptake by rice crop, in all cultivars during all years, got more than doubled with B fertilization (Table 4). However, fertilizer B use efficiency by the current rice crop (i.e., the fraction of applied B taken up by above-ground plant parts) was very low, i.e., 1.68–1.78% in 2002, 2.8–3.4% during 2003 (Tables 2, 4) and 1.59–2.38% during 2004 (Table 4). Though a major fraction of the unutilized fertilizer B may get fixed in the calcareous soils (Brady and Weil, 2002) or may get leached down the profile beyond root zone because of imperfect plow pan in flooded paddy fields, yet even a safe dose of 0.75–1.0 kg B ha-1 is affected leave a beneficial residual effect on succeeding crops in the rotation. Like rice, wheat is also categorized as less sensitive to B deficiency (Rerkasem and Jamod 1997), yet suffers with this nutritional disorder in Pakistan (Rashid et al. 2002 a, b) and elsewhere (Rerkasem and Jamod 2004). Thus, wheat in the rice–wheat system would benefit tremendously from residual effect of fertilizer B applied to rice. As rice is a cash crop and, thus, is relatively a high–input crop in the rotation, we suggest B application to rice rather than wheat.

Table 3. Impact of boron application on rice (*cv. Super Basmati*) grain and cooking quality during 2004

Grain characteristic	Boron applied (kg ha^{-1})				LSD (0.05)
	0.0	0.5	1.0	1.5	
Total milled rice (%)	70.4c	71.6b	72.3a	71.7b	0.5
Head rice (%)	51.6c	54.6b	57.0a	56.4a	0.6
Quality index (L/ BT)	2.96b	2.96b	3.0a	2.97b	0.03
Elongation ratio upon cooking	1.88d	1.94c	2.00a	1.98b	0.02
Bursting-upon-cooking (%)	9.9a	9.0b	8.0c	8.4c	0.5
Alkali spreading value (Score 1–7)[1]	4.5c	4.8b	4.9a	4.8b	0.1

1 Alkali spreading value: 4–5 score = Intermediate G.T. type rice
* Means followed by different letters are statistically different at LSD (0.05)

Economics of boron use in rice

Considering paddy yield increases alone, B application in rice proved highly cost-effective, with value–cost ratio (VCR) of 32–42:1 in *cv.* Super Basmati (Table 4), 55:1 in *cv.* Basmati-385, and 36:1 in cv. IR-6. Improvements in milling return, head rice recovery, and cooking quality traits were added advantages of great economic significance. Also, fertilizer B leaves an appreciable residual effect as well as improves efficiency of other farm inputs including major nutrient fertilizers. As the recommended B dose in rice is 0.75 kg ha^{-1}, farmer-level VCRs are expected to be better than the ones observed in our research investigations. Thus, B fertilizer use in rice crop will increase farm productivity, farmer profitability, national economy and soil resource sustainability.

Table 4. Efficiency and economics of fertilizer boron use in rice (*cv. Super Basmati*) in Punjab

B applied (kg ha^{-1})	Yield (t ha^{-1})		B uptake (g ha^{-1})	B use efficiency (%)	Value: Cost ratio
	Paddy	Straw			
		2003 (6 Sites)			
0	2.80c	6.84c	17.6	—	—
0.5	3.20b	7.98b	31.2	2.73	46:1
1.0	3.51a	8.97a	45.1	2.75	42:1
2.0	3.28	8.29b	42.4	2.48	14:
LSD (0.05)	0.05	0. 55			
		2004 (8 Sites)			
0	2.88c	5.0	13.4	—	—
0.5	3.25b	5.3	25.3	2.38	38:1
1.0	3.51ab	5.6	31.6	1.82	32:1
1.5	3.61a	5.8	37.2	1.59	25:1
LSD (0.05)	0.27	N.S.			

Conclusions

In B deficient soil situations, increased incidence of panicle sterility and less number of productive tillers per plant are major causes of yield reduction in rice. Deterioration in grain cooking quality is an additional serious set-back caused by B deficiency. Thus, rice, a major staple crop in the Indo-Gangetic Plains and elsewhere in the world, appears to be suffering significant yield and quality losses with soil B deficiency. Luckily, its remedy is simple and effective as soil B application enhances crop productivity as well as improves milling return and cooking quality of rice. Use of B is a highly cost-effective solution to the problem. As a single dose of B may last for several crop seasons, rice growers are encouraged to include B in their fertilizer use program. However, experience in other part of the world dictates that B use can get adopted only through B-fortified major nutrient fertilizers. Therefore, fertilizer industry must supply an appropriate concentration of B in major nutrient fertilizer formulation for rice crop.

Acknowledgements

We thank Chaudhry Abdul Ghaffar, Director General, Agricultural Extension and Adaptive Research, Punjab and M/S Engro Chemical Pakistan Ltd. for support in field experimentation, Mr. M. Asif Ghumman for statistical analysis, and Mr. M. Jawad and Mr. M. Irfan for word processing, graphics and composing of the manuscript .

References

Aslam M., Mahmood I.H., Qureshi R.H., Nawaz S., and Akhtar J., 2002 Salinity tolerance of soil as affected by boron nutrition. Pakistan J. Soil Sci **21**: 110–118.

Berger K.C. and Truog E., 1944 Boron tests and determination for soils and plants. Soil Sci **57**: 25–36.

Bingham F.T., 1982 Boron. In: Page A.L., (ed) Methods of Soil Analysis, Part 2: Chemical and Microbiological Properties edited by American Society of Agronomy: Madison, W.I., USA, pp. 431–448.

Brady N.C. and Weil R.R., 2002 The Nature and Properties of Soils, 13th ed. Prentice Hall, New Jersey, USA.

Byerlee D., Ali M., and Siddiq A., 2003 Sustainability of the rice-wheat system in Pakistan's Punjab: how large is the problem? In: Ladha J.K., Hill J.E., Duxbury J.M., Gupta R.K., Buresh R.J. (eds) Improving the Productivity and Sustainability of Rice-Wheat Systems: Issues and Impacts. ASA Special Pub. 65. American Society of Agronomy, Madison, Wisconsin, USA pp. 77–96.

Chaudhry F.M., Latif A., Rashid A., and Alam S.M., 1977 Response of rice varieties to field application of micronutrient fertilizer. Pakistan. J. Sci. Indus. Res. **19**: 134–139.

Dunn D., Stevens G., and Kending A., 2005 Boron fertilization of rice with soil and foliar applications. Online Crop Management. doi: 10.1094/CM–2005–0210–01–RS.

Gaines T.P., and Mitchell G.A., 1979 Boron determination in plant tissue by the azomethine-H method. Commun. Soil Sci. Plant Anal. **10**: 1099–1108.

GOP 2004 Pakistan Economic Survey 2004–2005. Ministry of Food, Agriculture and Livestock, Government of Pakistan, Islamabad.

Jones J.B. Jr., Wolf B., and Mills H.A., 1991 Plant Analysis Handbook: A Practical Sampling, Preparation, Analysis, and Interpretation Guide. Micro–Macro Publishing: Athens Georgia, pp. 312.

Rafique E., Rashid A., Ryan J., and Bhatti A.U., 2006 Zinc deficiency in rainfed wheat in Pakistan: Magnitude, spatial variability, management, and plant analysis diagnostic norms. Commun Soil Sci Plant Anal **37**: 181–197.

Rashid A., Rafique E., Muhammad S., and Bughio N., 2002d Boron deficiency in rainfed alkaline soils of Pakistan: Incidence and genotypic variation in rapeseed-mustard. In: H.E. Goldbach et al. (ed.), Boron in Plant and Animal Nutrition. Kluwer Academic/Plenum Polishers, New York pp 363–370.

Rashid A. and Fox R.L., 1992 Evaluating the internal zinc requirements of grain crops by seed analysis. Agron J **84**: 469–474.

Rashid A., Kausar M.A., Hussain F., and Tahir M., 2000 Managing zinc deficiency in transplanted flooded rice grown in alkaline soils by nursery enrichment. Tropical Agriculture (Trinidad) **77**: 156–162.

Rashid A., Rafique E., and Bughio N., 2002a Boron deficiency in rainfed calcareous soils of Pakistan. II. Incidence and boron requirement of wheat In: Goldbach H.E., Brown P.H., Rerkasem B., Thellier T., Wimmer M.A., Bell R.W., Eds. Boron in Plant and Animal Nutrition, Kluwer Academic Publishers: New York pp. 339–348.

Rashid A., Rafique E., and Bughio N., 1994 Diagnosing boron deficiency in rapeseed and mustard by plant analysis and soil testing. Commun. Soil Sci. Plant Anal. **25**: 2883–2897.

Rashid A., Rafique E., and Ryan J., 2002b Establishment and management of boron deficiency in field crops in Pakistan. A country report. In Goldbach H.E., Boron P.H., Rerkasem B., Thellier T., Wimmer M.A., Bell R.W., (eds.) Boron in Plant and Animal Nutrition, Kluwer Academic Publishers: New York, pp. 339–348.

Rashid A., Muhammad S., and Rafique E., 2000 Genotypic variation in rice susceptibility to boron deficiency. International Rice Research Notes (IRRI, Philippines) 25. **3**: 29–30.

Rerkasem B. and Jamod S., 2004 Boron deficiency in wheat: a review. Field Crops Res. **89**: 173–186.

Rerkasem B. and Jamod S., 1997 Boron deficiency induced male sterility in wheat (*Triticum aestivum* L.) and implications for plant breeding. Euphytia **96**: 257–262.

Reuter D.J., Edwards D.G., and Wilhelm N.S., 1997 Temperate and tropical crops. In: Reuter D.J., Robinson J.B., (eds.) Plant Analysis: An Interpretation Manual, 2nd ed; CSIRO Publishing: Collingwood, Victoria, Australia, pp. 81–284.

Savithri P., Perumal R., and Nagarajan R., 1999 Soil and crop management technologies for enhancing rice production under micronutrient constraints. In: Balasubramanium V., et al. (eds) Resource Management in Rice Systems: Nutrients. Kluwer Academic Publishers, Dordrecht, pp. 121–135.

Shorrocks V.M., 1997 The occurrence and correction of boron deficiency. Plant Soil **193**: 121–148.

Yasin M., Rashid A., and Rafique E., 2002 Boron status of cotton soils: Relationship with salinity. In: Abstracts, 9th Congress of Soil Science, 18–20 March 2002, Faisalabad. Soil Science Society of Pakistan, Islamabad, p. 25.

Sillanpaa M., 1982 Micronutrients and the Nutrient Status of Soils: A Global Study. Food and Agriculture Organization of the United Nations, Rome.

Yau S.K., 1997 Differential responses of barley, durum and bread wheat to high levels of soil boron. In: Ryan J. (ed), Accomplishments and Future Challenges in Dryland soil Fertility Research in the Mediterranean Area. ICARDA, Aleppo, Syria. pp. 209–218.

Hidden Hunger for B in Tropical Cauliflower: Evaluation of B Sources and Methods for Correction

M. Edward Raja
Indian Institute of Horticultural Research, Bangalore, India.
E-mail:medward@iihr.ernet.in

Introduction

India is a predominantly vegetarian country but per capita vegetable consumption is not up to the desired national standards due to low productivity. Though introduction of F-1 hybrids in recent times has resulted in substantial yield increase, the full genetic yield potential could not be realized, due to several yield limiting factors including nutrient deficiencies. Though cauliflower is temperate in its climate requirement, introduction of heat tolerant tropical cauliflower F1 hybrids have transformed periurban cauliflower production around Bangalore, situated in semi arid tropics of India. But the yield and quality are not optimum due to nutrient deficiencies especially. B Visible deficiency symptoms (like browning of curds, hollow stem) and yield response to B have been recorded in local varieties (Kotur 1994) in traditionally B deficient regions like Chotanagpur and Assam in North India (high rainfall, coarse soils with high humidity) but not in semi arid regions of southern India like Bangalore. But yields are low at 10-15 t ha^{-1} and quality of the curds (compactness and color) is affected resulting in low market acceptability. Since sub clinical deficiency or hidden hunger is a major problem in a comparatively low input field crop production in developing countries like India (Wallace 1982) identifying them is a challenge. Nutrient deficiency expression is genotype dependent and in some crops like wheat and barley plants result in total sterility due to copper and Manganese deficiencies without visible expression of any symptoms. Hence breeding nutrient efficient varieties is a better way of solving the hidden hunger (Graham, 1984). Since breeding for such nutrient efficiency is a very time consuming task, correction of the probable deficiencies by properly standardized input is necessary. Since Cauliflower is known to be highly susceptible to B deficiency a study was undertaken in the field to evaluate different sources and methods of B correction of B for eliminating hidden hunger for high yield and quality in F1 hybrids. Since the F1 hybrids have a fast, growth pattern (Srinivas and Bhatt 1994) the established soil and plant critical levels of nutrients for the low yielding open pollinated varieties (not F1 hybrids) is redundant. Hence there

F. Xu et al. (eds.), Advances in Plant and Animal Boron Nutrition, 117–124.

is a need to follow new critical levels for these F1 hybrids (Anonymous 1989). Hence in this study though the available B (hot water soluble B) is at 0.48 mgkg^{-1} which is nearer to the critical level of 0.5 mgkg^{-1}, the response study was initiated so that the full yield potential and best quality can be realized.

Materials and methods

A green house and field experiments were conducted to evaluate the B sources and methods of correction of the hidden hunger in F1 hybrid Cauliflower in Bangalore region

Green house experiment

Two cauliflower cultivars Pusa Katki a local open pollinated cultivar and F1 hybrid 'Suvasini' both identified as late season cultivars (Nov – March) were selected for the study. The open pollinated local variety Pusa Katki had a slow growth rate and B uptake, whereas the F_1 hybrid showed a faster growth rate and higher B uptake. The soil samples were analyzed by procedure enlisted in Jackson (1976) and plant B by method of Gains and Mitchel (1979). The experimental farm soil of IIHR, Bangalore (Typic paleustalf) was filled up in plastic pots and the 25 days old seedlings of the above cultivars were transplanted at the rate of 2 plants/pot and the plants were grown till the final harvest. The B uptake was studied at 2 stages: Before button initiation and at harvest. The dry matter gain and B, Zn, K and P uptake at two stages were recorded and plotted in a graph to establish the nutrient requirement at two important stages which will decide the treatments for the field experiment. The Zn was analyzed by atomic absorption spectrometry, K by flame photometry and P by colorimetric.

Field experiment

The following treatments were followed for a field experiment (2003 Nov to March 2004) to evaluate B sources and methods of application. There were 3 sources of B: Borax, (11% B) Diocta borate (DOB 20% B) for foliar spray treatment and sodium tetra borate (STB 16% B) and borax (11% B) for soil application. Two soil levels (1.0 and 1.5 kg B/ha) as basal application and 2 foliar spray concentration (1g L^{-1} and 2g L^{-1} for DOB (solubor) and 1.9g L^{-1} and 3.8 g L^{-1} for Borax) two times at 30 and 60 days after planting and in all there were nine-treatment combination with a No B control. The randomized block design was followed for statistical requirement. The B content, the yield and quality parameters like compactness, color at different post harvest periods were recorded, statistically analyzed by ANOVA and results are presented and discussed below.

Results

The soil at experimental station of IIHR Bangalore (Table-1) is a member of fine, mixed Isohyperthermic family of typic Paleustalf. They have neutral sandy loam to loam surface soil developed on weathered gneiss and nearly level to gently sloping land at an elevation of 900 feet above MSL. They are hard setting soil with a bulk density of 1.45 g cc^{-1}which is a constraint in water and nutrient management. They are low in organic matter (0.62%) but the hot water soluble B is 0.46 mg kg^{-1} in the surface (0-15 cm) soil and subsoil (15-45cm) have higher B than surface soil and according to the critical level fixed (Berger and Truog, 1939) it can be described B-sufficient soil. The climate is semi arid (Fig-1) with an annual rainfall of 750-850 mm (bulk of it is distributed within 4 months) with low mean humidity (30-40%) and a mean annual temperature of 28^0C which is ideal for the tropical cauliflower production.

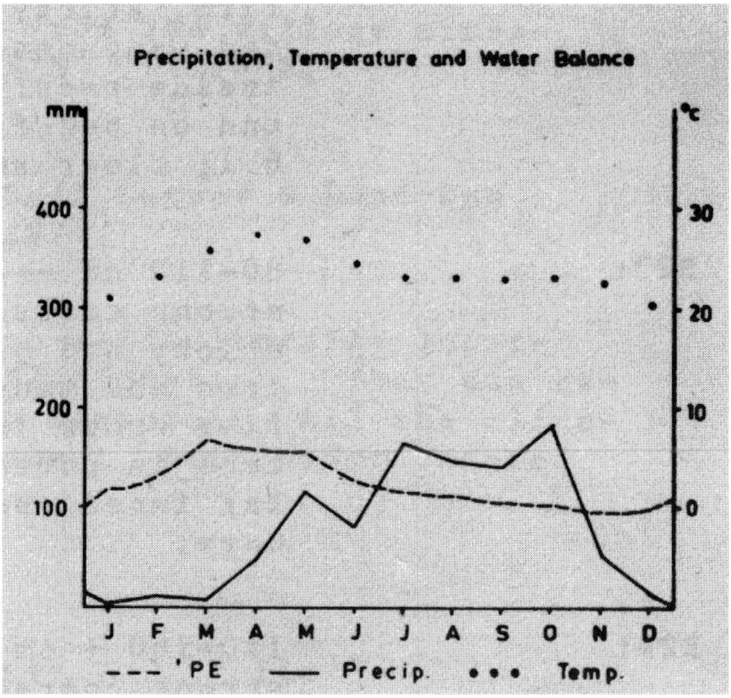

Figure 1. Weather parameters of IIHR farm Bangalore

Table 1. Physical chemical properties of the experimental farm soil at IIHR Bangalore

Soil Depth cm	pH (1:2.5)	EC dSm^{-1}	Texture	Clay %	CEC Cmol $(+)kg^{-1}$	OC%	HWS B	DTPA Zn	DTPA Fe	DTPA Mn	Exchangeable K	Base saturation	Bulk density gcc^{-1}
							mg/Kg						
0-15	6.2	0.78	Loam	18	9.2	0.62	0.46	0.62	12.1	14.1	184	75	1.45
15-30	6.4	0.94	Clay Loam	30	14.2	0.73	0.52	0.41	13.4	12.1	144	81	1.58
30-45	6.8	0.93	Clay Loam	34	13.8	0.28	0.47	0.33	14.3	13.1	133	84	1.68

Table 2. Nutrient uptake pattern of two cauliflower cultivars at two growth stages

Particulars	Dry Matter (g) 60 d	Mean Nutrient uptake(mg/plant) (Button initiation stage 60 d after planting						Dry matter (g) 120d	Mean nutrient uptake(mg/plant) at harvest					
Cultivars		Fe	B	Zn	P	K	Ca		Fe	B	Zn	P	K	Ca
Pusa Katki (open pollinated)	85	3.82	2.21	2.38	18.7	195..5	200	145	7.45	2.9	3.194	40.6	275.5	261
		(45)	(26)	(28)	(0.22%)	(2.3%)			(52)	(20)	(22)	(0.28%)	(1.9%)	(1.8%)
Suvasini F1	165	5.45 (33)	31.35 (19)	3.63 (0.18%)	29.7 (1.9%)	313.5 (1.3%)	1.7 (1.3%)	425	11.05 (26)	11.9 (28)	7.65 (28%)	89.5 (0.21%)	722.5 (1.7%)	382.5 (0.9%)

Values in( ) Indicate Concentration

Table 3. Comparative studies on the effects of different boron sources on quality and yield in cauliflower cv. F_1 hybrid: Suvasini

Treatment	Mean Curd weigh (kg)	Mean curd diameter (cm)	Mean Yield (kg) / plot(size=19 m^2)	Yield (t/ha)	Compactness (ratio of Wt. in Kg/Dia meter in m)	Stability of curd Colour at room Temp. * Rating in hours. Visual Score. 24 hrs	48 hrs	72 hrs
T1	0.95	17.6	44.57	23.58	5.36	2.5	2	1.5
T2	1.7	18.8	57.86	30.61	9.03	4.5	3.8	2.5
T3	1.24	17.8	51.48	27.23	6.96	4.3	3.9	2.4
T4	1.15	18.2	48.76	25.8	6.33	3.8	3.5	2.8
T5	1.17	18.5	49.11	25.98	6.33	3.7	3.5	2.5
T6	1.68	19	57.45	30.39	8.44	4.4	4	3.5
T7	1.65	17.6	55.75	29.49	9.37	4.3	3.9	2.8
T8	1.11	17.93	47.95	25.36	6.2	3.9	3.4	3
T9	0.96	18.26	46.38	24.54	5.23	3.8	3.5	3.2
	CD@5%=0.063		CD@5%=0.98	CD@5%=0.541	CD@5%=0.62			
	SEM=0.02		SEM=0.329	SEM=0.180	SEM=0.207			

Note : T1 , Control
T2, Two foliar sprays of Diocta borate @ 1.0 g/L
T3, Two foliar sprays of Diocta borate@ 2.0 g/L
T4, Two foliar sprays of Borax @ 1.9 g/L
T5, Two foliar sprays of Borax@3.8 g/L
T6, Soil application of Sodium tetraborate @ 1.0 Kg B/ha
T7, Soil application of sodium tetraborate@ 1.5 Kg B/ha.
T8, Soil application of Borax@ 1.0 Kg B/ha
T9, Soil application of Borax @ 1.5 Kg B / ha

Effect of different B sources and methods of application on Cauliflower

The results of field experiment on comparative effect of different B sources on quality and yield of Cauliflower (Table-3) indicated two foliar sprays of DOB (solubor) at 0.1% has given significantly higher yield of 30.6 t ha^{-1} which is on par with yield by sodium tetra borate (STB) soil application at 1.0 and 1.5 kg B/ha^{-1} over 23.58 t ha^{-1} recorded in no B control. Both foliar and soil application of Borax though recorded higher yield than no B control was very significantly less than the DOB spray and STB soil application. The DOB and STB also resulted in better quality of curds in terms of compactness and stability of curd colour when compared to other treatments.

Table 4: Effect of B sources and methods on B concentration of cauliflower F_1 hybrid *cv*. Suvasini (mg kg^{-1})

Particulars	Samples drawn from field experiment at 90 d after planting			
Treatment	Leaf	Curd	Stem	Root
No B control	28	28	30	28
0.1% DOB spray	107	47	31	31
0.2% DOB spray	231	48	37	44
0.19% Borax spray	90	32	25	36
0.38% Borax spray	170	45	22	28
1Kg /ha B-STB	40	42	34	28
1.5 B/ha B-STB	48	43	31	37
1.0Kg B/ha as Borax	41	36	21	25
1.5 Kg B/ha as Borax	42	38	29	28
CD 5%	11	9	7	9

Boron status of cauliflower

The effect of B sources and methods of application on B status of cauliflower (Table-4) indicated the foliar spray of DOB maintained highest B in leaf and curds followed by foliar spray of Borax. It is significant to note that the curd maintained B content of 28 mg kg^{-1} B in no B control where as the local cultivar had only 20mgkg^{-1} in the curds indicating the higher B requirement of F1 hybrids compared to local cultivars.

Discussion

With advent of heat tolerant F-1 hybrid cauliflowers the tropical cauliflower production has a received tremendous boost since it is being cultivated in non traditional areas like semi arid tropics of India. But the low yield and poor quality resulted in reduced market acceptability and profitability to farmers. This study has helped in identifying hidden hunger (sub clinical deficiency) of B as one of the important reasons for this poor performance of the F1 hybrids and this is attributed to several reasons. The soils of semi arid regions have some in built constraints for cauliflower due to low organic matter and high bulk density and hard setting nature (Mullin et al. 1990). The hard setting nature of these soil results in restriction on root proliferation and moisture availability resulting in inadequate exploration of the sub soil for nutrient mobilization. Since B is taken up by mass flow soil restriction on root limit the availability of B.

The fast growth rate of F-1 hybrid especially after button formation in terms of dry matter increase as curds from 165g to 425g (a 157% increase) and B uptake by 144% (Table-2) is another reason for the inability of the soil to meet the B requirement though the soil is having enough available B in the surface and sub soil (0.46 and 0.52 $mgkg^{-1}$ respectively). Cultivar difference in susceptibility to B deficiency has been recorded as earlier 1962 by Wall & Andrus in tomato and in oilseed rape by Fangsen Xu et al. (2002). He observed short growth period avoided B deficiency in some cultivars of oilseed rape. Response of B in a soil sufficient in B is due to the fast growth rate of the curds in a F1 hybrid and consequently enormous B requirement. Graham and Nambiar (1981) observed in the same soil wheat, oats and barley showed deficiency for micronutrient whereas the rye did not show and he attributed this phenomena to genetic deficiency in the plant rather than a deficiency in the soil. The hidden hunger observed in cauliflower is aggravated by the high temperature (Fig.1) during curd formulation (25^0C), which accelerates the growth rate, resulting in the soil's inability to supply very high amount of B for a short duration. The long hours of sunlight in tropical condition also contribute to B deficiency (Shorocks, 1997) so this is an important aspect of B nutrition in tropical cauliflower and hence external supply with B fertilizers as spray is very essential for high productivity and quality or higher soil B level has to be maintained throughout growth.

The superiority of foliar spray of DOB (solubor) at 0.1% and 0.2% over other treatments also reinforces the observation made above that high amount of B is needed at short notice and duration and foliar spray of DOB with high B content (20%) has amply met the requirement and hence it has proved better. Hanson (1991) has recorded a better mobility for foliar sprayed B and hence superiority of foliar spray over soil application. The reason for STB being equally efficient may be because it is less water-soluble than borax and has resisted the leaching loss suffered by the borax due to coarse texture and flooding type of irrigation practiced. The better quality of curds also is related to B status of curds.

Conclusion

This study indicated that in tropical cauliflower production the soil related constraints like hard setting nature, low organic matter and the heavy dry matter production during curd formulation at short notice resulting in high B requirement lead to in hidden hunger for B deficiency. Hence adequate moisture supply with foliar spray of high B sources like DOB (solubor) can result in correction of hidden hunger and result in high productivity and market acceptable quality. The earlier critical B level fixed for these soils have to be modified based on crop requirement.

Acknowledgement

The help given by Mr. Mazhaer Jameel in conduct of the field experiment and Lab analysis. The Author also acknowledges the financial help given by M/s Rallis India Bangalore in the conduct of this experiment. The support given by Director IIHR Bangalore is also gratefully acknowledged.

References

Berger K.C. and Truog E., 1939. Boron determination in soils and plants. Ind. Eng. Chem., Anal. Ed. 11:540-545.

Xu F.S., Wang Y.H., Ying W.H., and Meng J.L., 2002 Inheritance of boron nutrition efficiency in Brassica. J. Plant nutrition. 25: 901-902.

Hanson E.J., 1991 Movement of boron out of tree fruit leaves. Hort. Science. 26: 271-273.

Gaines T.P. and Mitcheel G.A., 1979 Boron determination in plant tissue by Azomethine-H method. Commun. Soil Sci Plant Anal. 10: 1099-1108.

Graham R.D., 1984 Breeding for nutritional characterization in cereals. Adv. Pl. Nutr. 1: 57-102

Graham R.D. and Nambiar E.K.S., 1981. Advances in research on copper deficiency in cereals. Aust. J. Agric. Res. 32:1009-1037.

Kotur S.C., 1993 Response of cauliflower to lime and boron in a boron deficient soil. Indian J. Hort. 50: 344-349.

Mullins C.E., McLeoo D.A., North Cote K.H., Tisdale J.M., and Young L.M., 1990 Hard setting soils: Behaviour, Occurrence and Management. Adv. Soil Science. 11: 37-108.

Srinivasa Rao, N.K. and Bhatt, R.M. 1992. Leaf area, growth and photosynthesis in relation to heterosis in tomato Photosynthetic 26(3):449-453

Wall J.R. and Andrus C.R., 1962. The inheritance and physiology of boron response in the tomato. Amer. J. Bot. 49:758-762.

Absorption of Foliar Sprayed Boron and its Translocation in the Citrus Plants When Applied at Different Phenological Phases

Rodrigo Marcelli Boaretto, Takashi Muraoka Maria Fernanda Giné and Antonio Enedi Boaretto

Centro de Energia Nuclear na Agricultura – Universidade de São Paulo Lab. Nutrição Mineral de Plantas C.P. 96, CEP 13400-970, Piracicaba - São Paulo, Brasil

E-mail: aeboaret@cena.usp.br

Introduction

The application of micronutrients in citrus plants has usually been done by foliar spraying. The citrus plants are exigent in boron, zinc, manganese, iron and deficiency of these micronutrients is common in worldwide citriculture. In Brazilian citriculture, the B and Zn deficiencies are most frequent (Quaggio et al. 2003). For this reason, these micronutrients are routinely applied as foliar fertilizers (Boaretto et al. 1997).

Although the B mobility in the phloem is generally limited, its redistribution can occur in some circumstance (Hanson 1991). Plant species differ dramatically in B mobility, and may be classified into species with restricted B mobility and those in which B is highly mobile (Brown and Shelp, 1997). If the B mobility in citrus plants is restricted, it is important that the nutrients are applied every new vegetation flush, as the nutrient applied before former vegetation does not redistribute to new tissues.

The B redistribution occurs in plant species producing polyols (*Pyrus*, *Malus* and *Prunus* genera), as they form polyol-B-polyol complexes in photosynthetic tissues and are transported through the phloem to active drains, as vegetative and reproductive meristems (Brown and Hu 1996; Hu et al. 1997). The polyols are single sugars, as sorbitol, manitol and dulcitol present in many plants, as detected by Zimmerman and Ziegler (1975), but not present in several citrus species. In these species, sucrose is present in phloem, which as it does not have cis-diol configuration, does not form stable borate complexes (Marchner, 1997).

The objective of this work was to study the foliar sprayed B uptake time course and also to verify the absorption and translocation of boron sprayed to the leaves at different phonological phases of orange plants.

F. Xu et al. (eds.), Advances in Plant and Animal Boron Nutrition, 125–131.

Materials and methods

Experiment 1

The experiment with 3 replications was carried out in green house of CENA-USP, using young sweet orange plants (*Citrus sinensis*) (L) Osbeck) grafted on sour orange (*C. aurantium* L.) conducted with two twigs and had been transplanted to plastic pots containing substrate proper for growing orchard plants. In each plant an average of 18.7 g of a solution of 255 mg L^{-1} containing ^{10}B (boric acid enriched with 92.64% ^{10}B) was applied to two of recently developed twigs.

The whole plant, except the sprayed part, was protected accordingly with plastic sheet, in order to avoid any ^{10}B contamination of non-sprayed part of plant. The amount of sprayed solution which effectively deposited to the plant was obtained by weighing the pot containing the plant, before and immediately after the application of ^{10}B with analytical balance with 0.1 g sensitivity.

The plants were harvested 3, 6, 12, 24 hours and 5, 15, 30, 75, 120 and 240 days after ^{10}B application, and separated into shoot and graft stock. In the plant shoot the different growing flushes were separated into leaves and stems. The plans harvested at 30 day had not emitted new growing flushes, but plants harvested after 75 days. The citrus plants were collected and separated in: "old" flush (stem and leaves from central stem existing at the moment of ^{10}B application); "applied" flush (stems ad leaves from the developing flushes which received ^{10}B); and "new" flush (stems and leaves developed after ^{10}B application), found only in the treatment 75, 120 and 240 days. Second developing flush had occurred in the plants and was collected at 240 days.

Experiment 2

This experiment was carried out in a four years old orchard of sweet orange (*C. sinenis* (CL) Osbeck) grafted on sour orange (*C. aurantium* L.), grown in a B deficient sandy soil, "São Paulo" State, Brazil. The foliar spraying was done at first hours of morning with manual sprayer, covering whole plant, but avoiding the applied solution to drain to soil. The ^{10}B solution was applied to whole plant, with exception to 3 branches per plant, which were protected covering them with plastic bag at the moment of spraying. The treatments (^{10}B solution spraying) were as follow: T1: before blooming; T2: during blooming; T3: at the beginning of fructification (2-3 cm diameter fruits). The sampling was done when the fruits were totally developed, for each treatment collecting 3 branches, which were protected when the B solution was sprayed and also 3 sprayed branches.

Chemical and isotopic analysis: After sampling and separation of the leaves and stems, they were washed with distilled water to remove unabsorbed B, dried in an oven at 65℃, and analyzed for total B content by azomethine-H method by spectrophotometry method (Malavolta et al. 1997) and $^{10}B/^{11}B$ isotopic ratio by ICP-MS.

Results and discussion

Experiment 1

In order to B be absorbed by leaves it is necessary the nutrient to penetrate through the cuticle and epidermal cells. In the mature citrus leaves, the adaxial cuticles thickness is 4.2 μm and the abaxial cuticle is 3.9 m. Peach and apple cuticles (1.6 and 2.1 m adaxial thick cuticle and 2.0 and 2.9 m abaxial cuticle) are thinner than orange cuticle (Leece, 1976), the B absorption is therefore limited in the citrus leaves. As shown in our work, with less than 9% of the sprayed B absorbed by plant. The B uptake through the leaves occurred mainly in the first day after foliar application (Figure 1).

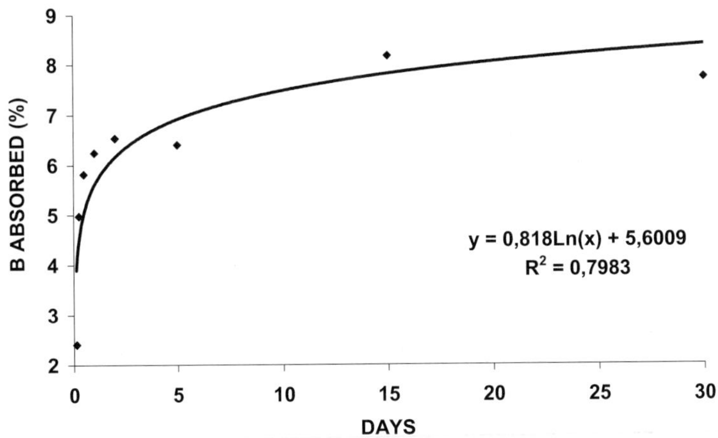

Figure 1. Time course of foliar sprayed B uptake (in %)

The B content increased in the leaves which received foliar spraying, when compared to control treatment leaf (table 1). The value of %B in the plant derived form the fertilizer (% Bdff) increased until 15th day after the spraying and then started to decrease, but the B content in the tissue continued increasing, as the plant continued to absorb B through the roots, diluting the total amount of B taken up by the leaves.

As does not occur polyols (sorbitol, manitol and dulcitol) in the phloem of citrus trees is expected that the B-phloem mobility in the plant will be limited, and the absorbed B remained mainly where it was applied. It may be noticed that only 0.3% and 3.2% of B taken up by the citrus leaves was translocated in the plant, at 75 and 240 days respectively (Table 2). The B concentration increased in the new flush due to B sprayed was lower than 1 mg kg^{-1}. As the B redistribution in the orange tree is restricted, its application in leaves will not be efficient in nourishing the plant parts that developed after fertilization, being necessary spray the nutrient whenever the plant develop new shoots.

Table 1. Boron contents in different plant part, and % B in the plant derived from the fertilizer (% Bdff) at different times after spraying (means of 3 replicates)

	Old leaf	Sprayed leaf	Sprayed leaf
	---------- mg kg^{-1} ----------		% B dff[(1)]
Control	111 a	54 c	0
3 Hours	96 a	60 c	15 ± 2
6 Hours	101 a	75 b	20 ± 2
12 Hours	102 a	79 ab	22 ± 2
24 Hours	110 a	82 ab	28 ± 4
2 Days	94 a	78 ab	29 ± 3
5 Days	107 a	87 ab	30 ± 4
15 Days	102 a	86 ab	32 ± 4
30 Days	102 a	90 a	27 ± 5

[(1)] % B in the plant derived from fertilizer Means values followed by the same letters do not differ significantly within the columns ($P > 0.05$).

Table 2. Localization of absorbed B in the plant at different times after spraying

	30 Days	75 Days	120 Days	240 Days
	---------------------- % [(1)] ----------------------			
% Young part	---	0.3	0.5	2.5
% Sprayed part	100	99.7	99.4	96.8
% Old part	0	0	0.1	0.7
% Total translocated	0	0.3	0.6	3.2

[(1)] Quantily absorbed = 100%

Unlike in the leaves, the older is the plant stem lower is the B concentration (Figure 2). This probably occurred, because of low evapo-transpiration water loss by stems, which practically does not occur in the citrus plant, and the B taken up by the roots is transported to the above ground part of the plant by the transpiration stream (xylem), running mainly to sites of highest water loss. Brown and Hu, 1996 reported that then was B mobile plants (almond, peach and plum) exposed to high B, higher accumulation of B in the young stems than in the mature leaves.

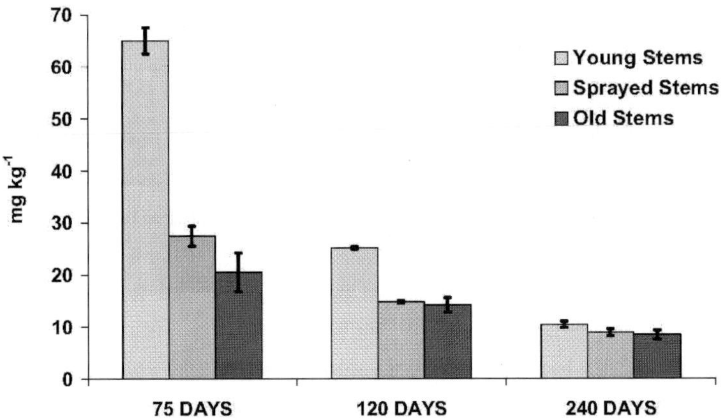

Figure 2. B concentration in the different plant part derived from the fertilizer, 75, 120 and 240 days after B application

Experiment 2

The foliar applied B increased the leaf B content compared to control. However, did not occur difference in the fruit B concentration. In contrast, Wójcik et al. 1999, in study with apple trees (B-mobile plant) found significant increase in the fruit B concentration. This suggests that in species which B is immobile, the fertilization does not change the nutrient fruit content.

The phenological stage of the citrus tree influenced the B uptake, as show in Figure 3. The more advanced the developing flush stage of the tree at the fertilizer spraying, the higher was the absorption of B by the fruits. Also, the larger the leaf area the larger is the amount of fertilizer deposited in the leaf and fruit surface and consequently higher is the nutrient uptake by the plant organs. However, there was no increase in the %Bdff when the fertilizer was sprayed at fruiting stage, probably due to the intense raining (38 mm) during three days following the spraying. Wójcik et al. 1999 also found a bigger increase of B content when the foliar fertilization was done after blooming than before blooming.

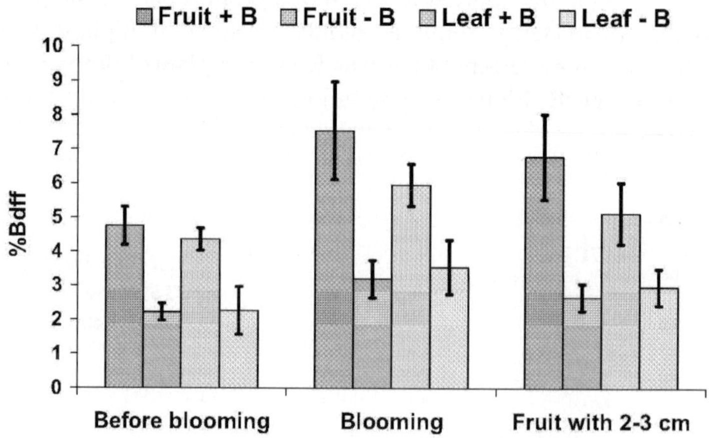

Figure 3. Percentage of B in the different plant part derived from the fertilizer as affected phenological stage of B application

The % Bdff in the branches which received the B spraying was 2 to 3 times higher than those in the stems covered during the spraying. Around 2 to 4% of B present in the fruits of branches, which were covered at the moment of spraying, was derived form the fertilizer, i.e., mobilized from the others parts of the plant. It represented an increase of 2 mg kg^{-1} in B derived from fertilizer for the B concentration in these flushes.

Conclusion

The foliar B fertilization increased the leaf B content. However, the B content did not alter in the leaves developed after spraying. The phenological phase of the citrus tree affected the B absorption. The more advanced the plant developing flushes at the spraying, higher was the fruit content on B derived from the fertilizer.

Acknowledgements

The research was supported by the funds from FAPESP (Fundação de Amparo a Pesquisa no Estado de São Paulo) and CAPES (Coordenação de Aperfeiçoamento de Pessoal de Nivel Superior) – Brazil.

References

Boaretto A.E., Tiritan C.S., and Muraoka T., 1997 Effects of foliar applications of boron on citrus fruit and on foliage and soil boron concentration. In: Boron in Soils and

Plants. Eds. R.W. Bell and B. Rerkasem. pp. 121–123. Kluwer Academic Publishers, Dordrecht, the Netherlands.

Brown P.H. and Hu H., 1996 Phloem mobility of boron in species dependent evidence of boron mobility in sorbitol-rich species. An. Bot. **77**, 497–505.

Brown P.H. and Shelp B.J., 1997, Boron mobility in plants. In: Boron in Soils and Plants. Eds. R.W. Bell and B. Rerkasem. 85–101. Kluwer Academic Publishers, Dordrecht, the Netherlands.

Caetano A.A., Moreira C.S., Negri J.D., Figueiredo J.O., Teófilo Sobrinho J., and Hiroce R., 1986 Study of efficiency of different source of zinc, manganese and boron sprayed in Valencia orange tree Citrus sinensis (L.) OSBECK (In Portuguese). Proc. Mtg. Sociedade Brasileira de Fruicultura. Brazil **8**, 14–31.

Hanson E.J., 1991 Movement of boron out of tree fruit leaves. Hortscience. **26**(3), 271–273.

Hu H., Penn S.G., Lebrilla C.B., and Brown P.H., 1997 Isolation and characterization of soluble boron complexes in higher plants. Plant Physiol. **113**, 649–655.

Leece D.R., 1976 Composition and ultraestructure of leaf cuticles from fruit trees, relative to differential foliar absorption. Aus. J Planta Physiol. 3, 833–847.

Malavolta E., Vitti G.C., and Oliveira S.A. (eds) 1997 Evaluation of nutritional state of plants: principles and aplications. (in Portuguese) Associação Brasileira para Pesquisa da Potassa e do Fosfato, Piracicaba, SP, Brazil.

Marschner H. (eds) 1997 Mineral nutrition of higher plants. Academic Press Ltd., London.

Maurer M. and Truman J., 2000 Effect of foliar boron sprays on yield and fruit quality of navel oranges in 1998 and 1999. In 2000 Citrus and Deciduous Fruit and Nut Research Report, College of Agriculture and Life Sciences, the University of Arizona, Tucson, Arizona, USA.

Perica S., Bellaloui N., Greve C., Hu H., and Brown P.H., 2001 Boron transport and soluble carbohydrate concentrations in olive. J Amer. Soc. Hort. Sci. **126**(3), 291–296.

Picchioni G.A., Weinbaum S.A., and Brown P.H., 1995 Retention and the kinetics of uptake and export of foliage-applied, labeled boron by apple, pear, prune and sweet cherry leaves. J. Amer. Soc. Hort. Sci. **120**(1), 28–35.

Quaggio J.A., Mattos Junior D., Cantarella H., and Tank Junior A., 2003. Fertilization with boron and zinc in complementation of foliar sprayed in rangpur orange tree. (In Portuguese). Pesq. Agropec. Bras. **38**, 627–634.

Santos C.H., Duarte Filho J., Modesto Junior C., Grassi Filho H., and Ferreira G., 1999 Foliar chelates fertilizers and salts in the absorption of boron, manganese and zinc in rangpur orange trees (in portuguese). Sci. Agric. **56**, 999–1004.

Shu Z.H., Oberly G.H., Carry E.E., and Rutzke M., 1994 Absorption and translocation of boron applied to aerial tissues of fruiting 'Reliance' peaches trees. Hortscience. **29**(1), 25–27.

Wójcik P., Cieslinsk G., and Mika A., 1999 Apple yield and fruit quality as influenced by boron applications. J. Plant Nutrition. **22**(9), 1365–1377.

Zimmermann M.H. and Ziegler H., 1975 List of sugars and sugar alcohols in sieve-tube exudates. In: Encyclopedia of plant physiology. VI: Transport in plants, I: Phloem transport (M.H. Zimmermann and J.A. Milburn, eds.) Spring-Verlag, Heidelberg, NY, USA. pp. 480–503.

Cotton and Wheat Responded to Boron on High-pH Calcareous Soils in Pakistan

Ibrahim M.[1], A. Niaz and M.Y. Nadeem

Soil Chemistry Section, Ayub Agricultural Research Institute, Faisalabad, Pakistan

[1] Corresponding Author: soilchem@fsd.paknet.com.pk

Introduction

The earlier workers (Kausar et al. 1979; Sillanpaa, 1982; Rashid, 1995) reported the micronutrients deficiency in soils and plants. The observation of farmers regarding the premature shedding of cotton squares and bolls provided the enough background to initiate the B research to determine its response by two major crops in the region. The high soil pH and presence of $CaCO_3$ in soil are the main factors influencing the availability of B to plants (Barrow 1989; Lehto 1995; Mezuman and Keren 1981; Goldberg and Forster 1991; Elseewi, 1974; Elseewi and Almalky 1979). Keeping in view the importance of cotton and wheat in Pakistan, this study was initiated to determine the response of these crops to B on high-pH-calcareous soils.

Materials and methods

Initially this study was conducted in pots on loamy sand and sandy loam soils. Both soils were calcareous and have pH above 8.0. After getting positive response of cotton to B in greenhouse experiments, the study was shifted to field at two different locations. At location-1 soil was loam, pH 8.2, $CaCO_3$ 4% and B 0.4 $\mu g\ g^{-1}$ and at location-2, soil was clay loam, pH 8.1, $CaCO_3$ 3.9%, and B 0.3 $\mu g\ g^{-1}$. On cotton treatments tested were, 0, 1, 2, 3, 4 and 6 kg B ha^{-1}. After cotton, study continued on wheat at three different locations for three years. For wheat experiments, soil at location-1 was clay loam, pH 8.1, $CaCO_3$ 3.90% and B 0.28 $\mu g\ g^{-1}$, at location-2, sandy loam, pH 7.6, $CaCO_3$ 4.70% and B 0.60 $\mu g\ g^{-1}$ and at location-3, sandy clay loam, pH 7.9, $CaCO_3$ 4.95% and B 0.32 $\mu g\ g^{-1}$. On wheat B treatments tested were, 0, 1, 2, 3 and 4 kg B ha^{-1}.

Results and discussions

Cotton

Soil used in greenhouse for cotton experiments was calcareous and high in pH. Boron was lower than the critical level (0.50 $\mu g\ g^{-1}$). The greenhouse results (not

F. Xu et al. (eds.), Advances in Plant and Animal Boron Nutrition, 133–136.

reported here) showed that 2 μg g^{-1} B treatment increased number of bolls, plant height and seed cotton yield. After greenhouse experiments, study was shifted to field. Field results showed that B application significantly increased seed cotton yield in all the treatments containing B during 2000 (Table 1). The yields of all B treatments were considerably higher than control treatment although statistically non-significant. The cotton yield in control was lowest (1.76 t ha^{-1}) and highest (2.70 t ha^{-1}) in 2 kg ha^{-1} B treatment. At higher B rates there were slight decreases in yields. The percent increases in yield over control due to B application varied from 25-53%, the highest (53%) was in 2 kg B ha^{-1}(Table 1).

Table 1. Cotton yield as affected by Boron application in field

Boron Treatments Kg ha^{-1}	Seed cotton Yield t ha^{-1}		Increase due to B %
Year 2000			
0	1.76	b	-
1	2.37	ab	35
2	2.70	a	53
3	2.28	ab	30
4	2.20	ab	25
6	2.37	ab	38
Year 2001			
0	1.88	b	-
1	1.94	b	3.2
2	2.04	a	53
3	2.20	a	30
4	2.12	a	25
6	2.10	a	38

Next year (2001), experiment with same treatments was repeated on clay loam soil having B 0.30 μg g^{-1}. This time cotton yield was significantly higher in all B treatments than control except of 1 kg ha^{-1} B treatment. The yield increases over control varied from 3.2 to 17.0%. Again the highest increase in yield was from 2 kg ha^{-1}. During this year increases due to B were much lower than the last year, 2000 (Table 1).

Wheat

Total nine B experiments were conducted on wheat during three years (2002-2004) at three different locations. During first year (2002) effect of B on grain was significant at all three locations, although it varied (Table 2). At location 1 and 2, treatment having 4 kg ha^{-1} B gave the highest yields, e.g. 5.01 and 4.77 t ha^{-1} respectively. At location three, the highest yield (4.43 t ha^{-1}) was obtained from 2 kg B ha^{-1} treatment.

During 2003, yields were also higher due to B application at all three locations but significant at location-1. At location-2 and 3 the effects were non-significant. The yields were higher at location-1 that followed by location-2 and then by location-3 (Table 2).

Unlike previous years (2002 and 2003) yields during 2004 at location two were lower than at location-1 and 3, perhaps it was due to less availability of irrigation water during the season at this location (Table 2). This year (2004) wheat grain yield increase due to B became significant at location-3 at 2 kg ha^{-1} B. No doubt yields were increased due to B application at location-1 and 2 but increases were non-significant. Again it was observed that B at the rate of 2 kg ha^{-1} gave reasonably good yield. Looking at overall results of both the crops (cotton and wheat) it is evident that yields were reasonably higher at 2-3 kg ha^{-1}B application. Therefore, on an average 2.5 kg ha^{-1} B application became the most appropriate dose for both the crops.

Table 2. Wheat grain yield as affected by Boron application

Boron Treatments (Kg ha^{-1})	Locations 1	2 (t ha^{-1})	3
2002			
0	4.32 b	3.50 c	3.70 b
1	4.69 a	4.00 b	3.70 b
2	4.62 a	4.13 b	4.43 a
3	4.89 a	4.15 b	4.33 a
4	5.01 a	4.77 a	3.88 b
2003			
0	4.60 b	4.20 a	3.20 a
1	5.20 ab	4.20 a	3.30 a
2	5.40 a	4.40 a	3.40a
3	5.90 a	4.90 a	3.30 a
4	5.60 a	4.50 a	3.30 a
2004			
0	3.99 a	1.72 a	3.60 a
1	4.35 a	1.70 a	4.00 a
2	4.58 a	2.41 a	4.20 a
3	4.49 a	2.27 a	4.40 ab
4	4.09 a	1.78 a	4.50 a

Conclusions

Both cotton and wheat responded to B application at all the locations on soils that are calcareous having pH in the range of 7.5-8.3. The degree of response varied from non-significant to significant. Moreover, the native B was low in all these soils, which necessitates the application of boron on these crops to get better yields. Therefore, 2.5 kg ha^{-1} B is recommended for both cotton and wheat.

Acknowledgements

The work was financially supported from the funds of the Soil Chemistry Section provided by the Province of Punjab. The authors are thankful to all the colleagues who helped during the study in the field and to Mr. Mohsin Ali Shah for typing the manuscript.

References

Barrow N.J., 1989 J. Soil Sci. **40**, 427–435.

Elseewi A.A., 1974 In Egypt, Alex. J. Aric. Res. **22**, 463–473.

Elseewi A.A. and Almalky A.E., 1979 Soil Sci. Soc. Am. J. **43**, 297–300.

Goldberg S. and Forster H.S., 1991 Soil Sci. **152**, 304–310.

Kausar M., Alam S.M., Sharif M., and Pervaiz 1979 Pakistan J. Sci. Indus. Res. **22**, 156–161.

Lehto L., 1995 For. Ecol. Manage. **78**, 11–20.

Rashid A., 1995 Micronutrient Project Annual Report 1993–94, National Agricultural Res. Center, Islamabad.

Sillanpaa M., 1982 FAO Soils Bulletin No. 48, FAO, Rome, Italy.

Mezuman U. and Keren R., 1981 Soil Sci Soc. Am. J. **45**, 722–726.

Potato Responses to Boron and Zinc Application in a Calcareous Udic Ustochrept

A. Rashid[1]*, M. Masud Mahmood[2], E. Rafique[1] and K. Farooq[2]

[1]Land Resources Research Program, National Agricultural Research Center (NARC), Islamabad-45500, Pakistan;

[2]Potato Research Program, NARC, Islamabad-45500, Pakistan,

*Corresponding author: abdul.rashid@comsats.net.pk

Introduction

Deficiencies of boron (B) and zinc (Zn) are well established in many agronomic and horticultural crops grown in calcareous soils of Pakistan (Rashid 2006). As crop responses to B as well as to Zn are appreciable and use of their fertilizers is highly cost–effective, application of these micronutrients is now recommended in the country. The history of Zn use in the country is about three decades old with rice being the first crop affected by its deficiency, and now Zn use is a regular practice in many crops including citrus, deciduous fruits and maize (Anonymous 1998; Rashid 1996). Establishment of field scale B deficiency in crops is a relatively a recent development and its formal use in the country started with cotton during late 1990s (Rashid and Rafique 2002). However, with the passage of time the magnitude and severity of B deficiency is increasing, primarily because of enhanced research information duration the past decade (Rashid et al. 2002).

Though potato *(Solanum tuberosum)* is known to be highly susceptible to the deficiency of B as well as of Zn (Shorrocks 1997; Anonymous 1998), local research information on this aspect remains scarce. Therefore, field experiments were carried out to study the response of potato crop to B and Zn application in a calcareous Udic Ustochrept.

Materials and methods

Field experiments were carried out in a B and Zn deficient calcareous Udic Ustochrept (HWE B, 0.40 mg kg^{-1}; AB-DTPA Zn 0.76 mg kg^{-1}) during spring and autumn 2000 crop seasons at National Agricultural Research Centre, Islamabad, Pakistan (latitude 33° 43' N, longitude 73° 5' E) to study the response of potato (*cv*. Cardinal) to B and Zn applications. Experimental treatments were: (i) control; (ii) three foliar sprays of 0.05% B solution; (iii) three foliar sprays of 0.1% Zn solution; (iv) three foliar sprays of 0.05% B + 0.1 Zn solution; and (v) soil application of 1.5 kg B ha^{-1} + 5.0 kg Zn ha^{-1}. The source of B was boric acid

F. Xu et al. (eds.), Advances in Plant and Animal Boron Nutrition, 137–142.

(17% B) for foliar sprays and borax (11% B) for field application. The source of Zn was $ZnSO_4.7H_2O$. A surfactant, 0.05% detergent powder, was added to the foliar spray solutions, and foliar sprays were performed 20 days after germination (DAG) and at 15–20 days intervals thereafter. Surface soil (0–20 cm) of the experimental field, sampled prior to planting was analyzed for physico-chemical properties including HWE B (Berger and Truog 1994). Salient soil characteristics are presented in Table 1.

Table 1. Soil characteristics of the field site

Soil series	Miani
Soil family	Fine-silty, mixed, hyperthermic Udic Ustochrept
pH (1:1)	8.1
EC (1:1)	0.25 dS m^{-1}
$CaCO_3$ equiv.	2.1%
Organic matter	0.9%
AB-DTPA extractable:	
NO_3–N	9.0 mg kg^{-1}
P	2.8 mg kg^{-1}
K	101 mg kg^{-1}
Zn	0.76 mg kg^{-1}
HWE B	0.40 mg kg^{-1}

The field experiments were laid out in a randomized complete block design with four replications. The plot size was 12 m^2. Basal fertilizer use comprised of 250 kg N ha^{-1} as urea (applied in 2 equal installments), 54 kg P ha^{-1} as DAP and 104 kg K ha^{-1} as sulfate of potash. The spring crop was planted in early January and harvested during first week of May. The autumn crop was planted at end September and harvested in mid of January of the subsequent year. The crop received standard irrigation and other production practices for the area. All fertilizers were broadcast applied prior to planting. Uniform field broadcast of small quantities of borax and $ZnSO_4.7H_2O$ was attained by premixing the micronutrient fertilizers with about 5 times volume of well pulverized soil.

Youngest mature leaves of potato, sampled 45 days after germination (DAG), 60 DAG (i.e., at flower initiation stage) and 75 DAG (i.e., when tubers half grown; Reuter et al. 1997), were analyzed for B and Zn content. Boron content in leaf tissue was determined by dry ashing (Gaines and Mitchell 1979) and colorimetry using azomethine-H (Bingham 1982). Zinc in plant tissue was measured by HNO_3-$HClO_4$ digestion and atomic absorption spectroscopy. Tuber yield was recorded by harvesting whole plots.

Results and discussion

Boron as well as Zn use increased tuber yield, almost equally, during both crop seasons ($P \leq 0.05$; Table 2).

Table 2. Effect of boron and zinc application on potato yield

Treatment	Tuber yield (t ha^{-1})	
	Spring	Autumn
Control	10.75c	11.22c
Foliar-applied B	14.52b	14.70ab
Foliar-applied Zn	14.09b	14.05ab
Foliar-applied B+Zn	16.22a	16.38a
Soil-applied B+Zn	13.38ab	13.80ab
LSD (0.05)	1.62	1.80

Potato yield increases were substantial and consistent, i.e., upto 35% with B, upto 31% with Zn, and upto 51% with B+Zn. Foliar feeding of micronutrients proved more effective than their soil application. Also, the response to micronutrients was better during spring season compared with autumn season. The economics of B and Zn use was highly attractive; value cost–ratio (VCR) being 260:1 for foliar feeding and 35:1 for soil application.

Micronutrient application, both by foliar spray and soil broadcast, substantially increased B and Zn concentrations in diagnostic leaves of potato (Tables 3, 4).

However, concentration of B as well as of Zn decreased with advancement of crop age. For example, B concentration in youngest mature leaves of control plants was 21–23 mg kg^{-1} at 45 days after germination (DAG), 19–20 mg kg^{-1} at 60 DAG and 17–18 mg kg^{-1} at 75 DAG. As the decrease in concentration was consistent and irrespective of micronutrient treatments, it is attributed to dilution effect, i.e., a consequence of biomass increase with plant growth. The suggested adequate B concentration range in youngest mature leaves of potato is 30–60 mg kg^{-1} at early–late flowering stage (Huett et al. 1997; Westerman 1993) and 30–40 mg B kg^{-1} when tubers are half grown (Walsh and Beaton 1973). Thus, in our experiments, the potato plants grown without B application were deficient in B.

Table 3. Boron concentration (mg kg^{-1}) in youngest mature leaves of potato as affected by boron and zinc application

Treatments	Days after germination		
	45	60	75
	Spring crop		
Control	21	19	17
Foliar-applied B	41	39	36
Foliar-applied Zn	22	21	18
Foliar-applied B+Zn	39	37	34
Soil-applied B+Zn	30	27	25
	Autumn crop		
Control	23	20	18
Foliar-applied B	44	39	37
Foliar-applied Zn	21	22	20
Foliar-applied B+Zn	38	36	35
Soil-applied B+Zn	34	31	28

Table 4. Zinc concentration (mg kg^{-1}) in youngest mature leaves of potato as affected by boron and zinc application

Treatments	Days after germination		
	45	60	75
	Spring crop		
Control	18	19	17
Foliar-applied B	21	20	18
Foliar-applied Zn	49	45	42
Foliar-applied B+Zn	40	36	34
Soil-applied B+Zn	30	28	25
	Autumn crop		
Control	21	22	18
Foliar-applied B	24	25	22
Foliar-applied Zn	49	44	40
Foliar-applied B+Zn	43	42	37
Soil-applied B+Zn	28	26	27

With foliar feeding, B concentration increased to 38–44 mg kg^{-1} at 45 DAG, 36–39 mg kg^{-1} at 60 DAG and 35–37 mg kg^{-1} at 75 DAG (Table 3). The tuber yield increases with foliar sprays of B were 1–35% over control during both crop seasons. Soil applied B also increased B concentration in leaves; however, the magnitude of increase was lesser compared with the impact of foliar feeding, i.e., 30–34 mg B kg^{-1} at 45 DAG, 27–31 mg B kg^{-1} at 60 DAG and 25–28 mg B kg^{-1} at 75 DAG during both crop seasons. The potato yield increases with soil applied B+Zn were 23–24% over control yields. Much lesser yield increases with soil-applied micronutrients compared with those observed with foliar feeding indicate that soil applied B+Zn proved insufficient for bringing micronutrient nutrition of potato plants to adequate levels.

In the leaves of control plants, Zn concentration was 18–21 mg kg^{-1} at 45 DAG, 19–22 mg kg^{-1} at 60 DAG and 17–18 mg kg^{-1} at 75 DAG (Table 4). As the generally suggested Zn deficiency critical level in potato leaves is 20 mg Zn kg^{-1} (Walsh and Beaton 1973; Weir and Cresswell 1993; Westerman 1993; Huett et al. 1997), the potato plants grown without Zn application were deficient in Zn to varying degrees. Leaf Zn concentration increased with foliar feeding as well as with soil applied Zn; however, concentration increases were much greater with foliar sprays (Table 4). Similar to the B nutrition scenario, lesser yield increase with soil applied Zn (Table 2) is attributed to ineffectiveness of this treatment to fully cure the deficiency.

Acknowledgments

We thank Mr. Muhammad Irfan and Mr. Muhammad Jawad for word processing and composing of the manuscript.

References

Anonymous 1998 Micronutrients in agriculture: Pakistani perspective. National Fertilizer Development Center, Islamabad, Pakistan. pp. 51.

Berger K.C. and Truog E., 1944 Boron tests and determination for soils and plants. Soil Sci. **57**: 25–26.

Bingham F.T., 1982 Boron, In: (Page A.L. ed), Methods of Soil Analysis, Part 2: Chemical and Microbiological Properties American Society of Agronomy, Madison, W.I., USA, pp. 431–448.

Gaines T.P. and Mitchell G.A., 1979 Boron determination in plant tissue by azomethine-H method. Commun. Soil Sci Plant Anal **10**: 1099–1108.

Huett D.O., Maier N.A., Sparrow L.A., and Piggot T.J., 1997 Vegetable crops. In: Reuter D.J., and Robinson J.B., (eds), Plant analysis: an interpretation manual, Second Edn, CSIRO., Collingwood, Victoria, Australia, pp. 383– 464.

Rashid A., 1996 Secondary and micronutrients, In: Rashid A. and Memon K.S. (Managing Authors), Soil Science. National Book Foundation, Islamabad, Pakistan. pp. 341–385.

Rashid A., 2006 Incidence, diagnosis and management of micronutrient deficiencies in crops: success stories and limitations in Pakistan, Proc. IFA Int Workshop on Micronutrients, 27 Feb 2006, Kunming, China.

Rashid A. and Rafique E., 2002 Boron deficiency in cotton grown in calcareous soils of Pakistan. II. Correction and criteria for foliar diagnosis, In: Goldbach H.E. et al. (ed), Boron in plant and animal nutrition. Kluwer Academic/Plenum Publishers, New York, pp. 357–362.

Rashid, A., Rafique E., and Ryan J., 2002 Establishment and management of boron deficiency in field crops in Pakistan: a country Report, In: Goldbach H.E. et al. (ed), Boron in Plant and Animal Nutrition. Kluwer Academic/Plenum Publishers, New York. pp. 339–348.

Reuter D.J., Robinson J.B., Penerill K.I., Price G.h., and Lambert M.J., 1997 guidelines for collecting, handling and analyzing plant materials. In: Reuter D.J., and Robinson J.B., (eds), Plant analysis: an interpretation manual, Second Edn, CSIRO, Collingwood, Victoria, Australia, pp. 55–70.

Shorrocks V.M., 1997 The occurrence and correction of boron deficiency. Plant and Soil **193**: 121–148.

Walsh L.M. and Beaton J.D., 1973 Soil testing and plant analysis. Soil Sci Soc Am Publication No. 368.

Weir R.G. and Cresswell G.C., 1993 Plant nutrient disorders 3. Vegetable crops. NSW Agriculture, Inkata Press, Melbourne.

Westermann D.T., 1993 Fertility management in potato health management (ed Rowe R) American Psychopathological Society, St Paul, Minnesota, USA, pp. 77–86.

Occurrence and Correction of Boron Deficiency in Wheat and Mustard in Bangladesh

M. Jahiruddin[1*], M.U. Ahmed[1], M.A. Hossain[1], M.R. Islam[1] and M.F. Islam[2]

[1] Department of Soil Science, Bangladesh Agricultural University, Mymensingh 2202, Bangladesh.

[2] Plant nutrition specialist, SFFP, Khamarbari, Dhaka, Bangladesh.

* Corresponding author: m_jahiruddin@ yahoo.com

Introduction

Boron deficiency is a major reason for lower yield of wheat and mustard in Bangladesh. This element deficiency has arisen mainly due to continuous mining of soil nutrients for increased cropping intensity without adequate replenishment. Boron deficiency induces grain sterility. Again, crop species and varieties may differ in their sensitivity to boron deficiency (Rerkasem and Jamjod, 1997; Brown et al. 1999; Kataki et al. 2001). Wheat, next to rice, is the most important cereal crop and mustard ranks first among the oilseed crops in Bangladesh. These two crops are found to be much affected by B deficiency (Jahiruddin et al. 1995; Islam et al. 1997; Haque et al. 2000). The present study was undertaken to find out a suitable dose of B for these crops and also to screen out the crop varieties tolerant or sensitive to B deficiency.

Materials and methods

The field trials were conducted in three locations: Bangladesh Agricultural University (BAU) farm, Mymensingh; Regional Agricultural Research Station (RARS) farm, Jamalpur and On-Farm Research Division (OFRD) farm, Rangpur.

The experiments with wheat (cv. Kanchan) were carried out at those three locations in order to determine a suitable dose of boron. The B doses were 0, 0.5, 1, 1.5, 2 and 2.5 kg ha^{-1}. The layout was kept undisturbed for two years. Transplant Aman rice (second or last crop of a year) was grown on the same plots with and without added B to evaluate direct and residual effects of boron. Further, twelve varieties of wheat, ten from Bangladesh and two from Thailand, were tested at BAU farm, Mymensingh under two boron treatments (0 and 1 kg B ha^{-1}). The Bangladeshi varities were Kanchan, Sourav, Gourab, Protiva, Akbar, Aghrani, Shatabdi, Pavon, Ananda, and Barkat, and the Thai varieties (used as checks) being Fang 60 (B efficient) and SW 41 (B inefficient).

F. Xu et al. (eds.), Advances in Plant and Animal Boron Nutrition, 143–148.

The mustard experiments were conducted in RARS, Jessore and OFRD, Rangpur. Boron was applied at the rates of 0, 0.5, 1, 1.5, 2 & 2.5 kg ha^{-1}. The soil was calcareous having pH 8.2 and B 0.13 mg kg^{-1}. Further an experiment with mustard was conducted at RARS, Jessore to evaluate the response of different varieties to boron application (0 and 1 kg B ha^{-1}). Eight varieties representing three species: *Brassica campestris*, *Brassica napus* and *Brassica juncea* were tested.

In each experiment, every plot received an equal dose of N, P, K, S and Zn from urea, TSP, MOP, gypsum and zinc oxide, respectively. Boron was added as boric acid (17% B). Intercultural operations viz. weeding and irrigation were done as and when necessary. Data on the yield and yield contributing characters have been recorded.

Results and discussion

The grain yield of wheat was significantly influenced by boron application. Generally, the yield increased as the B dose increased (Figure 1), however successive increments of added boron gave successively smaller increases in grain yield indicating that yield response function is quadratic ($y = a + bx + cx^2$), not linear. Thus, the yields due to 1, 1.5, 2 and 2.5 kg B ha^{-1} were statistically identical.

Again, application of boron at 0.5 kg B ha^{-1} did not influence grain yield significantly over control. Transplant Aman rice responded a little to B application indicating that 1 kg B ha^{-1} yr^{-1} is sufficient to correct B deficiency for the two crops (wheat and rice) (Figure 2).

Except Fang 60, all the varieties of wheat responded significantly to boron application (Figure 3). The highest grain yield (61% yield increase over control) was recorded in Gourab followed by Shatabdi (47%) and Kanchan (42%). Shourav had the lowest yield increase (9% only). On the other hand, the Thai varieties, Fang 60 (B efficient) gave only 3% yield increase while SW41 (B inefficient) produced 34% increased yield due to B application (1 kg ha^{-1}). This result suggests that Sourav, among the Bangladeshi varieties, was the most B in responsive (B efficient) variety which, thus, can be used as a breeding material for development of high yield potential varieties of wheat having a unique quality of tolerance to B deficiency. For an area of widespread low B soils where B fertilizer is not used by the farmers, evaluation of B efficiency would be essential before selection materials go for on-farm trials.

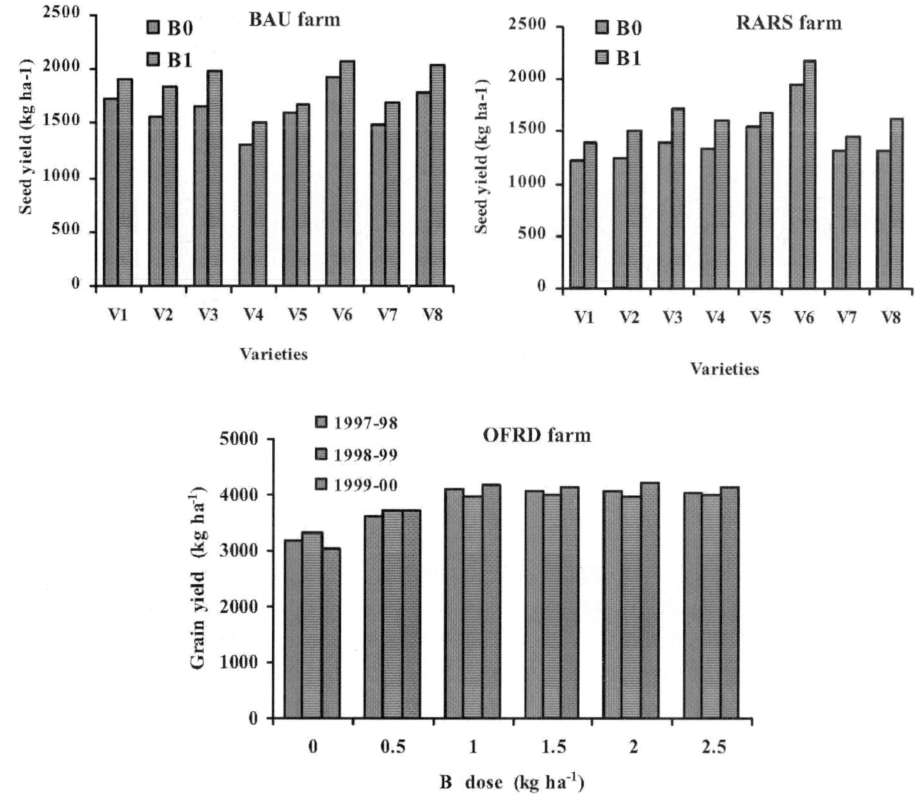

Figure 1. Effect of added B on grain yield of wheat

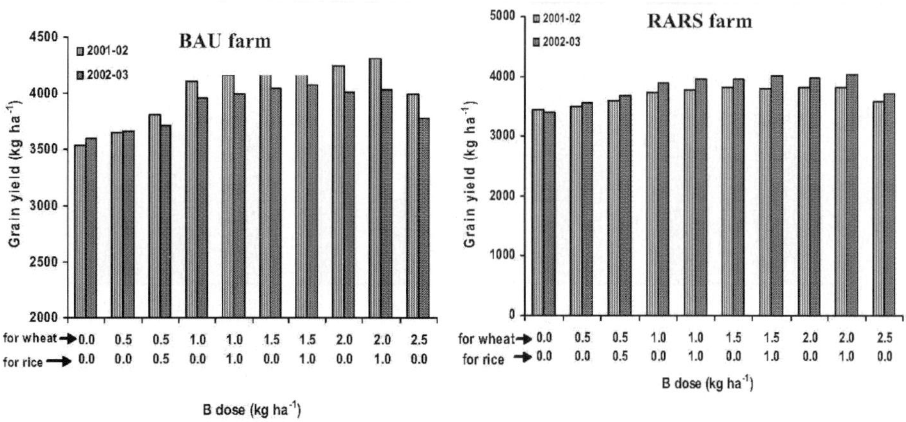

Figure 2. Direct and residual effects of added B on grain yield of T. Aman rice

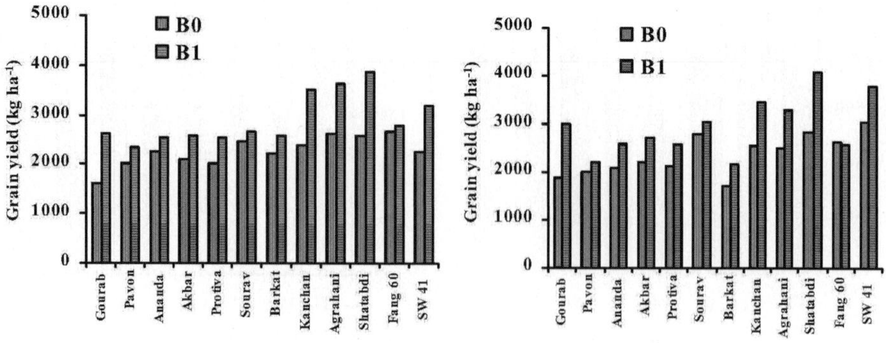

Figure 3. Response of different varieties of wheat to B application

There was a significant positive effect of boron on the seed yield of mustard (cv. BARI Sarisha 8), the yield did not vary significantly between 1 and 2 kg B ha^{-1} (Figure 4). Concerning varietal response to boron application, it appeared that the *Brassica napus* varieties (BARI Sarisha 2, 3 & 8) responded the highest, *Brassica juncea* (BARI sarisha 5 & 6) responded medium and *Brassica campestris (*BARI Sarisha 1, 4 & 7) had the lowest response (Figure 5). The optimum dose of boron was found to be 1 kg B ha^{-1}. Seed yield was positively correlated with the number of pods $plant^{-1}$ as well as the number of seeds pod^{-1}.

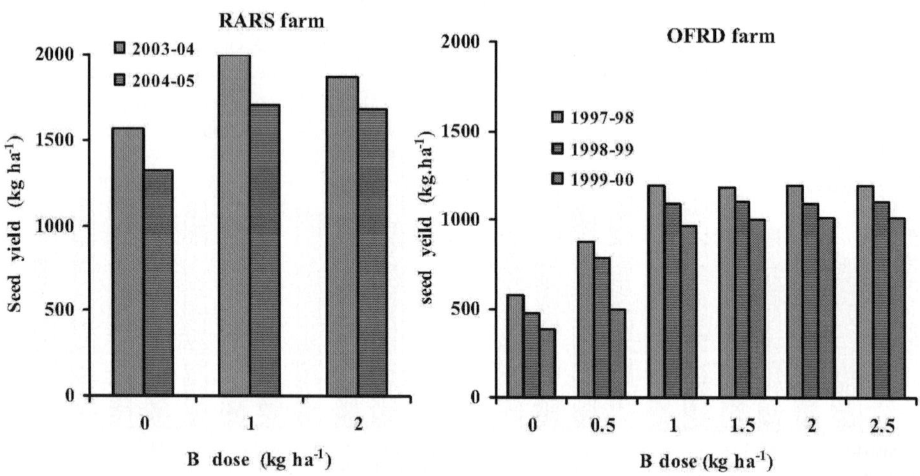

Figure 4. Effect of added B on seed yield of mustard

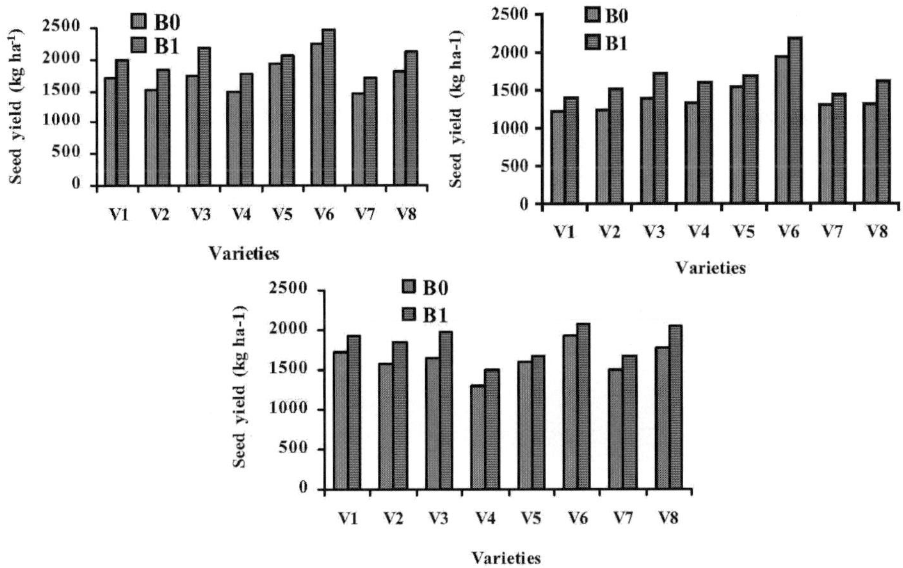

Figure 5. Response of different varieties of mustard to B application (V1, V4&V7 represent *B. campestris*, V2, V3&V8 *B. napus* and V5&V6 *B. juncea*)

Acknowledgement

This work was supported by DANIDA funded Integrated Soil Fertility & Fertilizer Management Project (SFFP) of the Department of Agriculture Extension (DAE), Dhaka.

References

Brown P.H., Bellaloui N., Hu H., and Dandekar A., 1999 Transgenically enhanced sorbitol synthesis facilitates phloem boron transport and increases tolerance of tobacco to boron deficiency. Plant Physiol. **119**, 17–20.

Haque M.A., Jahiruddin M., and Islam M.R., 2000 Effect of sulphur and boron on seed yield of mustard (Brassica napus). Bangladesh J. Seed Sci. & Tech. **4**(1&2), 7–11.

Islam M.R., Riasat T.M., and Jahiruddin M., 1997 Direct and residual effects of S., Zn and B. on yield and nutrient uptake in a rice-mustard cropping system. J. Indian Soc. Soil Sci. **45**(1), 126–129.

Jahiruddin M., Ali M.S., Hossain M.A., Ahmed M.U., and Hoque M.M., 1995 Effect of boron on grain set, yield and some other parameters of wheat cultivars. Bangladesh J. Agric. Sci. **22**, 179–184.

Kataki P.K., Upreti H.K., and Bhatta M.R., 2001 Soil boron deficiency induced wheat sterility in Nepal: respone to boron and nitrogen application. J. New Seeds **3**(4), 23–39.

Rerkasem B. and Jamjod S., 1997 Genotypic variation in plant response to low boron and implications for plant breeding. Plant and Soil **193**, 169–180.

Boron Deficiency and its Nutrition of Groundnut in India

A.L. Singh, Vidya Chaudhari and M.S. Basu

National Research Centre for Groundnut, P. B. 5, Junagadh-362 001, India. E-mail: alsingh@nrcg.res.in and alsingh_ad1@sancharnet.in

Introduction

The groundnut (*Arachis hypogaea* L.) is an important food legume and oilseeds crop of India grown in about 8 m ha land, the largest area in the world, but stand second in production due to its low productivity. Presently, the average productivity of groundnut, in India, is around 1300 kg ha^{-1} which is very low as compared to USA and China mainly because, the crop is mostly grown as rain fed in dry lands, under low fertility and low input management, often subject to the vagaries of the weather conditions. However, in recent years, combination of improved genotypes and nutrient management practices has increased the productivity.

Presently, Gujarat, Andhra Pradesh, Karnataka, Maharashtra and Tamil Nadu states contribute about 90% of the groundnut production in India, of these Gujarat, Tamil Nadu and Maharashtra are the major areas producing export quality groundnut. However due to deficiencies of boron (B) and calcium (Ca), there is poor seed filling and hence low quality produce is obtained from these areas (Singh et al. 2004). Thus looking to the export market, which requires high quality and well-filled seed, it is essential that these aspects be looked carefully. After much work Ca fertilizer is recommended, but there are only few reports on B nutrition of groundnut in India. Soil parent material and texture are considered to be major soil factors associated with the occurrence of B deficiency that can readily be prevented and corrected by soil and foliar applications (Shorrocks, 1997). Thus the study on B nutrition of groundnut was initiated at this research center about a decade ago and some of the findings are summarized here.

Materials and methods

Field, pot and micro-plot experiments were conducted at the National Research Centre for Groundnut, Junagadh, India in a medium black, calcareous (19 - 21% $CaCO_3$), clayey (35% sand and 44% clay) soil containing 7.6 - 7.8 pH, 0.71- 0.81% organic carbon, 660 - 710 mg kg^{-1} total N, 6.0-6.2 mg kg^{-1} available P, 10 - 11 mg kg^{-1} heat soluble S, 3.02, 4.9, 0.61 and 0.60 mg kg^{-1} DTPA extractable Fe, Mn, Zn, and Cu, respectively and 0.20 - 0.36 mg kg^{-1} hot water soluble B.

F. Xu et al. (eds.), Advances in Plant and Animal Boron Nutrition, 149–162.

The crop was grown under recommended package of practices and proper care was taken to protect it from weeds, insects, pests and diseases during the entire cropping season. The crop was harvested at maturity, dried in sun for a week and pod and haulm yields, shelling percent, percentage sound mature seeds and 100-seed mass recorded. The B in plant samples were analyzed colorimetrically using curcumine. All these data were analyzed statistically. The experiment-wise details of the methodologies are given below.

Pot experiments: diagnosis of B deficiency symptoms and requirement

Soil and sand culture pot experiments were conducted under various levels of B to find out the deficiency symptoms and role of B in the nutrition of groundnut. Fifteen kg of soil or sand, as per the experimental requirement, was filled in a number of pots of 25 cm diameter with a bottom hole plugged with glass wool. The groundnut genotypes were sown at a rate of four seeds per pot in 4 replicates.

A series of sand culture pot experiment were conducted under various deficiency and toxicity levels of B, during 1993-1995, to find out the correct levels of B for best growth and yield of groundnut and to diagnose the B deficiency symptoms under low B conditions. The rate of photosynthesis, growth and pod yield recorded.

In sand culture experiments Steinberg's nutrient solutions, was used. For different doses of B, the minus solution containing all nutrients except B was taken as base solution and as per the treatment, the required quantity of B from the stock solutions (Boric acid) was added. The pots were initially irrigated with one liter of complete solution in each pot on the day it was sown. After 7 DAE (days after emergence) the pots were flushed with water and the treatment of different doses of macronutrient was started by providing 250 mL of nutrients solution pot^{-1} day^{-1} from 7-30 DAE, and 500 mL pot^{-1} day^{-1} there after. The nutrient solution was stopped 5 days before harvest. The pots were flushed with water at weekly interval. After maturity, the plants from each pot were harvested, dried and observations on the plant height and pod numbers and yield recorded. The oven dried plant and seed samples were ground to a fine powder and analyzed for B.

Soil culture pot experiments were conducted at various doses of B to find out the deficiency and toxicity levels of B for growing groundnut in calcareous soil.

Field experiments

The field was prepared by ploughing and labeling and 10 cm deep furrows were opened at 30 cm spacing. The field was then divided into small plots of

25 m_2 (5 m x 5 m) by raising bunds. A basal dose of 20 kg and 22.5 kg P ha^{-1} as diammonium phosphate and 40 kg K ha^{-1} as muriate of potash were mixed in the soil before sowing. The groundnut was sown at 30 cm x 10 cm spacing in the furrows and covered with soil.

Micro-plot experiment

Experiments were conducted, in polythene-lined micro-plots in field and supplied with various doses of B, to find out the sufficiency and deficiency levels in both plant and soils. The soil and plant samples from the experiments were analyzed and based on the soil analysis, the critical sufficiency level of B in soil was fixed.

Assessment of yield losses due to B deficiency

Field experiments were conducted during three (wet 1993, dry 1994 and dry 1995) seasons in a randomized block design with three treatments as detailed in Table 1 and six replications. Groundnut genotype FeESG 10-1 was grown and observations on yield and related attributes recorded.

Effectiveness of soil and seed dressing of boron

A field experiment was conducted during two wet seasons in a factorial randomized block design with two modes of application through seed and in furrows as the main plots and the B fertilizers in the sub-plots with three replications.

The boron containing salts at 5 kg ha^{-1} (which contained 1 kg B ha^{-1}) were applied through seed and directly in soil using furrows (Table 2). To apply through seed, these salts were mixed with the water soaked seed and shaken so as to make a uniform coating of these salts around the seed and then sown in the furrows. In the other method the nutrient salts were applied in the furrows. The groundnut genotype, FeESG 10-1 was grown and effect of B application on seed and seedlings and percent germination were recorded in the field. The crop was harvested at maturity, and observations on yield and related attributes recorded.

Comparison of soil, foliar and drip application

Field experiments were conducted, using groundnut variety GG 2, during Rabi-summer 1995 (February to May) and Rabi 1995-96 (Mid October to February) seasons in a randomized-block design with five treatments as in table 5 and three replications. For the first time flood irrigation was given to all plots to maintain adequate soil moisture for uniform germination and from the second irrigation onward, as per treatment, the drip or flood irrigation was given. The seed germinated within a week. The B (1 kg ha^{-1} B as boric acid) was applied through soil application (in furrows), through foliar sprays (foliar application of

0.2% aqueous solution) and through drip irrigation (Feeding 0.5% aqueous solution through ventury) thrice at 30, 50 and 70 DAS (Days after sowing).

The chlorophyll content of leaves was measured at 40 DAE (Days after emergence). Five plants from each plot were uprooted randomly both at 60 DAE and at maturity, washed, separated into leaves, stems and nodules, weighed after drying in an oven at 60°C for about a week, and subjected to dry wt. measurement.

Nutrition of bold seed groundnut

Field experiments were conducted during rabi-summer 1995 (February to May) and rabi 1995-96 (Mid Oct.-Feb.) seasons in a factorial randomized-block design with two groundnut genotypes (TKG 19A, a large seeded and FeESG 10 as an ordinary groundnut) in main plot and eight treatments, as in table 7, as sub plots with three replications. The treatments 100 kg ha^{-1} Ca as gypsum, 100 kg ha^{-1} K as muriate of potash and 2 kg ha^{-1} B as borax were applied in soil 50% as basal and 50% 50 days after emergence (DAE).

Results and discussion

Deficiency and toxicity symptoms and requirement

From the sand and soil culture pot experiments at various levels of B, the deficiency and toxicity levels of B and their symptoms in groundnut were identified and the requirement of B was worked out.

The B deficiency caused retarded growth in groundnut plant particularly of the apical portion, death of the stem apex, and regeneration from the lateral bud, malformation of the leaf vein, chlorosis, necrosis of basal margins in emerging leaves are commonly observed. Boron deficiency also causes thickened and stiff growing bud at terminal end, young leaves turn pale green starting at basal end, leaves deformed, twisted, and brittle, shortened internodes and bushy or rosette appearance of the plant (Singh et al. 2004). However, these visual symptoms on leaves were not observed in field grown groundnut crop, even on B deficient soils. But, as B deficiency symptoms invariably occur in the young tissues and apical meristems, shortened growth and bushy appearance was common. The B deficiency was similar to that of Ca, except that in B the necrotic areas are localized near leaf margins but in Ca they are distributed over the entire surface. Roots become blackened and growth of root nodules is suppressed. Goldberg and Forster (1991) reported that B can be adsorbed by Calcium carbonate and a positive correlation was found between amount of adsorbed B and Calcium carbonate.

The deficiency of B causes low pod filling and hollow darkening or off-colour area develop in the center of the seed known as `hollow heart' of groundnut reducing the quality of seed. Many a times there were shriveled seeds due to B deficiency. In groundnut, boron facilitates translocation of sugar and fat synthesis and is important for RNA synthesis, cell division, differentiation, and maturation and pollen germination. It is transported primarily in xylem, but is relatively immobile in phloem and hence commonly occur in neutral to alkaline and weathered light texture soils where groundnut is mainly grown in India. The most common B deficiency symptom is hollow-heart of kernel in which the inner faces of the cotyledons are depressed and discolored reducing the quality of seed (Cox and Reid, 1964). Groundnut kernels containing 6.7-17.3 mg B kg^{-1} showed the incidence of hollow-heart. Deficiency of B is more commonly observed in groundnut grown in coarse textured and hilly soils of India particularly in Tamil Nadu, northern Bihar and part of Assam, West Bengal, northern Orissa, north east India, Karnataka and Gujarat (Singh and Basu, 2005). Though the clear-cut symptoms of B deficiency are observed in Tamil Nadu, Karnataka, Bihar and a few pockets of Gujarat, the response of B is reported from most of the groundnut growing areas in India causing considerable yield losses.

Results of the soil culture pot and field experiments in polythene-lined microplots at various B doses clearly demonstrated the sufficiency and deficiency levels of B in both plant and soils. The soil and plant samples when analyzed the sufficiency level of B (hot water soluble B) in soil was found to be 0.35-0.5 ppm. The B deficiency occurred when the B of the soil was less than 0.2 ppm, however, depending upon the soil and groundnut genotypes, the critical limits of B may vary from 0.2-0.4 ppm (Singh, 1994). There is a strong correlations between B and organic carbon contents of soils and the risk of B deficiency increases when organic matter declines as it has the ability to complex large amount of B (Yermiyaho et al. 1988). Boron is an important both at deficient as well as toxic level in soil. It is required in very small quantity and application of 5-10 kg ha^{-1} B showed toxicity symptoms in leaves. The concentration above 5 ppm of B in soil was toxic to the groundnut plant (Singh, 1994).

In polythene-lined micro-plot, in the field, a good correlation between the amount of B applied, soil B and pod yield was obtained, however, the B requirement of groundnut varied with cultivars. On an average, the healthy plant showed 40 ppm B, however, the sufficiency level of B in the leaf during flowering and fruiting (40-70 DAE) was 25-60 ppm. The critical level of B was observed to be 20 ppm, but the clearcut deficiency symptoms were observed only when the leaf B concentration fell below 15 ppm. However, Hill and Morrill (1974) reported 26 ppm of B at 30-60 DAP, and Gopal (1968) found 20 ppm as the critical level in leaves. The groundnut kernels containing less than 17 ppm B showed the incidence of hollow-heart.

The toxicity of B caused stunted growth and interveinal chlorosis leading to iron deficiency. Excess of B accumulation caused imbalance of other nutrient, the leaflets tips become yellow, interveinal chlorosis followed by necrosis. The B toxic leaves showed curling with scorch sign, the chlorotic areas were golden yellow at margin. Leaves assumed a scorch appearance and fall off. The rate of photosynthesis, growth and pod yield of were maximum at 0.5 ppm of B in nutrient solution, indicating that this is the optimum concentration of B. The respiration rate in groundnut roots was more at the toxic levels (5 ppm) of B.

Assessment of yield losses due to B deficiency

In groundnut, severe mineral deficiencies are commonly observed in most part of the country, however the visual symptoms of B are rarely observed in spite of its good response. Thus, presuming that the B deficiency is hidden in groundnut, though many a time express in kernel, an experiment was conducted to assess the loss of yield due to B deficiency where encouraging results were obtained (Table 1).

Table 1. Yield losses due to B deficiency in groundnut in calcareous soils

Details of the Treatments	Pod yield (kg ha^{-1})			Yield losses over T2, a full package of macro-and micro-nutrients (%)		
	1993	1994	1995	1993	1994	1995
T1-Control	979	1292	2005	-	-	-
T2-All macro- and micro-nutrients	1903	2131	2795	-	-	-
T3- T2 minus B	1626	1576	2328	14.5	26.0	16.7

Note: the macronutrients N, P, K, Ca, S and Mg at 40, 40, 60, 100, 30 and 10 kg ha^{-1}, respectively were applied 50% as basal in the furrows and 50% at 30 DAE in all the treatments except control. The micronutrients Fe, Mn, Zn , Cu, B and Mo at 10, 10, 5, 2, 2 and 1 kg ha^{-1}, respectively were used 50% as basal in thefurrows along with macronutrients and 50% applied at 30 DAS.

The perusal of data show that application of macro- plus micro-nutrients produced 39.7-94.4% more pod yield than the untreated control and withholding of B caused considerable decrease in pod yield. Depending upon the year the yield-losses due to B deficiency were 14.5-26%. Being a legume crop groundnut requires B for partitioning, and hence shortage of this element showed yield reduction.

Boron (B) is a constituent of cell membrane and essential for cell division, nitrogen metabolism and protein formation and acts as a regulator of K/Ca ratio in the plant. It helps in nitrogen absorption and translocation of sugars and carbohydrates in the plants. It is important in pollination and seed production. As boron is highly essential to complete the plant life cycle, the deficiency of boron caused yield reduction in groundnut. Total soil B content ranged from 0.01 to 10 ppm. However, only a small fraction of this amount is available to the crop. Much of the total soil B is present as a component of tourmaline a highly insoluble mineral.

In Tamil Nadu, based on random soil sample analysis, the B deficiency accounts for 21% and depending upon the soils conditions, B application at 5-25 kg ha^{-1} increased yield of groundnut, however under irrigated conditions 15 kg ha^{-1} borax with NPK as basal is used (Muthusamy and Sundararajan, 1973). In field trials borax at 1.25 kg B ha^{-1} increased 25 to 30% yield of groundnut at Pollachi in Coimbatore district and 15-30% in South Arcot district (Chitdeswari and Poongothai, 2003). However, application of 2.5 kg ha^{-1} B at Tindivanam increased 57% yield of groundnut hybrid over control.

Effectiveness of soil and seed dressing of boron fertlilizers

The borax and boric acid are the two commonly used boron fertilizers in India and their effectiveness and methods of application, as seed dressing and in the furrows in the soil, at 1 kg B ha^{-1} were judged in groundnut crop. In general, these salts showed their positive response to groundnut grown in calcareous soil and increased the number of pods, pod yield, shelling per cent, oil content and seed size of groundnut over control (Tables 2 and 3). However, there was interaction among the methods of applications on the yields and other parameters. When borax and boric acids were applied as soil, in the furrows, showed positive response with good germination and increased the pod yield, pod number and oil content. But, as seed dressing these was detrimental to groundnut seedling. Boric acid was comparatively more detrimental than borax. In a study on the effects of nutrient salts on the germination reveals that, in general, the groundnut seed took 6 days for their germination, but seed coating with borax and boric acid delayed field emergence for 3 and 4 days respectively and reduced germination (Table 4).

Seed dressing with boric acid caused damage to seed initially, as a result reduction in field germination was observed and hence these could not cope up with growth and field cover to produce significant increase in yield and other parameters. Thus, these should not be used as seed dressing. However, as soil application, both boric acid, and borax were excellent and hence these should be used.

The increase in oil content, shelling percent and 100-seed weight of groundnut was observed due to application of borax either through seed or in soil indicating that B has played an important role in increasing these parameters and hence its application is essential in calcareous soil. Though role of micronutrients in increasing the yields and yield attributes, in calcareous soil, is well known

(Singh, 1999; Singh et al. 1990; Singh and Chaudhari, 1997), the present study has clearly demonstrated the effectiveness of these B fertilizers and its essentiality for groundnut in getting good quality produce.

Table 2. Influence of various Boron fertilizers and their modes of application on the pod and haulm yields, number of pods, oil content, percent shelling and 100-seed weight of FeESG 10 groundnut during Wet 1995

Treatment	Pod yield (kg ha^{-1})			Haulm Yield (kg ha^{-1})			Pods $plant^{-1}$		
	Seed	Soil	Mean	Seed	Soil	Mean	Seed	Soil	Mean
Control	950	950	950	3167	3167	3167	6.3	6.3	6.3
Borax	1211	1399	1305	3000	3433	3217	7.7	8.0	7.9
	(27.5)	(47.3)	(37.4)						
Boric acid	862	1460	1161	3522	3356	3434	7.0	8.6	7.8
	(-9.2)	(53.7)	(22.2)						
Mean	1008	1270	1139	3230	3319	3273	7.0	7.6	7.3
LSD (0.05)									
Application mode		64			82			0.3	
Treatments		183			234			0.9	
Interaction		259			332			1.2	
	Oil %			Shelling %			100-seed wt.		
Control	48.5	48.5	48.5	61.0	61.0	61.0	24.4	24.4	24.4
Borax	50.6	49.5	50.1	66.2	62.8	64.5	26.5	29.1	27.8
	(4.3)	(2.0)							
Boric acid	49.7	49.9	49.8	63.7	62.5	63.1	24.7	25.9	25.3
	(2.3)	(2.8)							
Mean	49.6	49.3		63.6	62.1		25.2	26.4	
LSD (0.05)									
Application mode		0.46			NS			NS	
Treatments		1.31			2.8			2.0	
Interaction		1.86			3.9			2.8	

Note: figures in parenthesis are the percent changes over control

Table 3. Influence of Boron fertilizers and their mode of application on the pod and haulm yields, oil content, percent shelling, sound mature kernels (SMK), and 100-seed weight of groundnut genotypes FeESG-10 during Wet, 1996

Treatment	Pod yield (kg ha^{-1})			Haulm yield (kg ha^{-1})		
	Seed	Soil	Mean	Seed	Soil	Mean
Control	985	1042	1013	1333	1600	1467
Borax	918 (-6.8)	1293 (24.1)	1106	1233	1800	1516
Boric acid	697 (-29.2)	1300 (24.8)	998	1300	1833	1567
Mean	867	1212		1289	1744	
LSD (0.05)						
Application mode		58			97	
Treatments		137			227	
Interaction		194			321	
		Shelling %			100 kernel wt (g)	
Control	64.9	65.1	65.0	28.1	28.0.	28.0
Borax	66.5	68.1	67.3	25.8	27.1	26.5
Boric acid	65.9	67.6	66.7	25.6	27.8	26.7
Mean	65.8	67.0		26.5	27.6	
LSD (0.05)						
Application mode		0.87			0.71	
Treatments		2.03			1.65	
Interaction		2.88			2.35	
		SMK %			Oil %	
Control	79.0	78.7	78.9	49.0	49.2	49.1
Borax	82.5	84.7	83.6	50.3	52.5	51.4
Boric acid	83.3	84.0	83.7	52.8	53.4	53.1
Mean	81.6	82.5		50.7	52.4	
LSD (0.05)						
Application mode		NS			0.64	
Treatments		3.1			1.52	
Interaction		4.7			2.15	

Note: figures in parenthesis are percent changes over control

Table 4. Influence of various salts and their modes of application on field emergence of FeESG 10 groundnut genotypes

Treatments	% Germination with various application modes		Time taken for 50% germination in seed treatment	
	Seed	Soil	Days for germination	Delay (days)
Control	90	90	6	
Borax	75	>80	9	3
Boric acid	45	>80	10	4
Iron sulphate	80	>80	8	2
Manganese sulphate	80	>80	8	2
Zinc sulphate	80	>80	8	2

Comparison of boron application through soil, foliar and drip

The micronutrients due to their lesser quantity are generally applied on foliage in solution form, but due to dry weather the leaves do not effectively absorb them fully. Now-a-days, looking to the water economy in semi-arid and arid regions which are the main groundnut area of India, the drip system of irrigation is becoming popular. Therefore, a field experiment conducted to reveals that application of B through drip irrigation, increased the yield and yield parameters, fertilizer use efficiency and was superior over their soil and foliar application in increasing all these parameters (Table 5 and 6).

Boron when applied in soil or as foliar spray, though could not influence much on pod and haulm yields and shelling percentage, it increased percent SMS and 100-seed weight during both the seasons. However drip application of B could increase all these parameters significantly.

Depending upon the season and year, soil, foliar and drip application of B increased pod yield by 5.6-7.7, 6.1- 6.4, and 14.8-20.8% respectively over control. Drip application of B increased, 12-33% 100-seed mass, 7% shelling out-turn and 11% SMS over control which were also superior over their soil and foliar applications. In B deficient soils, application of 0.5-1.0 kg ha^{-1} B as borax or boric acid recover the deficiency, but for better response the B should be applied prior to bloom stage. Foliar application of 0.05-0.1% aqueous solution of boric acid is effective in alleviating B deficiency of groundnut in the standing crop in the field (Singh et al. 1993). The response of boronated SSP in calcareous soil is also promising. Golakia and Patel (1986) the maximum yield of groundnut observed at 2 ppm of soil B and Ca:B ratio of 218-224.

Table 5. Boron and their methods of application on plant height, chlorophyll, pod weight, percent sound mature seeds (SMS) and oil in GG 2 groundnut

Treatments	Plant height at harvest		Chlorophyll content (mg g^{-1} dry wt.) in leaves at 40 DAE			Pod wt. ($g\ plant^{-1}$)		% SMS	% Oil	
	1995 RS	1995-96 R	Chl. a	Chl b	Total Chl	1995 RS	1995-96 R	1995-96 R	1995 RS	1995-96 R
Control	27.6	22.9	4.94	1.61	6.55	4.5	4.1	70.0	49.9	49.0
B, soil	30.5	23.8	5.22	1.73	6.95	5.7	5.5	73.7	49.8	49.9
B, foliar	29.3	24.1	4.76	1.60	6.36	4.9	5.8	76.7	49.1	49.9
Drip Water	32.0	25.2	5.95	1.88	7.83	6.2	4.9	82.0	49.0	49.2
B, drip	32.6	25.3	5.47	1.65	7.12	6.9	6.0	84.7	49.5	49.7
LSD (0.05)	4.0	2.1	0.80	0.45	1.02	0.9	1.03	8.3	NS	NS

Note: where R is Rabi (Mid Oct.- Feb.) and RS is Rabi-summer (Feb.-May) crop.

Table 6. Influence of boron and their methods of applications on the pod and haulm yields, shelling percent and 100-seed wt of groundnut variety GG 2

Treatments	Pod yield (kg ha^{-1})		Haulm yield (kg ha^{-1})		Shelling (%)		100-seed wt (g)	
	1995 RS	1995-96 R	1995 RS	1995-96 R	1995 RS	1995-96 R	1995 RS	1995 96 R
T1-Control	2158	2329	3360	2567	65.1	63.4	35.0	34.2
T4-B, soil	2278 (5.6)	2509 (7.7)	3778	2867	67.7	66.0	39.5	35.9
T7-B, foliar	2297 (6.4)	2470 (6.1)	3658	2606	69.5	64.6	43.1	37.6
T8-Drip Water	2523 (16.9)	2521 (8.3)	3925	2900	71.3	65.1	44.2	38.2
T1-B, drip	2607 (20.8)	2673 (14.8)	3904	3033	72.8	67.9	46.6	38.4
LSD (0.05)	475	283	529	286	3.45	2.8	5.1	1.8

Note: where R is Rabi and RS is Rabi-summer crop. Figures in parentheses indicate percent increase over control

The major advantages of B application through drip were precise application at appropriate times with desired concentration, uniform distribution, less damage to crop and soil and ultimately higher yield. The development of root system is more in a restricted volume of soil in drip irrigation and application of B through drip can efficiently place it in this zone of highest root concentrations resulting in increased nutrient use efficiency over soil and foliar applications. In Tamil Nadu B deficiency is more common and foliar application of 0.5 kg ha^{-1} borax recorded higher plant height, dry matter, pod yield, shelling %, oil content in groundnut (Shanker et al. 2004). Saxena and Mehrotra (1984) found that groundnut variety "Type 28" gave maximum increase in pod yield with application of 11.2 kg borax ha^{-1}. A net profit of Rs.4,328 ha^{-1} was obtained with application of 2.5 kg ha^{-1} borax in soil + 0.25% foliar with a cost benefit ratio of 1.94 in Tamil Nadu (Sudharsan and Ramaswami,1993). Tripathy et al. (1999) reported that borax 10 kg ha^{-1} increased the groundnut pod yield significantly.

Nutrition of bold seeded groundnut

Being bigger in size and good looking, the large-seeded (bold) groundnuts and also the hand picked and selected (HPS) seeds from the ordinary genotypes attract export for table consumption and Saurashtra and Kutch, in India, is the main area for the production of such export groundnut. But the large seeded groundnuts having more than 65 g 100 seed-mass, under ordinary management condition show 5-10% lower shelling out-turn and 100 seed mass than its potential, indicating that these are not getting proper nutrition and need extra input of Ca, K and B for better seed filling. A field experiment on pod filling and yield of groundnut, in large-seeded (TKG, 19A) and an ordinary (FeESG 10-1) genotypes reveals that application of Ca (100 kg ha^{-1}) as gypsum, K (100 kg ha^{-1}) as muriate of potash, and B (2 kg ha^{-1}) as boric acid alone or in combinations improved the pod filling and pod and seed yields (Table 7).

Application of Ca, K, and B increased 34.6, 23.4 and 15.8% pod yield over control, respectively, however combined application of Ca+K+B increased 44.6% pod yield over control. Application of these nutrients improved kernel filling, as a result the shelling percent and 100 seed mass increased. In Thailand, application of 0.25-0.5 kg B ha^{-1} increased the number and weight of pod and seed and proportion of large seed to small seed and decreased percentage of hollow-heart from 13-49% to less than 1% (Keerati-Kasikorn and Panya, 1988a).

Table 7. Effects of Ca, K and B on pod yield, shelling (%) and seed-mass of groundnut genotypes

Treatments Details	Pod yield (kg ha^{-1})		Shelling (%)		100 seed wt (g)	
	O	L	O	L	O	L
Control	1944	1064	60.9	63.2	27.2	55.9
B2	2172	1312	62.4	64.3	27.6	56. 1
K100	2352	1360	63.5	67.2	31.2	59.3
K100+ B2	2434	1420	64.6	67.7	30.4	61.1
Ca100	2455	1594	64.8	67.1	29.9	59.8
Ca100+B2	2669	1397	65.7	66.6	29.2	59.9
Ca100+K100	2745	1555	65.3	67.4	30.7	60.8
Ca100+K100+B2	2814	1534	66.1	68.0	31.7	61.3
LSD (0.05)						
Treatments	256		1.95		2.45	
Genotypes	128		0.98		1.23	
Interactions	NS		NS		NS	

Note: L and O are large-seeded (TKG 19A) and ordinary (FeESG 10-1) genotypes

Conclusions

Thus, soil application of 1.0 kg B ha^{-1}, either as basal or at 20-30 days after emergence, is invariably recommended for production of well-filled and quality seeds of groundnut in India. As large-seeded groundnut requires more B for pod filling it must be applied. Further as the application of B through drip irrigation was more beneficial over their soil and foliar applications, this practice is recommended in the areas wherever drip irrigation facility exists in semi-arid and arid regions.

References

Chitdeshwari T. and Poongothai S., 2003 yield of groundnut and its nutrient uptake as influenced by Zinc, Boron and Sulphur. Agric. Sci. Digest.**23**(4): 263–266.

Cox F.R. and Reid P.H., 1964 Calcium-boron nutrition as related to concealed damage in peanuts. Agron. J. **56**: 173–176.

Golakiya B.A. and Patel M.S., 1986 Effect of Calcium carbonate and boron on yield of groundnut. Indian J. Agric. Sci. **56**: 41–44.

Goldberg S. and Forster H.S., 1991 Boron sorption on calcareous soils and reference calcite. Soil Sci. **152**, 304–310.

Gopal N.N., 1968 Boron deficiency in groundnut (*Arachis hypogaea* L.) Indian J. Agric. Science **38**: 832–834.

Keerati-Kasikeran P. and Panya P., 1988a Effect of boron on yield of groundnut cultivar in Tainan 9. In proceedings of the 7th Thailand National Groundnut Meeting. Pattaya Chonburi Thailand pp. 384–389.

Saxena H.K. and Mehrotra O.N., 1984 Effect of Boron and Molybdenum, in presence and absence of phosphorus and calcium on groundnut. Indian J. Agric. Res. **18**(2): 91–94.

Shanker M.A.G.N., Gajanana H.K., Mohan Kumar, Ramappa Jakkanur, Ravi M.V., and Roopadevi V.D., 2004 IFA International symposium on micronutrients. Effect of Zinc and Boron on growth, yield and quality of finger millet – groundnut based cropping system for Alfisols of Karnataka. Dryland Agriculture Project, UAS, India.

Shorrock V.M., 1997 The occurrence and correction of boron deficiencies. Plant and Soil **193**: 121–148.

Singh A.L., Joshi Y.C., Chaudhari V., and Zala P.V., 1990 Effect of different sources of iron and sulphur on leaf chlorosis, nutrient uptake and yield of groundnut. Fert. Res. **24**: 85–96.

Singh A.L., Chaudhari V., and Koradia V.G., 1993 Spray schedules of multi-micronutrients to overcome chlorosis of groundnut. Indian J. Plant Physiol. **34**: 35–39.

Singh A.L., 1994 Micronutrient nutrition and crop productivity in groundnut. In Plant Productivity Under Environmental Stress (Eds. K. Singh and S.S. Purohit), pp. 67–72. Agro Botanical Publishers India, Bikaner, India.

Singh A.L., 1999 Mineral Nutrition of Groundnut. In Advances in Plant Physiology (Ed. A. Hemantaranjan). pp. 161–200. Scientific Publishers (India) Jodhpur.

Singh A.L. and Basu M.S., 2005 Integrated nutrient management in groundnut-a farmer's manual. National Research center for groundnut (ICAR), Junagadh India p. 54.

Singh A.L. and Chaudhari V., 1997 Sulphur and micronutrient nutrition of groundnut in calcareous soil J Agronomy and Crop Sci. **179**: 107–114.

Singh A.L., Basu M.S., and Singh N.B., 2004 Mineral Disorders of Groundnut. National Research center for groundnut (ICAR), Junagadh ,India p. 85.

Sudharsan S. and Ramaswami P.P., 1993 Micronutrient Nutrition in Groundnut – Blackgram cropping system. Fertilizer News. **38**(2): 53–55.

Tripathy S.K., Patra A.K., and Sameri S.C., 1999 Effect of Micronutrients on nodulation, Growth, yield and nutrient uptake by groundnut. Indian J. Plant physicol.Vol. 4. **3**: 207–209.

Yermiyaho U., Keren R., and Chen Y., 1988 Boron sorption on composted organic matter. Soil Sci. Soc. Amer. Proc. **53**, 1309–1313.

Poor Tree Form in *Eucalyptus nitens* Linked to Boron Deficiency

Timothy E. Smith

Horticulture and Forestry Science, Department of Primary Industries and Fisheries, Gympie, Queensland 4570, Australia.

Email: tim.smith@dpi.qld.gov.au

Introduction

Apical dominance is the suppression of lateral shoots by a robustly growing shoot apex. The mechanism for this apical dominance is generally accepted to be through the polar flow of auxins, most notably indoleacetic acid (IAA), synthesized in the shoot apex and young leaves (Li et al. 2001).

Boron deficiency has caused loss of apical dominance in a number of species (Marschner, 1995; Smith 2004), but the direct role of B in apical dominance is less certain. However, it has been shown that B performs a specific role in forming borate-diester bonds in the rhamnogalacturonan II (RGII) region of the pectin network in the primary cell wall (Ishii and Matsunaga, 1996; Kobayashi et al. 1996; O'Neill, et al. 1996) and that when B supply was withheld in pea (*Pisum sativum*) plants, there was a dramatic decrease in IAA export out of the shoot apex and a decline in IAA concentrations in the shoot and root apices, prior to a decline in elongation of the apex (Li et al. 2001).

A problem with B deficiency was anticipated when eucalypts were grown on Ferrosol soils (Australian Soil Classification (Isbell, 1996) approximate to Oxisol soils under Soil Taxonomy (Soil Survey Staff, 1975)). Smith (2004) found that at least 12 times higher concentrations of hot 0.01 M $CaCl_2$ extractable soil B levels were required in Ferrosol soils to sustain critical leaf B concentrations in avocados compared to sandier textured Chromosol, Kandosol and Kurosol soils (Alfisol, Soil Survey Staff, 1975). Ferrosol soils are high in sesquioxides, such as iron and aluminium oxides, which have a high affinity for B adsorption (Hatcher and Bower, 1967; Sims and Bingham, 1968; McPhail et al. 1972).

The objective of this experiment was to determine if B deficiency occurred in *Eucalyptus nitens* trees grown on Ferrosol soils and if a loss of apical dominance (LAD) then resulted as a consequence of low B supply.

F. Xu et al. (ed.), Advances in Plant and Animal Boron Nutrition, 163–169.

Materials and methods

A nutrient omission field experiment was established on a Red Ferrosol soil at Nowendoc, New South Wales, Australia (31° 32' 46 S, 151° 46' 34 E). The Ferrosol soils are acidic, deep, well structured and free draining light to medium clay soils with >5% free iron (Isbell, 1996). They have a 1:1 (silica: aluminium) kaolinite based clay lattice. The Red Ferrosol used in the experiment had 47% clay in the soil surface and 54% clay in the B horizon, with a soil pH (1 soil : 5 water) of 5.7 and 5.5, respectively. The organic carbon content of the 0–10 cm soil was 6.1%. The initial 0–10 cm hot 0.01 M $CaCl_2$ extractable soil B was 0.62 mg kg^{-1}, total N was 0.28%, total P was 421 mg kg^{-1} and available P was 12.6 mg kg^{-1}. The site had no fertiliser history.

The experiment was a randomised complete block design with nine treatments and three replicate blocks (27 plots). Each plot comprised 72 trees, at a stocking of 1201 stems/ha and the trial area was 2.158 ha. The experiment was planted in May 2001 with open pollinated *Eucalyptus nitens* seedlings.

The nutrient omission treatments (applied June 2001) were: 'complete' with soil surface applications of 200 kg ha^{-1} N, 60 kg ha^{-1} P, 100 kg ha^{-1} Ca, 50 kg ha^{-1} K, 10 kg ha^{-1} B, 10 kg ha^{-1} Zn and 5 kg ha^{-1} Cu; complete –N; complete –P; complete –Ca; complete –K; complete –B; complete –Cu and –Zn; 200 kg ha^{-1} N and 60 kg ha^{-1} P only; and nil fertiliser.

Boron was applied as Ulexite ($NaCaB_5O_9.8H_2O$, 10% B). Nitrogen was applied over a two–year period as ammonium nitrate (NH_4NO_3, 34% N), with 27.5 kg N ha^{-1} applied in June 2001, 72.5 kg N ha^{-1} applied in February 2002 and 100 kg N ha^{-1} applied in January 2003. The fertiliser forms of other elements were triple super phosphate ($CaH_4(PO_4)_2.H_2O$ & $CaHPO_4$, 20.7% P) for P, gypsum ($CaSO_4$, 18.5% Ca) for Ca, muriate of potash (KCl, 50% K) for K, copper sulphate pentahydrate ($CuSO_4.5H_2O$, 25.3% Cu) for Cu, and zinc sulphate monohydrate ($ZnSO_4.H_2O$, 35% Zn) for Zn.

Tree growth measures of total height and diameter at breast height (dbh; at 1.3 m above ground level) were measured at ages 2 and 3 and basal area was calculated. Tree volume index was calculated using the equation: tree volume index = 1/3 tree height × basal area.

Tree form was assessed using a scoring system for loss of apical dominance (LAD). The scoring system was: 1 = no LAD; 2 = minor LAD in the top of the crown, no stem defects; 3 = moderate LAD in the top of the crown, with few minor defects in the upper third of the stem; 4 = moderate to severe LAD in the crown and few moderate to severe defects due to LAD in the upper stem, with minor defects in the lower third of the stem; and 5 = several severe LAD defects along the entire stem and the crown.

Statistical analysis was conducted using GenStat®, sixth edition. Statistical significance was determined using analysis of variance and differences between treatments were assessed using a Fishers protected least significant difference (LSD) test at the 5% level.

Results

Tree growth

Trees had grown to an average tree height of 9.6 m and average dbh of 10 cm at three years from planting. There were significant treatment effects in both years two and three for tree height ($p < 0.01$), stem diameter ($p < 0.01$), and stem volume ($p < 0.01$). In year three, the complete treatment resulted in increases of 12%, 15.5% and 45.3% in tree height, dbh and volume, respectively, compared to the nil fertiliser treatment. Treatment effects on stem volume in years two (2003) and three (2004) are shown in Figure 1. Tree height and stem diameter followed similar patterns.

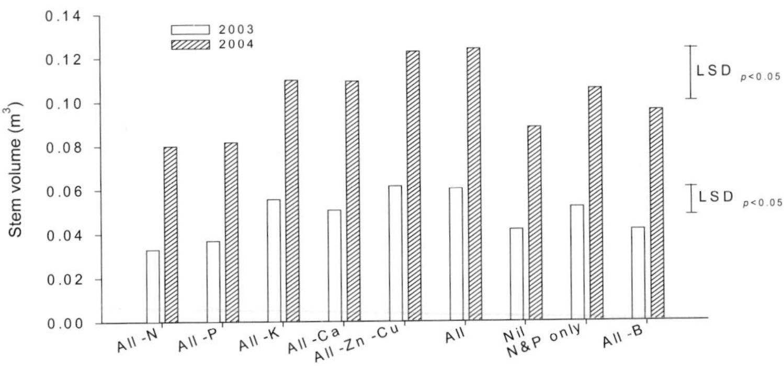

Figure 1. Effects of nutrient omission treatments on stem volume of *Eucalyptus nitens* at age two (2003, white bars) and age three (2004, hatched bars). Complete fertiliser represented by 'All'.

At three years, the omission of N, P or B independently resulted in: decreases in tree height of 15.9%, 14.4% and 10.4%, respectively; decreases in dbh of 20.2%, 17.6% and 12.2%, respectively; and decreases in stem volume index of 45.4%, 39% and 31.5%, respectively, compared to the complete treatment. The omission of Ca, K, Zn or Cu had no effects on tree height, dbh, or volume of *E. nitens* on a Ferrosol soil. When N and P were applied on their own, there was no significant decrease in tree growth measures compared to the complete treatment.

Loss of apical dominance

Fertiliser treatments had a significant effect on LAD severity (p <0.01; Figure 2). Nitrogen fertilisation exacerbated the severity of LAD. The complete –N treatment had the lowest LAD being significantly less than the complete, but this effect was at the expense of tree growth (decreased stem volume by 45.4%). The N and P only and complete –B had significantly higher LAD scores than the complete, with the complete –B treatment having the highest LAD score.

The LAD scores in the complete –P, complete –Ca, complete –K, complete –Cu and –Zn, and nil treatment were not significantly different from the complete treatment.

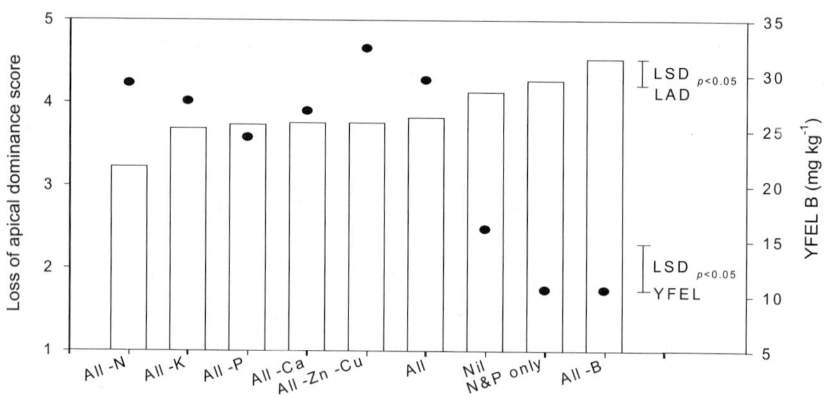

Figure 2. Effects of nutrient omission treatments on loss of apical dominance (bars) and youngest fully expanded leaf B (dots) of *Eucalyptus nitens*. Complete fertiliser represented by 'All'.

Foliar boron

Youngest fully expanded leaf B concentrations in *E. nitens* responded to fertiliser treatments (p <0.01; Figure 2). The complete –P (24.3 mg kg^{-1}), nil (16.1 mg kg^{-1}), N and P only (10.6 mg kg^{-1}) and the complete –B treatments (10.6 mg kg^{-1}) had lower foliar B concentrations than the complete treatment (29.6 mg kg^{-1}).

Loss of apical dominance was negatively correlated (R^2 = 0.73, p <0.01) with foliar B concentrations (Figure 3).

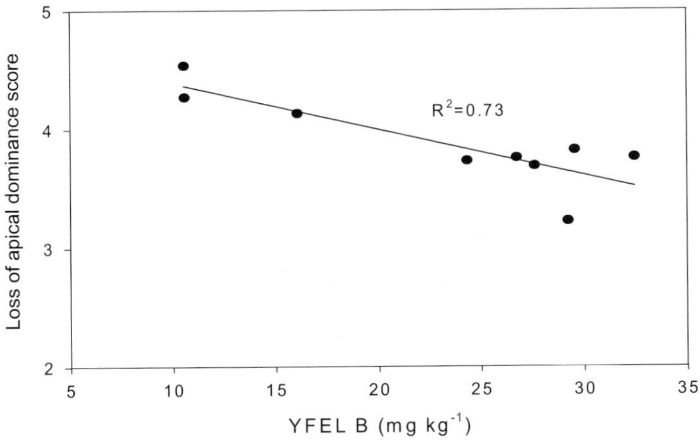

Figure 3. Correlation between loss of apical dominance scores and youngest fully expanded leaf (YFEL) *Eucalyptus nitens*

Discussion

Loss of apical dominance was negatively correlated with foliar B concentrations in this experiment, indicating that B nutrition played a role in maintenance of apical dominance, and thus, stem form of *E. nitens*. Stem form declined in treatments that were not subject to B applications, compared to the complete treatment, corresponding to lower foliar B concentrations.

Loss of apical dominance has implications for the market value of trees. Trees with LAD scores of 4 were exhibiting severe defects in the upper stem, and those with a score of 5 had severe defects along the entire stem including the lower 1/3 of the stem or butt log, which is the most valuable. The treatments without B, namely the nil, N and P only, and the complete –B, exhibited LAD scores of 4.1, 4.3 and 4.5, respectively. These treatments had a high proportion of stems with severe defects, reducing their market value.

This experiment demonstrated that N and P increased tree growth of *E. nitens* on a Ferrosol soil. There was no loss in growth (compared to a complete fertiliser treatment) by applying N and P on their own, as is commonly practiced in the Australian hardwoods plantation industry. Thus, the tree growth results support a view that additional fertiliser inputs were unnecessary and simply add to the cost of production. However, this experiment also demonstrated that the saving made on fertiliser may come at the cost of stem form and, by inference, wood quality if the B supply to trees is limited. Thus fertiliser inputs of N, P and B are justified for *E. nitens* on a Red Ferrosol soil to sustain healthy tree growth, in the absence of previous fertiliser inputs.

The amount of B fertiliser required for Red Ferrosol soils to facilitate a significant shift in leaf B levels in hardwoods is higher than that required for other soil types due to a greater B adsorption capacity of Ferrosols. These soils contain kaolinite clays, which have lower B adsorption rates than other clay types such as illite or montmorillonite (Hingston, 1964; Keren and Mezuman, 1981; Keren and Bingham, 1985), but Ferrosols also contain relatively high amounts of iron and aluminium oxides which have a high affinity for B adsorption (Hatcher and Bower, 1967; Sims and Bingham, 1968; McPhail et al. 1972). The B rate applied to the Red Ferrosol in this experiment (10 kg B ha^{-1}) was sufficient to increase foliar B concentrations in *E. nitens* from 10.6 to 28.3 ($\pm$ 2.8) mg kg^{-1}, but this rate would probably result in B toxicity in a sandy textured soil.

In conclusion, apical dominance of *E. nitens* on Red Ferrosols was improved with B applications. The application of B, in addition to the standard N and P fertiliser additions, was justified on the basis of improved stem form and increased market value of timber. The next step in this research will be the evaluation of treatment effects on wood quality and harvest yields of *E. nitens*.

Acknowledgements

I acknowledge Forest Enterprises Australia for the provision of land for the field experiment, and Paul Keay and John Huth for technical assistance they provided for this experiment. I also acknowledge John Simpson, Dr Valerie Debuse and Dr Stephen Trueman for their constructive criticism of this manuscript.

References

Hatcher J.T. and Bower C.A., 1967 Adsorption of boron by soils as influenced by hydroxyl aluminum and surface area. Soil Sci. 104, 422–426.

Hingston F.J., 1964 Reaction between boron and clays. Aust. J. Soil Res. 2, 83–95.

Isbell R.F., 1996 The Australian Soil Classification. CSIRO Publishing, Collingwood, Victoria.

Ishii T. and Matsunaga T., 1996 Isolation and characterisation of a boron-rhamno-galacturanon II complex from cell walls of sugar beet pulp. Carbohyd. Res. 284, 1–9.

Keren R. and Bingham F.T., 1985 Boron in water, soil and plants. In: Advances in soil Science, Vol. 1 (B.A. Stewart, ed.), Springer-Verlag, New York, pp. 229–276.

Keren R. and Mezuman U., 1981 Boron adsorption by clay minerals using a phenomenological equation. Clays and Clay Min. 29, 198–204.

Kobayashi M., Matoh T., and Asuma J., 1996 Two chains of rhamnogalacturanon II are cross-linked by borate-diol ester bonds in higher plant cell walls. Plant Phys. 110, 1017–1020.

Li C., Pfeffer H., Dannel F., Römheld V., and Bangerth F., 2001 Effects of boron starvation on boron compartmentation, and possibly hormone-mediated elongation growth and apical dominance of pea (*Pisum sativum*) plants. Physiol. Plant. **111**, 212–219.

McPhail M., Page A.L., and Bingham F.T., 1972 Adsorption interactions of monosilicic and boric acid on hydrous oxides of iron and aluminum. Soil Sci. Soc. Am. Proc. **36**, 510–514.

Marschner H., 1995 Mineral Nutrition of Higher Plants. (2nd Edition) Academic Press, San Diego, USA, pp. 379–396.

O'Neill M.A., Warrenfeltz D., Kates K., Pellerin P., Doco T., Darvill A.G., and Albersheim P., 1996 Rhamnogalacturanon II, a pectic polysaccharide in the walls of growing plant cells, forms a dimer that is covalently cross-linked by a borate ester. J. Biol. Chem. **271**, 22923–22930.

Sims J.R. and Bingham F.T., 1968 Retention of boron by layer silicates, sesquioxides and soil materials: II. Sesquioxides. Soil Sci. Soc. Am. Proc. **32**, 364–369.

Smith T.E., 2004 Boron nutrition of Hass avocado (*Persea americana* Mill.). PhD Thesis, University of Queensland, p. 196.

Soil Survey Staff 1975 Soil Taxonomy – a Basic System of Soil Classification for Making and Interpreting Soil Surveys. USDA, Washington, D.C.

Response of 'Schattenmorelle' Tart Cherry Trees to Drip Boron Fertigation

Pawel Wojcik

Department of Fruit Crop Management and Plant Nutrition, The Research Institute of Pomology and Floriculture, Pomologiczna 18,

96-100 Skierniewice, Poland

E-mail: pwojcik@insad.pl

Introduction

Tart cherry (*Prunus cerasus* L.) is commonly grown in Poland because of favorable soil and climate conditions. During the last 10 years, tart cherry fruit production in Poland varied from 140 to 180 thousand tons, accounting for ca. 20% of the world production (Kubiak, 2000). Introduction of mechanical harvest has increased an interest in culture of this species (Wawrzynczak, 2000). The export of tart cherry fruit to EU markets in the recent years has also stimulated production of this species in Poland. It is predicted that tart cherry production will gradually increase since fruit of this species are valuable for some products such as yoghurt, cream cheese, jam, juices, and marmalades (Kubiak, 2000).

Boron (B) is an essential element required for growth and development of higher plants (Marschner, 1995). Boron deficiency is most frequently observed on course-textured soils with low status of organic matter (Gupta, 1979; Shorrocks, 1997) and such soils prevail in Poland. It is estimated that approximately 70% of the agricultural land in Poland has low B status (Grzeskowiak, 1996).

Irrigation is beneficial treatment in regions in which drought periods frequently occur (Pacholak, 1986). In Central Poland, where sandy loam soils predominate and annual precipitation seldom excess 600 mm, irrigation of fruit trees is required to reach high productivity and marketable fruit quality. In many fruit-growing regions mineral nutrients are applied through an irrigation system (fertigation) which gives possibility to control precisely both water and nutrient supplies to crops and to maintain the desired concentration and distribution of ions and water in the soil (Dasberg et al. 1988; Komosa et al. 1999; Treder, 2003). Fertilizers in the drip fertigation system are applied at lower rates compared to broadcast fertilization (Haynes, 1985). Reduced fertilizer rates under the drip fertigation result from improved fertilizer utilization efficiency and limited leaching of nutrients beyond tree rooting zone (Bravdo and Hepner,

F. Xu et al. (eds.), Advances in Plant and Animal Boron Nutrition, 171–177.

1987). A possibility to regulate nutrient amounts and the frequency of their applications according to plant needs and weather conditions in the growing season is an additional advantage of the drip fertigation (Bar-Yosef, 1999). However, there is little information concerning the efficiency of the drip B fertigation in tart cheery orchards. Therefore, the aim of this experiment was to examine response of tart cherry trees to application of B via the drip irrigation system.

Materials and methods

Plant material and growth conditions

The experiment was carried out in 2004 at a commercial orchard in Central Poland on mature 'Schattenmorelle' IR-7 tart cherry (*Prunus cerasus* L.) trees grafted on Mahaleb (*Prunus mahaleb* L.) seedlings. The experimental site is located 136 m above sea level, and receives annually ca. 500 mm of a rainfall. Tart cherry trees were planted on a sandy loam soil (Albic Luvisols). Some physico-chemical properties of the soil from the surface horizon (0-20 cm) are given in *Table 1.* Composite soil sample for analysis was taken in 2003 along tree rows. Soil pH was determined potentiometrically in 1 : 2.5 soil/1 M KCl suspensions after shaking for 24 h; particle contents of sand, silt and clay by using the aerometric method according to the procedure of Casagrande and Proszynski (Ostrowska et al. 1991); organic matter according to the chromic and titration procedure (Allison, 1965); exchangeable potassium (K), magnesium (Mg), and calcium (Ca) by means of 1 M NH_4-acetate solution (Ostrowska et al. 1991); available phosphorus (P) by means of Ca-lactate solution (Ostrowska et al. 1991); oxides of aluminium (Al) and iron (Fe) using 0.175 M NH_4-oxalate (Jin et al. 1987); manganese (Mn) oxides with 0.1 M $NH_2OH{\cdot}HCl$ solution (Chao, 1972); and available B by means of hot water (Berger and Truog, 1944). Phosphorus, K, Mg, Ca, Al, Fe, Mn, and B were determined using an inductively coupled plasma spectroscopy (Thermo Jarrell Ash, Franklin, USA).

Table 1. Some physico-chemical properties of the soil

Sand	Silt	Clay	pH	Organic C	Exchangeable			P	Fe	Al	Mn	B
					K	Mg	Ca					
%				g kg^{-1}	cmol kg^{-1}			mg kg^{-1}	mg 100g^{-1}		mg kg^{-1}	
70	15	15	5.4	12	4.1	6.6	47	73	323	120	15	0.31

The experimental tart cherry trees were planted in 1997 at a spacing of 4 m (between rows) x 3 m (within row) and trained as a spindle system. The trees were drip irrigated from early May until mid August when water potential in the soil dropped below –0.03 MPa. Pressure-compensating emitters were placed in a row line every 0.5 m, delivering 1.75 L of water per h. Time of water application was regulated by tensiometers placed in the soil at a depth of 40 cm at a distance of 20 cm from a dripper. Samples of the water used for irrigation were periodically analysed and B concentrations varied from 0.01 to 0.03 mg L^{-1}. Herbicide strips (1.5 m wide) along tree rows were maintained with Azotop (50% simazine), Basta (15% NH_4-gluphosynate), and Nabu (12% sethoxydim). The sod in the interrows was regularly moved. In 2004 nitrogen (N), and K were applied via the drip irrigation system. Nitrogen was delivered at a rate of 20g $tree^{-1}$ in form of calcium nitrate (15-0-0) weekly over 12 weeks commencing mid-May. Potassium was provided at a rate of 15 g $tree^{-1}$ as potassium chloride (0-0-50), weekly over 8 weeks in June-July. The rates and the terms of application of fertilizer N and K in this experiment were consistent with the recommendations given by Treder (2003). Protection against pathogens and pests was made according to the standard recommendation for commercial orchards (Olszak and Bielenin, 1999).

Boron treatments and the experiment design

The drip B fertigation was applied over 4 weeks at 3-d intervals beginning at the stage of bud break at a rate of 1g $tree^{-1}$ which corresponds to 25% of the recommended per-hectare broadcast B rate (Wojcik, 2003). Such early the B fertigation treatment resulted from an assumption that under conditions of low B availability in a soil, flower B status will be a critical factor determining tree yielding. The above assumption seems to be true since it is known that B plays a key role in the reproductive process (Dell and Huang, 1997). For the B fertigation Borvit material (liquid, 8% B as boric acid; Intermag, Olkusz, Poland) was used. The trees unfertilized with B but irrigated the same as the B-fertigated trees served as a control. The experiment was conducted using a randomised block design with three replications. Each replicate consisted of 10 trees.

Measurements and observations

(i) Soil B was determined at 28 days after the B fertigation was terminated. Soil samples were collected at a distance of 20 cm from the emitter along the tree row from a depth of 0-40 cm. They were air dried, ground with a wooden roller to pass a 1-mm sieve, and then homogenized. Soil B was extracted by means of hot water according to the procedure of Berger and Truog (1944), and determined by using an inductively-coupled plasma spectroscopy; (ii) tree vigor was assessed after ending the vegetation period as the total length of current season shoots per tree calculated according to the method of Jolly and Holland (1958); (iii) leaf B concentration was determined at 80 days after full bloom. Composite samples of 100 leaves per replicate were collected from the mid-portion of extension shoots of the current year's growth. The leaf samples were dried in a forced-draft oven at 75 °C, ground to pass through a 0.84-mm stainless steel screen, ashed in a muffle furnace at 480 °C for 12 h, and then dissolved in 0.5% HCl. Boron was determined by means of an inductively-coupled plasma spectroscopy; (iv) flower B was determined on a 100 flower sample per plot. Flowers were collected at full bloom without pedicles. Preparation of the flower samples and B determination were the same as for leaf samples; (v) total fruit yield was measured separately for each plot; (vi) mean fruit weight, soluble solids concentration (SSC), and titratable acidity (TA) of fruits were calculated/determined on a 200 fruit sample per plot. Soluble solids concentration was measured by means of an Abbe refractometer, and TA by titrating the fruit homogenate with 0.1 N NaOH to pH 8.1. The TA results represented citric acid content expressed as a percentage.

Statistical analysis

Analyses of variance were performed on all data. Differences between treatment means were evaluated using Duncan's Multiple Range Test at $P \leq 0.05$. Data of the total length of current season shoots per tree were transformed according to $y = \log(x)$ as outlined by Szczepanski and Rejman (1987).

Results and discussion

Water-extractable B concentration in the soil of the control plots was low (*Table 2*). The B fertigation increased soil B level up to an optimal range (0.45-0.70 mg kg^{-1}) proposed by Wojcik (2003) for the most fruit crops grown under temperate climate.

The total length of current season shoots per tree was not affected by the B fertigation, averaging 141.9 m. Leaf B concentration of the control trees was within an optimum range (25-45 mg kg^{-1}) proposed by Sadowski (1996) (*Table 2).* Thus, despite low soil B level, concentration of this trace element in leaf tissues was adequate. This indicates that B fertilizer requirements of tart cherry are low. The above statement appears to be true because B deficiency symptoms in tart cherry orchards have not been observed in Poland despite that a large part of the orchards is established on soils with low B availability. In our study the B fertigation increased leaf B concentration (*Table 2*). However, the B fertigation had no effect on flower B level (*Table 2*). A lack of effect of the drip B fertigation on status of this trace element in flower tissues indicates that the movement of soil-applied B into the reproductive tissues in the spring was limited. The reduced transport of B into flowers of tart cherry trees might be caused by a low transpiration rate resulting from a small leaf area developed before tree blooming. This explanation appears to be true since it is known that plant B distribution is primarily governed by the transpiration stream (Marschner, 1995).

Table 2. Effect of drip boron fertigation on hot water-soluble boron concentration in the soil and 'Schattenmorelle' tart cherry response

Treatment	Soil B conc. (mg kg^{-1})	Leaf B conc. (mg kg^{-1})	Flower B conc. (mg kg^{-1})	Yield (kg tree^{-1})	Mean fruit weight (g)	Soluble solids conc. of fruit (%)	Titratable acidity of fruit (%)
Boron fertigation	0.56b	48b	46a	19.6a	5.1a	12.5b	1.03a
Control	0.32a	34a	49a	19.2a	5.0a	11.3a	1.01a

Note: means within column with the same letter are not significantly different by Duncan's Multiple Range Test at $P \leq 0.05$.

Tree yielding was not affected by the drip B fertigation (*Table 2*). A lack of impact of the B fertigation on yield despite increased leaf B status indicates that B was not factor limiting the reproductive growth. Taking into consideration that the B fertigation did not affect flower B status we can speculate that under B deficiency conditions, foliar sprays rather than soil application of B should be recommended to improve the reproductive growth.

Mean fruit weight was not influenced by the B fertigation (*Table 2*). Soluble solids concentration in tart cherry fruits was increased as a result of the B fertigation (*Table* 2) which might result from enhanced photosynthesis rate. On the other hand, increased SSC in tart cherry fruits on the B-fertigated plots might be caused by improved transport of carbohydrates from leaves to fruit tissues. This explanation may be true because B facilitates the short- and long-distance transport of sugars in a plant via formation of borate-sugar complexes increasing consequently the export of products of photosynthesis from the leaves to developing organs such as buds, flowers and fruit (Starck, 1998). In our study the drip B fertigation did cause no significant changes in TA of fruits (*Table 2*).

In conclusion we can state that high yield of tart cherry trees can be reach even on a soil with water-soluble B concentration as low as 0.32 mg kg^{-1}. However, under these soil conditions, the drip B fertigation at a rate of 1g tree^{-1} increases soluble solids concentration in tart cherry fruits. Boron fertilization is not successful in improving vigor and yielding of 'Schattenmorelle' tart cherry trees if B concentrations in leaf and flower tissues are above 34 and 46 mg kg^{-1} DW, respectively.

References

Allison L.E., 1965 Organic carbon. In Methods of soil analysis (D.D. Black, J.L. Evans, I.E. White, I.E. Engminger, and F.E. Clark, eds.) American Society of Agronomy, Madison, WI, pp. 1367–1378.

Bar-Yosef B., 1999 Advances in fertigation. Adv. Agron. **65**: 1–77.

Berger K.C. and Truog E., 1944 Boron tests and determination for soil and plants. Soil Sci. **57**: 25–36.

Bravdo B.A. and Hepner Y., 1987 Irrigation management and fertigation to optimize grape composition and vine performance. Acta Hort. **206**: 49–67.

Chao T.T., 1972 Selective dissolution of manganese oxides from soils and sediments with acidified NH2OH·HCl. Soil Sci. Soc. Am. Proc. **36**: 764–768.

Dasberg S., Bar-Akiva S., Spazisky S., and Cohen, A., 1988 Fertigation versus broadcasting in an orange grove. Nutrient Cycling in Agroecosystems **15**: 147–154.

Dell B. and Huang L., 1997 Physiological response of plants to low boron. Plant Soil **193**: 103–120.

Grzeskowiak A., 1996 Nawozy mineralne we wspolczesnym rolnictwie. In Nawozenie mineralne roslin uprawnych (R. Czuba, ed.) Zaklady Chemiczne "Police", Police, Poland, pp. 17–56.

Gupta U.C., 1979 Boron nutrition of crops. Adv. Agron. **31**: 273–307.

Haynes R.J., 1985 Principles of fertilizer use for trickle irrigated crops. Nutrient Cycling in Agroecosystems **6**: 235–255.

Jin J., Martens D.C., and Zelazny L., 1987 Distribution and plant availability of soil boron fractions. J. Amer. Soc. Soil Sci. **51**: 1228–1231.

Jolly G.M. and Holland D.A., 1958 Sampling methods for the measurement of extension growth of apple trees. A.R.E. Malling Research Station, England, pp. 87–90.

Komosa A., Pacholak E., Stafecka A., and Treder W., 1999 Changes in nutrient distribution in apple orchard soil as the effect of fertigation and irrigation. II. Phosphorus, potassium and magnesium. J. Fruit Ornam. Plant Res. **2**: 71–80.

Kubiak K., 2000 Marketing owocow wisni i czeresni. National Conference "Intensyfikacja produkcji wisni i czeresni", ISK, Skierniewice, Poland, pp. 14–32.

Marschner H., 1995 Mineral nutrition of higher plants. Academic Press, London.

Olszak E. and Bielenin A., 1999 Program ochrony roślin sadowniczych. Agrosan, Sandomierz, Poland.

Ostrowska A., Gawlinski S., and Szczubialka Z., 1991 Metody analizy i oceny wlasciwosci gleb i roslin. Instytut Ochrony Srodowiska, Warsaw, Poland.

Pacholak E., 1986 Wpływ nawozenie i nawadniania na wzrost i plonowanie jabloni 'James Grieve'. Rocz. Akad. Rol. Poznan **160**: 1–79.

Sadowski A., 1996 Nawozenie roslin sadowniczych. Sad Karlowy **1**: 71–96.

Shorrocks V.M., 1997 The occurrence and correction of boron deficiency. Plant Soil **193**: 121–148.

Starck Z., 1998 Transport floemowy i dystrybucja substancji pokarmowych. In Podstawy fizjologii roslin (J. Kopcewicz and S. Lewak, eds.) PWN, Warsaw, Poland, pp. 387–412.

Szczepanski K. and Rejman S., 1987 Metodyka badan sadowniczych. PWRiL, Warsaw, Poland.

Treder W., 2003 Wpływ fertygacji nawozami azotowym i wieloskladnikowym na zmiany chemiczne gleby oraz na wzrost i owocowanie jabloni. Zesz. Nauk. ISK **97**: 1–97.

Wawrzynczak P., 2000 Zmechanizowanie zbioru owocow wisni w Europie. National Conference "Intensyfikacja produkcji wisni i czeresni", ISK, Skierniewice, Poland, pp. 115–128.

Wojcik P., 2003 Znaczenie boru w uprawie jabloni. In Integrowana produkcja owocow (A. Holewinski ed.) Graf-Sad, Skierniewice, Poland, pp. 34–38.

Effect of Micronutrients on Citrus-Fruit Yield Growing on Calcareous Soils

Ibrahim M[1]., N. Ahmad, S.A. Anwar and T. Majeed

Soil Chemistry Section, Ayub Agricultural Research Institute, Faisalabad, Pakistan.

[1]Corresponding Author Email: soilchem@fsd.paknet.com.pk

Introduction

Due to high pH and calcareousness of soil, the availability of micronutrients is affected (Talibudeed, 1981; Marschner, 1995) and growing citrus on such soils frequently shows deficiency symptoms like yellowing of young growth, interveinal chlorosis and dieback of branches. These are typical symptoms of iron (Fe) and zinc (Zn) deficiencies. Some short duration studies were conducted on micronutrients indicating their response specifically to Zn (Khattak, 1994; Siddique et al. 1994). The soil conditions are such that applied nutrients may not give immediate impact on yield in perennial plants like citrus due to their limited root system and hairs on roots absorbing nutrients from surface soil (Cohen, 19976). Moreover, impact of environmental factors is more affective on fruit plants as compared to field crops due to many reasons not mentioned here. Keeping in view the existing status of information and importance of citrus fruits in Pakistan, a long-term study regarding micronutrient effects on fruit yield is planned and two methods of nutrient application are tested in two separate experiments in farmers' field in citrus growing area famous for Kinnow Mandarin. Up to the end of 2004, two years fruit-yield data have been taken.

Methods and materials

For establishing the field experiments on soil and foliar application of micronutrients an eight yeas old citrus orchard of Kinnow Mandarin was selected. Five nutrients were applied in two separate experiments (soil and foliar application) in following combinations: (1) Control, (2) Zn, (3) Zn + Cu, (4) Zn + Cu + Fe, (5) Zn + Cu + Fe + Mn, (6) Zn + Cu + Fe + Mn + B. Nine plants were kept in each treatment. For soil application, 50 g Zn, 25 g Cu, 100 g Fe, 100 g Mn and 15 g B were applied on per plant basis.

Nutrients were broadcasted and mixed in soil by light hoeing followed by irrigation. For foliar application, 1000 mg l^{-1} solution of each Zn, Cu, Fe, Mn and 60 mg l^{-1} was sprayed during May and June each year. Fruits were picked by hand during December/January and weighed. Before treatments application soil

F. Xu et al. (eds.), Advances in Plant and Animal Boron Nutrition, 179–182.

and plants were analyzed for micronutrients. The soil used was clay loam having pH 8.0, B 0.5, Cu 4.0, Fe 6.2, Mn 2.2 and Zn 0.4 mg kg^{-1}.

Results and discussion

Fruit yield of soil application of micronutrients experiments

Individual fruit yield data of each plant in treatments are not given here to avoid the bulk but instead yields of nine individual plants during two years are summed and presented in Table 1. This was necessary because yields of individual plants varied from 0 to 250 kg per plant. Yield of every plant was different from other in the same treatment indicating the individual behavior of plants. It has been further observed that all the plants showed alternate bearing habit (heavy and light fruit bearing in alternate year) independently. The total yields of year 2003 and 2004 are summed which showed that Zn (2260 kg) and Zn + Cu (2527 kg) treatments gave higher than control (2073 kg). The yields were decreased by the application of Fe, Mn and B when included in the respective treatments.

Table 1. Effect of soil application of micronutrients on fruit yield of Kinnow Mandarin

Micronutrients Treatments	Yield of 9 plants 2003 (kg)	Yield of 9 plants 2004 (kg)	Two years yield of 18 plants (kg)
Control	870	1203	2073
Zn	866	1394	2260
Zn + Cu	1223	1304	2527
Zn + Cu + Fe	1113	1196	2309
Zn + Cu + Fe + Mn	971	1239	2210
Zn + Cu + Fe + Mn + B	788	1120	1908

The yields obtained were, 2309 kg from treatment Zn + Cu + Fe, 2210 kg from Zn + Cu + Fe + Mn, and 1908 kg from Zn + Cu + Fe + Mn + B. The yields are decreased when these nutrients are applied together. This might be due to the reason that when more than two nutrients are applied together these might have caused salinity in soil and affected the fruit yield. The citrus plant is known to be very sensitive to salinity. It has been reported in the literature than an irrigation water fit for field crops is unfit for citrus (Cohen, 1979). Therefore, the addition of all nutrients together may be avoided.

Fruit yield of foliar application of micronutrients experiments

Fruit yield results of different treatments during the 2003 and 2004 were summed and presented in Table 2. The treatment having only Zn gave the highest yield (1464 kg) that followed by control (1427 kg), Zn + Cu (1222 kg), Zn + Cu + Fe (974 kg), Zn + Cu + Fe + Mn (797 kg) and Zn +Cu + Fe + Mn + B (793 kg). Similar to the soil application Zn showed positive effect on yield and addition of Cu, Fe, Mn and B decreased yields in general. Since data are for two years and at this stage the reason for decrease in yields due to these nutrients seems to be salt effect because all salts were applied together in soil.

Addition of nutrient salts in spray solution certainly increased solution concentration that might have caused immature fruits drop and foliage injury. Therefore, it is suggested that all salts may not be mixed in the same tank at one time. This explanation may not be taken as a final word because in this situation two other abnormalities like uneven bearing of plants within treatments and alternate fruit bearing were common in citrus plants. Therefore long-term results are needed for better conclusion in such a situation.

Table 2. Effect of foliar application of micronutrients on fruit yield of Kinnow Mandarin

Micronutrients Treatments	Yield of 9 plants 2003 (kg)	2004 (kg)	Two years yield of 18 plants (kg)
Control	662	765	1427
Zn	614	850	1464
Zn + Cu	505	717	1222
Zn + Cu + Fe	388	586	974
Zn + Cu + Fe + Mn	466	331	797
Zn + Cu + Fe + Mn + B	457	336	793

Conclusions

The study has completed two years and fruit yield results indicated that plants in both experiments (Soil and foliar application) showed uneven fruit bearing within treatments. Moreover, all the plants showed the phenomenon of alternate fruit bearing. Due these two main reasons the treatment effects were not clear, therefore, long-term data are needed to draw definite conclusion. By pooling the two years yield data it was observed that when all five nutrients were applied together showed decrease in fruit yield. Starting from Control to the last treatment where all five nutrients were applied together, yields were decreased respectively with the addition of these nutrients in soil and as foliar application.. Yields were lower in experiment where nutrients were applied on foliage as compared to soil application.

Acknowledgement

The authors are highly thankful to Pakistan Agricultural Research Council (PARC) for providing the funds for this study under Agricultural Linkages Program. Moreover, the cooperation of Project Staff at Head Quarter is duly acknowledged. We also acknowledge the cooperation of Mr. Mohsin Ali for typing the manuscript.

References

Cohen A., 1976 Citrus Fertilization. In International Potash Institute (IPI) Bulletin No. 4.

Marschner H., 1995 Mineral nutrition of higher plants. 2nd Ed Academic Press, London, England.

Talibudeen O., 1981 Precipitation. pp. 81–144. In: D.J. Greenland and M.H.B. Hayes. Ed The chemistry of soil processes. John Wiley and Sons, New York, New York.

Khattak J.K., 1994 Effect of foliar application of micronutrients in combination with urea on the yield and fruit quality of sweet oranges. Final technical report 1991–1994. Pakistan Science Foundation Islamabad and NWFP Agric.Univ. Peshawar.

Siddique M.T., Rashid M., and Saeed M., 1994 Micronutrients status of soils and plants of citrus growing areas in Punjab. In proceedings of 4th National Congress of Soil Science, May 24–26 1994, Islamabad pp. 355–360.

Development of New Boron Micronutrient and Determination of Product Applications

Gaye Çakal[2], Ferhat Soylu[1,*], M. Engin Çelen[1], Sinem Şener[1], Süleyman Taban[3] and S.B. Shukla[4]

[1]National Boron Research Institute, Middle East Technical University Campus, 06531 Ankara/Turkey.

[2]Department of Chemical Engineering, Middle East Technical University, 06531 Ankara/Turkey.

[3]Department of Soil Science, Faculty of Agriculture, Ankara University, 06110 Ankara/ Turkey.

[4]Shukla Consultancy Co. Plot No. 25-A, Besant Street,Santacruz (W) Mumbai/India.

E-mail: ferhat@boren.gov.tr

Introduction

Turkey has world's largest boron mineral deposit, refined borate production capacity of about 800.000 t/y and borate ore production capacity of 1.800.000 t/y. The National Boron Research Institute "BOREN" was established under the authority of Ministry of Energy and National Resources. Main aim of BOREN is to increase the consumption of boron products and broaden the production range from refined boron products to commercial products. For this purpose new investigations concerning the uses of boron in glass, fiberglass, ceramics, building materials, detergents, flame retardants and wood preservation have been programmed. Boron products for health and agriculture are essentially in the scope of BOREN for the benefit of mankind. In this perspective, the role of BOREN will be forming a bridge between Public Institutions, Industrial Foundations, Universities, Research and Development Institutions, Independent Researchers and Consultants. From this point of view BOREN is associating and grouping the independent studies in related fields at all organizations mentioned above. Hence these studies will be able to lead to a useful product for technology. Since its establishment; BOREN has been forming a data collection system for studies about scientific and technological investigations as well as boron effects on plant nutrition and human health and is willing to cooperate with entrepreneurs all around the world to enlarge the use and benefit of boron minerals and enhance the market oriented product development.

One major scope of BOREN studies is the use of boron as a micronutrient for plant nutrition in agriculture. A systematic approach to the research topic has been applied as first determining the use of boron in agriculture, investigating

F. Xu et al. (eds.), Advances in Plant and Animal Boron Nutrition, 183–187.

the available products in the market and production trials with the products. In order to carry out the investigations BOREN has been cooperating with Department of Chemical Engineering Middle East Technical University, Department of Soil Science Ankara University and Shukla Consultancy Mumbai/India. For product development studies a laboratory and a laboratory scale pilot plant has been established with Department of Chemical Engineering, agricultural trials have been carried out mainly with Department of Soil Science and process details have been clarified in supervision with Shukla Consultancy.

Materials and methods

Boron is known to be an essential micronutrient for plants as organizing the glucose transfer in cell metabolism and photosynthesis and increasing the crop yield. In order to define the availability of boron for plants; sources of boron are examined. In nature, boron is available as minerals mainly Ulexite ($NaCaB_5O_9.8H_2O$), Colemanite ($Ca_2B_6O_{11}5H_2O$), Tincal ($Na_2B_4O_710H_2O$), Ascharite $MgBO_2OH$ etc. The nutrients from moisturized soils are used by the roots of plants. A uniform distribution of nutrients is necessary. Another point is that agricultural activities are performed on the top layer of soil where the minerals are not available to be used by the plants. Therefore boron is to be applied either by foliar or soil application. For these applications refined boron products such as Boraxdecahydrate ($Na_2B_4O_7.10H_2O$), Boraxpentahydrate ($Na_2B_4O_7{\cdot}5H_2O$), Boric acid (H_3BO_3) and Disodiumoctaboratetetrahydrate (DOT) come into consideration. DOT is the most commonly used available product in the market. Trials at BOREN have been firstly performed on DOT.

Spray drying method (dot and light powertarimbor production)

In these trials, a mixture of hot boric acid and borax solution is continuously fed to the spray dryer. Product with B_2O_3% of 67 and pH 8-8.5 were obtained. Trials with spray drying continued with varying amounts of boric acid and borax and an optimized composition which is the same as referred TARIMBOR is produced. Light powder spray dried TARIMBOR with B_2O_3% of 72 and pH 7.5-8 is produced.

As the result of spray drying trials, spray drying technology is found to be expensive therefore solution reaction systems are studied.

Solution reaction method (tarimbor production)

Solutions with certain amounts of boric acid and borax are prepared on laboratory scale. The reaction has been carried out in 8 lts of continuously stirred tank reactor at temperatures above the crystallization points. The process is composed of reaction, crystallization, filtration and drying steps. The reaction time (residence time), stirring rate and temperatures as process parameters have been optimized and TARIMBOR with B_2O_3 content of % 58-59, neutral pH and 650-750 (kg/m3) bulk density is produced.

In order to scale up this study to industrial scale, a laboratory scale pilot plant system has been designed. The pilot plant is composed of 80 lt stainless steel tank reactors full automation with a Programmable Logical Control System. In order to try different crystallization techniques granulation tanks and crystallization containers with immersed rods are designed. The process parameters are optimized with the guidance of laboratory test results. The solids are fed by automatic solid feeding system to the reactor while the reaction temperature is kept constant. The reactor is heated with hot oil jacket. After the reaction, two different forms of crystallization are studied. Granulation of the product by natural cooling in a continuously low rate stirred granulization tank yielded granular product namely TARIMBOR Granular with solubility of 16 of g solvent/100 g water at 20 °C, high rate of dissolution, B_2O_3 content of 58-59%, neutral pH and bulk density of 650-750 (kg/m^3). Crystallization of the product at the rod immersed crystallization container yielded TARIMBOR Macro Crystal with solubility of 16 of g solvent/100 g water at 20℃, low rate of dissolution, B_2O_3 content of % 58-59 and neutral pH. The macro crystals produced are about 4 cm thick and 15 cm long. TARIMBOR Granular will be applicable in both foliar and soil techniques. TARIMBOR Macro Crystals are tended to be applied in heavy rain climates. It is also possible to pulverize the TARIMBOR Macro Crystals to yield pulverized product.

Application of tarimbor

Foliar and soil applications of Granular TARIMBOR to Wheat (Triticum aestivum, Triticum durum), Maize (Zea mays) and Chickpea (Cicer arietinum) on laboratory conditions are performed in Department of Soil Science of Ankara University. On foliar applications 0.2% B for wheat (Triticum aestivum) and 0.1% B for maize (Zea mays) and on soil applications 2 ppm B for wheat (Triticum aestivum) and 1 ppm B for maize (Zea mays) resulted in a slight increase in crop yield.

Field trials for foliar and soil applications are performed on garlic fields. On Soil application 22.38% increase in crop yield is examined at doses of 200 g B / 1000 m^2 and on foliar application 7.64% increase in crop yield with 0.1 B% solutions is examined.

Results and discussion

The aim of this research is to develop a new boron micronutrient compatible to the available products in the market, study the production methods, develop a low cost production method and determine the application advantages by laboratory and field trials. Using boric acid and borax as feed materials, most commonly used product in the market Disodiumoctaboratetetrahydrate (DOT) with B_2O_3% of 67 and pH 8-8.5 and light powder TARIMBOR with B_2O_3% of 67 and pH 7.5-8 are produced by spray drying method. Studies continued on research for a lower cost production method. Therefore solution reaction systems are studied. The reaction of boric acid and borax on laboratory studies yielded TARIMBOR with B_2O_3 content of % 58-59 and neutral pH, the production is repeated on pilot plant scale. With different crystallization methods, TARIMBOR Granular with high solubility (16 of g solvent/100 g water at 20 °C), high rate of dissolution, neutral pH (7-7.5), B_2O_3 content of 58-59% and 650-750 (kg/m^3) and TARIMBOR Macro Crystal with high solubility (16 of g solvent/100 g water at 20 °C), low rate of dissolution, neutral pH (7-7.5), B_2O_3 content of 58-59% are produced. Solubilities and boron concentrations of available products in the market and TARIMBOR are compared in Figure 1, Table 1 and Table 2.

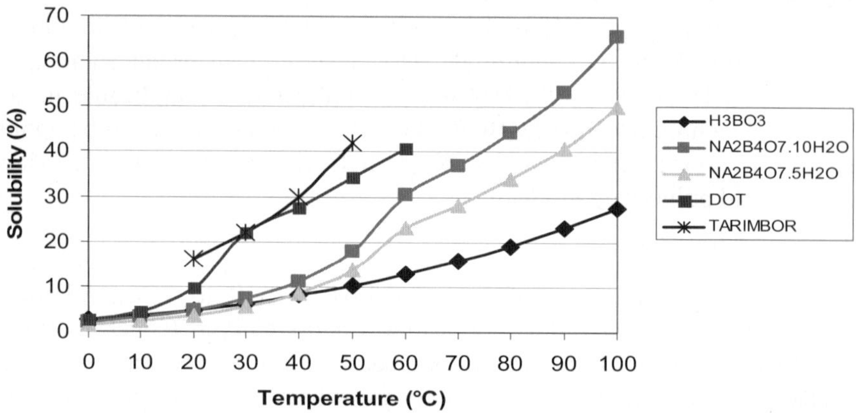

Figure 1. Solubilities of Boron Nutrient Products

It is examined that TARIMBOR is the high B containing micronutrient with high solubility. The Na/B ratio for TARIMBOR is found to be less than 0.43. The advantage of TARIMBOR Macro Crystal will be observed in heavy rain climates.

Application on Wheat (Triticum aestivum, Triticum durum), Maize (Zea mays) and Chickpea (Cicer arietinum) showed significant increase in crop yield on laboratory scale. Field studies of TARIMBOR Granular are performed on garlic and on Soil application 22.38% increase in crop yield are examined at doses of 200 g B / 1000 m^2 and on foliar application 7.64% increase in crop yield with 0.1 B% solutions is examined.

Table 1. Boron contents of boron nutrient products

PRODUCTS	% B	% B_2O_3	BULK DENSITY(kg/m^3)
TARIMBOR	18	59	750
TARIMBOR (SPRAY DRIED)	21	67	600
DOT (Disodiumoctaboratetetrahydrate)	21	67	650
Boric Acid	17	56	881
Borax Pentahydrate	15	48	992
Borax Decahydrate	11	37	850

Table 2. Na/B ratios of products

PRODUCTS	Na/B Ratios
BORAX	1.06
DOT	0.53
TARIMBOR	0.43

Boron Requirement and Distribution in the Oil Palm (*Elaeis guineensis* Jacq.) and Some Implications on Manuring Practices

K.J. Goh[1], H.H. Gan[1], K.K. Kee[1], P.S. Chew[2] and K.C. Teoh[3]

[1] Advanced Agriecological Research Sdn. Bhd., Locked Bag 212, Sg. Buloh P.O., 47000 Sg. Buloh, Selangor, Malaysia, aarsb@po.jaring.my

[2] 3, Lorong 12/15A, 46200 Petaling Jaya, Selangor, Malaysia, lmq@streamyx.com

[3] 11, Elbon Avenue, Epping, Sydney 2121, Australia

Introduction

Oil palm (*Elaeis guineensis* Jacq.) is grown on more than 12 million hectares in the humid tropics mainly between latitude 10°N and 10°S. It produces about 27% of the world's vegetable oils and fats. Oil palm is the most productive oil crop in the world, and it requires a relatively high amount of B to sustain its growth and production despite being a monocotyledon (Shorrocks, 1997). In fact, it is one of the 16 plants regarded as most sensitive to B deficiency and highly responsive to B application (Shorrocks, 1997).

In Southeast Asia, the oil palm is mainly cultivated on the highly weathered Ultisols and Oxisols derived from granite, sandstones and shales. These soils have low soil B contents (Shorrocks, 1997) and therefore, B deficiency symptoms on the oil palm in form of various types of malformed younger leaves are common particularly after drought. B deficiency causes premature lignification of the cell walls (Rajaratnam and Lowry, 1974) and under severe conditions; for example at the little leaf stage, yield may decline by about 83% (Rajaratnam, 1973a). Thus, water-soluble B fertilizer such as Fertibor (15% B) is regularly applied at the rate of 1 to 3 kg B ha^{-1} yr^{-1} in the first six years after planting to prevent B deficiency in the oil palm.

Ng et al. (1968), working with the older thick-shelled *Dura* planting materials, reported that B concentration in the oil palm canopy was similar to the stem. They did not find any trend in B concentrations between leaves of different ages and concluded that B may be mobile in the oil palm. In contrast, Rajaratnam (1972) using the current thin-shelled *Tenera* planting materials showed increasing B concentrations from the youngest to the oldest leaf suggesting that B is immobile in oil palm. However, they both agreed that B moves rapidly along the transpiration stream resulting in the accumulation of B at the tips of the oil palm leaf and leaflets.

F. Xu et al. (eds.), Advances in Plant and Animal Boron Nutrition, 189–202.

K has been demonstrated to inhibit B uptake by oil palm (Rajaratnam, 1973b) whereas N fertilization may increase palm growth leading to low or deficient B concentration through dilution effect (Shorrocks, 1997). Since then, N and K fertilizer applications for the oil palm have increased substantially (Goh and Kee, 2000). It is therefore important to determine the B requirement of oil palm under the current high fertilizer regime. Tinker and Smilde (1963) postulated that the nutrient requirement of oil palm might be established by analysing the elements in the palm tissues because the bulk of the crop, and consequently its nutrient content, is large, and the long-term nutrient-supplying power of the soil is poor. The fraction of the total available nutrients held in the crop itself can then be considerable, and its nutrient requirements will be related to the amounts immobilized by the plants (Tinker and Smilde, 1963). These conditions are approached by the oil palm growing on soils with poor B supplying capacity as found in most of Southeast Asia. Despite this simple concept, the B requirements and distribution in the oil palm plantation has received little attention in the last three decades and has not been investigated for the current *Tenera* planting materials except in their canopy. Hence, this study was conducted with the following objectives: 1) to determine the B requirement of oil palm on an Oxisol at different palm ages; 2) to investigate B mobility in oil palm using an indirect method; and 3) to examine the B distribution in an oil palm field at steady state on an Oxisol.

Materials and methods

Experimental sites

The oil palms selected for study came from the optimal NPK treatments of two factorial fertilizer response trials. Both experiments were conducted on Munchong series (Typic Hapludox) soil which was derived from shale. In the first experiment, destructive samplings of the palms were carried out at six palm ages to study the B requirements of growing palms. Two palms were sampled at 20, 37 and 71 months after field planting whereas three palms were sampled at 46, 57 and 82 months after field planting. At each month of sampling, the palms were chosen from different replicates and where possible, palms from the same replicates were taken across the months. Sodium borate was applied at an average rate of 3.1 kg B ha^{-1} yr^{-1}.

In the second experiment, two 16 years old palms, the interrow vegetation and the frond stacks were sampled to examine the B distribution in an oil palm field at the steady-state condition. Sodium borate was applied at an average rate of about 1 kg B ha^{-1} yr^{-1}.

In both experiments, the number of leaves produced by each sampled palm were measured at half yearly intervals, the fresh fruit bunch (FFB) yields at 10 day intervals and the male inflorescences at quarterly intervals. These data were then summarized on an annual basis to estimate the yearly B requirements of oil palm.

Palm destructive sampling and B analysis

The oil palm can be divided into unambiguous morphological components (Tinker and Smilde, 1963). These are: leaf, which can be further separated into leaflets, leaf rachis and leaf petiole, unopened spear leaves, growing point or "cabbage", stem, roots, male inflorescences and FFB. In this paper, the stem included the petiole bases, which were attached to it after pruning the leaves, and the root bole, which was the growing point for the roots.

Each fresh leaf was cut down ("fresh" being taken to mean that over half the total leaflet area was still green) and counted. Each leaf was then separated into the leaflets, petiole and rachis. The leaflets for nutrient analysis were sampled systematically by taking 1 in 10. The balance of the leaflets was then sampled for determination of fresh and dry weights. The rachis was cut into three equal sections. Each section was divided into three equal parts and a 10 cm section was cut from the middle of each part. One longitudinal half from each section was bulked for fresh and dry weight determination. Leaf petioles from leaves 1 to 6 were divided into two equal parts whereas the older ones into three equal parts. Leaf 1 was the youngest fully opened leaf. A 10 cm section was cut from the middle of each part. One longitudinal half was then taken from each section and bulked for fresh and dry weight determination. For B analysis, the leaflets, rachis and petiole were each bulked for various groups of leaves in the ratio of their dry weights. In the oil palm, each group or whorl of leaves consists of eight leaves i.e. leaves 1 to 8, 9 to 16, 17 to 24, 25 to 32, 33 to 40, 41 to 48 and 49 to 56. The unopened spear leaves were sub-sampled in the same manner as the leaf but their rachis and petiole were not separated because most of them had very short petioles.

The primary roots around the stem were cut to facilitate its felling at ground level. The cabbage was then cut off from the apex of the stem. Two 1/16th sections were sampled. The rest of the stem was divided into six equal sections. The middle 15 cm of each stem section was sampled and two 1/16th sections sub-sampled.

The roots were sampled using the trenching method. Since the palms were planted in an equilateral triangular pattern of 9.1 m and 8.9 m apart for experiment 1 and 2 respectively, the palm area was divided into three sections as follows: palm to palm of the same row, palm to palm of another row and palm to interrow area. In each section, a trench 30 cm wide and 90 cm depth was dug to the middle of the planting row or interrow area. The roots that were dug up were sieved out and collected for every 30 cm by 30 cm by 30 cm depth section and then bulked for each palm.

The male infloresecences on the palms at the time of sampling were separated into immature inflorescences, mature inflorescences and old inflorescences. The fresh fruit bunches for B analysis were collected from the second experiment only. A total of 26 FFB were collected over a three month period. In the laboratory, each bunch was weighed, stripped and divided into its major unique components of fruit, stalk, spikelets and parthenocarpic fruits. The fruit was further separated into the shell and kernel. Fresh weight of each component was taken. Two 100 g samples of each of the chopped and ground stalk and spikelets were collected for fresh and dry weight determination. The oil in the fruits were extracted using hexane following the modified Blaak's method (Rao et al. 1983).

The aerial parts of the interrow vegetation in each palm area were cut and weighed. The vegetation on the palm stem was also sampled, weighed and bulked with the interrow vegetation. Only one eighth of the fresh weight was taken for analysis.

All sub-samples of the vegetative and reproductive parts were sent to the laboratory for fresh and dry weight determination and B analysis. The latter followed the Azomethine-H method (John et al. 1975). Briefly, 1 g of plant sample was dry ashed at 530°C and digested with 10 mL of 1.4 M H_2SO_4. The solution was then filtered through Whatman No. 1 paper. 0.5 mL of 0.05 M EDTA, 1 mL of 0.5 M ammonium acetate and 1 mL of Azomethine-H solution were than added to 1 mL of the filtrate to prevent interferences and develop the colour. The B concentration in the filtrate was then read using a UV/VIS spectrophotometer.

Computing the annual B requirements from the data

In the first experiment, the time intervals were not fixed. Therefore, the cumulative B accumulation or uptake by the palm in each component at each sampling time has to be computed and the results modeled using non-linear regression. The gradient of the model at each yearly time period gave the annual B requirement for the component.

The B requirement of canopy between two sampling periods (t1 and t2) was: B content of new leaves produced between t1 and t2 plus B content accumulated by the older leaves between t1 and t2. It was also assumed that the B content in the canopy at the first sampling represented a continuous accumulation of B from the time of field planting since pruning of the leaves had not commenced then.

The B requirement of roots between two sampling periods (t1 and t2) was computed as follows:

B content at t2 - B content at t1 + α B content at t2

where α is the percentage of self-pruned roots. The α value for each time period was obtained from Jourdan and Rey (1997). It was also assumed that the root biomass continued to grow in the first 36 months after field planting with minimal self pruning following the logistic root model of Jourdan and Rey (1997).

The annual B requirements for FFB and male inflorescences were estimated based on their annual dry matter production multiplied by their average B concentration for the year.

Statistical analysis

Descriptive statistics, means and standard deviations, were calculated using Statistica Ver. 7.0 (StatSoft, 2004). Separate one factor Anova was computed for the B distribution within the vegetative components of the palm and the FFB. Non-linear regression was used to model the cummulative vegetative data using CurveExpert 1.38 (Hyams, 2001).

Results

B requirements of the oil palm

The B contents of the canopy and stem increased rapidly from 20 to 82 month after planting (Table 1). These increases were mainly due to palm growth since their B concentrations were relatively constant over the period of measurements. Root B concentration was 7.9 mg kg^{-1} at 20 months after planting but it quickly stabilized at an average 4.5 mg kg^{-1} from 37 months after planting resulting in a stable B content in the root (Table 1). The B requirements for FFB yields and production of male inflorescences between the sampling intervals increased exponentially before reaching a plateau at 71 months after planting (Table 1), again due to biomass increments rather than B concentrations in the palm components.

The cumulative B contents for each palm component followed a non-linear regression model (Table 2). The cumulative B contents for canopy, stem and male inflorescences fitted well to modified exponential models whereas those for roots and FFB yields to hyperbolic models. The latter implied that the B requirements were relatively constant after their maximum has been attained. The r-squares for all the equations exceeded 0.78.

Table 1. B contents (g ha^{-1}) in different vegetative components of the oil palm at the time of sampling and generative production between the time of sampling

	Canopy (g ha^{-1})		Stem (g ha^{-1})		Root (g ha^{-1})		FFB (g ha^{-1})		MI (g ha^{-1})	
Month	Mean	sd	Mean	sd	Mean	sd	Mean	sd	Mean	sd
20	22	5	10	0.3	NA	NA	0	0	NA	NA
37	62	11	50	6	44	15	28	2	16	8
46	82	20	52	14	36	2	47	14	38	15
57	133	16	113	8	27	5	86	7	29	12
71	178	15	174	16	37	NA	95	6	24	9
82	171	16	261	52	37	6	91	3	20	8

Note: FFB and MI denote fresh fruit bunches and male inflorescences, respectively. NA means not available and sd is standard deviation.

Table 2. Equations to calculate the cummulative B requirements of the vegetative and reproductive components of oil palm

Component	Equation	Number	r^2	Sy.x
Canopy	$Ln(y) = 15816 - (115/x)$	15	0.98	267
Stem	$y = 0.04x^{2.41}$	15	0.94	156
Root	$y = 721*Ln(x)-2288$	10	0.80	117
FFB	$y = (5.1x - 108)/(1 - 0.02x + 0.0002x^2)$	15	0.96	58
MI	$y = 887*Ln(x) - 3070$	15	0.78	147

Note: The variable "x" is time in month after planting and "y" is cumulative Brequirement in mg B per palm. The planting density was 138 palms per hectare

The annual B requirements of oil palm followed a sigmoidal curve with rapid increase from the second year after planting, reaching a maximum of 343 g B ha^{-1} yr^{-1} in the sixth year (Figure 1). The annual B requirement then declined to about 220 g B ha^{-1} yr^{-1} in the eighth year before stabilizing at 198 g B ha^{-1} yr^{-1} from the twelve year. The highest B requirement of oil palm came from its developing canopy, which peaked at the fifth year after planting when the canopies of neighboring palms were fully overlapped ("closed") and competition began to set in. At the peak, the B requirement for the canopy accounted for more than 40% of the total B requirement of oil palm.

The B requirement for stem development was also the highest in the sixth year at 67 g B ha^{-1} yr^{-1} before declining sharply as palm competition intensified and growth rate slowed down (Figure 1). In contrast, the B requirement for roots was not only the lowest but also attained its peak earlier at the third year after planting (Figure 1). Thereafter, it required about 9 g B ha^{-1} yr^{-1} to sustain root turnover or self pruning of roots.

The B requirement for FFB yield was the second largest among the palm components and accounted for about 33% of the annual B requirement at the sixth year and increasing to 37% as the B needs for stem decreased sharply (Figure 1).

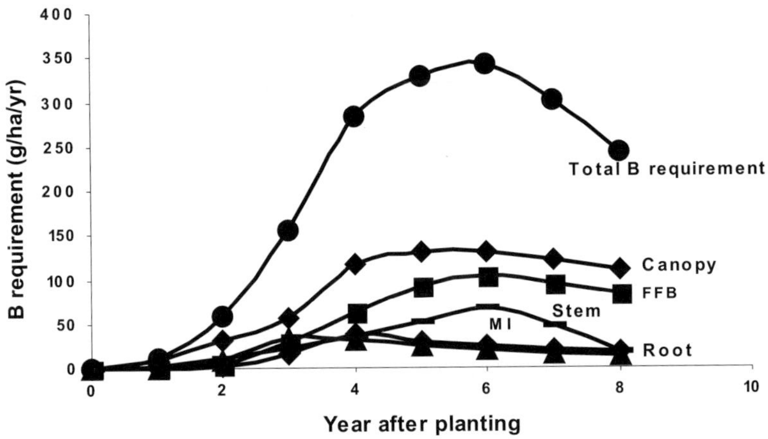

Figure 1. Annual B requirements for different components of oil palm on Munchong series (Typic Hapludox) soil

B distribution within the oil palm

The detailed B distribution in the oil palm was studied in the second experiment where the palms were 16 years old and in steady-state conditions (Tables 3 and 4). The cabbage, which composes mainly meristem cells, had the highest concentration of B at 12.3 mg kg^{-1} (Table 3). This was followed by the leaflets which were at the tail-end of the transpiration stream. Interestingly, the petiole, which supports the rachis and leaflets, and the petiole base, which are attached to the stem after the leaf is pruned off, had higher B concentrations than the rachis or stem. The stem and rachis had similar B concentrations of 3.3 and 3.6 mg kg^{-1}, respectively. Although the stem B concentration was low, it had accumulated a large quantity of B by the 16th years at 404 g ha^{-1}. The roots had the lowest B concentration of 1.4 mg kg^{-1}.

In FFB, larger B concentrations were found in the stalk, empty spikelets and mesocarp fibre (Table 4). These components have more rapid transpiration rate particularly during fruit development and oil formation stages (Jeje et al. 1978). B concentration in the kernel at 6.23 mg kg^{-1} was 2.25 times more than the shell (Table 4). The stalk and empty spikelets, which are the main components of empty fruit bunches after the milling process, contained about 43% of the B in FFB.

Table 3. B concentrations and contents in the vegetative components of 16 year-old oil palms on Munchong series (Typic Hapludox) soil

		B concentration (mg/kg)		B content (g/ha)	
Vegetative component	Number of samples	Mean	sd	Mean	sd
Leaflets	2	7.0	0.4	67.3	5.8
Petiole	2	6.2	1.7	51.6	27.2
Rachis	2	3.6	0.1	36.0	2.3
Cabbage	2	12.3	1.3	4.8	0.9
Stem	2	3.3	0.4	403.5	53.6
Root	2	1.4	0.6	23.3	6.3

Table 4. B concentrations and contents in the fresh fruit bunch (FFB) components of 16 year-old oil palms on Munchong series (Typic Hapludox) soil

	B concentration (mg kg^{-1})		B content (g ha^{-1})	
Bunch component	Mean	sd	Mean	sd
Stalk	12.87	0.23	3.91	0.60
Empty spikelet	11.57	1.50	27.18	3.55
Mesocarp fibre	11.30	1.15	27.46	3.10
Shell	2.77	0.67	4.16	0.76
Kernel	6.23	4.27	9.83	6.62
Parthenocarpic fruit	9.50	NA	0.15	NA
Total	2.39	0.41	71.58	12.39

Note: NA denotes not applicable because there was only one sample with parthenocarpic fruits

A detailed analysis of the B distribution in the canopy showed that the B concentrations seemed to be relatively constant from the spear leaves to leaf number 32 before decreasing in the older leaves (Table 5). In terms of B content, there was an initial increase from leaf 1 to leaf 9 due to increasing frond dry weight. The B contents between leaf 9 and leaf 32 were similar at an average 26.3 g B per leaf. The B contents then declined linearly between leaf 33 and leaf 49. This decrease was mainly due to the lower dry weights and B concentrations of both petiole and rachis (Table 5). There was no clear trend in B concentrations in the leaflets.

Table 5. Changes in B concentrations and contents of developing leaves of a 16 year-old oil palm on Munchong series (Typic Hapludox) soils

			B concentration (mg kg^{-1})		B content (g leaf^{-1})	
Leaf stage	Leaf position	No. of leaf	Mean	sd	Mean	sd
Developing leaf	Spear	NA	5.92	0.98	49.7	9.9
Maturing leaves	1-8	8	5.76	0.72	20.3	1.5
Matured leaves	9-16	8	5.96	0.92	26.3	5.0
Matured leaves	17-24	7.5	5.66	1.08	25.8	5.6
Matured leaves	25-32	7.5	5.65	0.87	26.8	6.3
Matured leaves	33-40	5.5	4.76	0.29	22.0	3.4
Old leaves	41-48	3.5	5.08	0.63	19.9	3.7
Oldest leaf	49+	2	5.21	0.92	17.7	3.7

Note: NA denotes not applicable. Leaf stage refers to the developmental stage of the leaves while leaf position indicates the physiological age of the leaf and position in the canopy. In the oil palm, each whorl of leaves comprises eight leaves but the lower whorls may haveless due to pruning during harvesting of fresh fruit bunches and canopy maintenance rounds.

B distribution in an oil palm plantation

The B distribution in the various components of one hectare of oil palms indicated that the largest store of B was in the stem at 404 g ha^{-1} followed by the canopy at 155 g ha^{-1} (Figure 2). The ground vegetation of mainly grasses and ferns contained very low B at only 6 g ha^{-1}.

In terms of B cycling, the pruned leaves would return 71 g B ha^{-1} yr^{-1} to the agroecosystem while the decaying male inflorescences about 15 g B ha^{-1} yr^{-1}. The pruned leaves were neatly stacked in the interrow area called leaf pile. The latter area stored about 47 kg B ha^{-1}, which was 68% of the annual B in the pruned leaves. The lower B value in the leaf pile might be attributed to the high decomposition rate of oil palm leaf where most of it would decay within six months.

In the oil palm agroecosystem, the FFB yields are transported to the oil palm mill for processing, and this component contained 52 g B ha^{-1} yr^{-1}. The oil palm also immobilized about 36 g B ha^{-1} yr^{-1} mainly for stem growth.

Discussion

The B requirement of oil palm in the first 82 months after planting was mainly driven by its growth rate, a process termed as growth demand for nutrients by

Tinker and Nye (2000). This is due to the relatively constant B concentration across palm age in each palm component, which is in agreement with the results of Ng et al. (1968). However, the current *Tenera* oil palm on infertile Oxisols requires about 27% more B than the older *Dura* planting material on fertile Inceptisols. This difference might be attributed to the larger canopy of the *Tenera* oil palm which is responsive to high fertilizer regime (Goh et al. 2003).

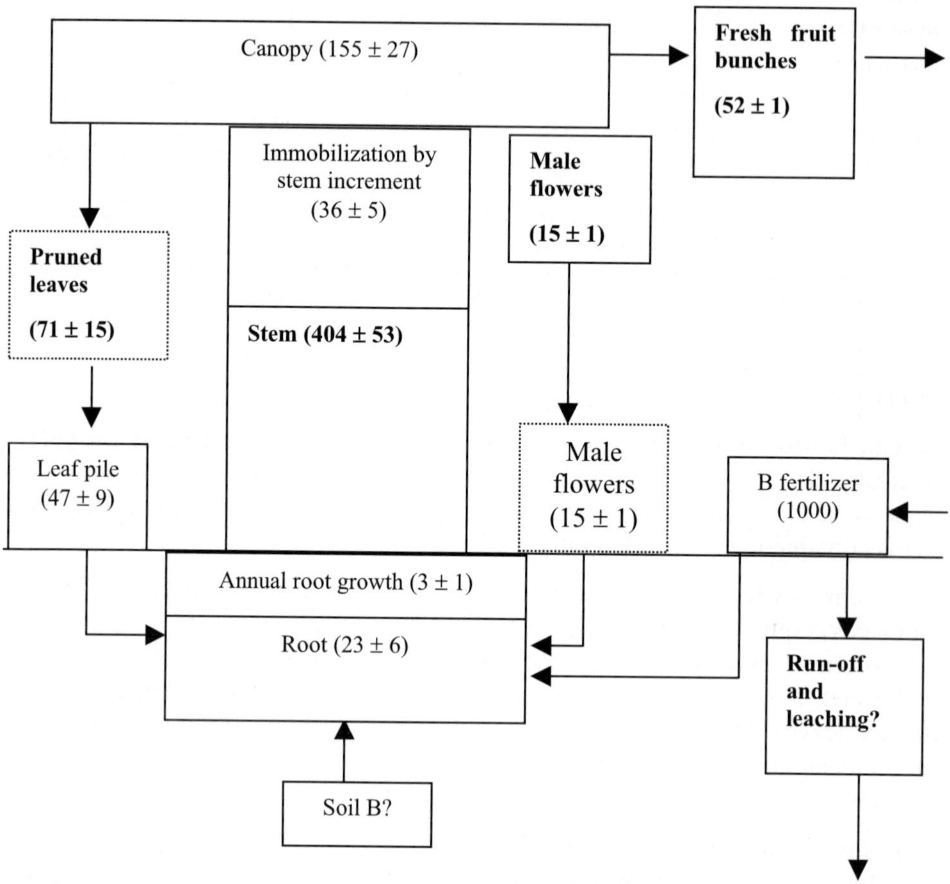

Figure 2. Total demands and sinks for and supplies of boron in an oil palm agroecosystem on Munchong series (Typic Hapludox) soil. Solid line boxes represent sinks for B (g ha^{-1}) and sub-box the annual demands for B (g ha^{-1} yr^{-1}) for each palm component. Dotted line boxes represent B supplies or recycled B (g ha^{-1} yr^{-1}) to the oil palm. Question marks denote unknown values for B. Numbers in each box show the mean ± standard deviation

The B content of the roots had not been studied previously. Initially the oil palm partitions a large proportion of its biomass to the roots to maximize the exploitation of soil water and nutrients. This results in relatively large B requirement for root development. But as the other vegetative matter develop and production commences, the proportion of B for roots declines to less than 6% (Figure 1).

The annual B requirement decreased from the sixth year after planting due mainly to the declining growth rate of the stem (Figure 1). Extrapolating the model for the annual B requirement shows that the steady-state B uptake of 198 g ha^{-1} yr^{-1} as estimated using the 16 years old palms in the second experiment would be reached at the twelve year after planting. This agrees with the findings of Gerritsma and Soebagyo (1999). They found that the vegetative growth of oil palm reached its maximum at the 12th years after planting, and FFB yields were relatively stable thereafter. Therefore, the nutrient requirements including B can be anticipated to be in a steady-state too.

The *Tenera* FFB contained 2.39 mg B kg^{-1} which was only 12% greater than Dura bunches reported by Ng et al. (1968). This was despite its bigger mesocarp and thinner shell and the former containing more B. This small difference could be partially attributed to the higher B concentrations in the stalk and empty spikelets of *Dura* bunches although strong conclusion cannot be made since Ng et al. (1968) analyzed only two bunches compared with 26 bunches in this study.

The B distribution in the vegetative components and FFB suggests that B is mainly transported by the xylem. In the oil palm, B is considered to be phloem immobile based on the work of Rajaratnam (1972). Moreover, B toxicity symptoms of oil palm are exhibited first in the tips and margins of leaflets of older leaves suggesting that it is a non-polyol producing plant (Brown et al.1999) and therefore, its B has restricted phloem mobility. However, a detailed analysis of the B concentrations in the canopy implies that B may be translocated in the phloem also. The young developing (spear) leaves had B concentration similar to the matured leaves (leaves 9 to 16) and higher than the maturing leaves (leaves 1 to 8). But the older leaves (leaf 33 and older) tended to have the lowest B concentration (Table 5). The declines in B concentration occurred in the petiole and rachis but not the leaflets. This indicates that B may be mobiled in the phloem of petiole and rachis only. Both Ng et al. (1968) and Rajaratnam (1972) did not investigate the B concentrations in the spear leaves and leaves older than 33. Thus, they did not observe the above and the latter concluded that B was phloem immobile while the former contended that there was no consistent trend. Rajaratnam (1972) further illustrated that B could be lost from the leaves through guttation in order to explain the differences in B contents in the leaves at different time of sampling and leaf age. However, this postulation could not explain our results since guttation can occur in all the leaves.

The oil palm is mainly planted on highly weathered Ultisols and Oxisols with generally low soil B content. Thus, increasing rates of soluble B fertilizer in the first six years after planting are usually required to match the annual B requirements (Figure 1). Apart from the first year, the B rates should range between 2.25 and 4.5 kg B ha^{-1} yr^{-1} due to the low fertilizer use efficiency of less than 15% as obtained in this study. Subsequently, the B rates may be decreased to between 2 and 3 kg B ha^{-1} yr^{-1} for palms between seven and twelve years old. At the steady-state of 12 years or older, only occasional B applications are probably necessary since B exported out of the system through FFB and immobilized in the stem is relatively low.

About 43% of the B in FFB is found in the stalk and empty spikelets which can be returned to the fields via empty fruit bunches. It is probably also worthwhile to build-up B in the oil palm canopy in the early years to increase the B content in the leaf pile, which is the dominant source of organic matter in the plantations. This is because organic matter can be an effective source of B by mineralization when the soil B is low (Bell et al. 2002).

The oil palm accumulates substantial B in its vegetative components and by the time of replanting at 25 years old, it should reach about 750 g B ha^{-1}. This is sufficient to meet the B requirements of oil palm in the first four years. The current zero-burn replanting technique where the palms are felled, chipped and pulverized, and the organic residues spread throughout the field may be able to return the palm B to the soils although the next generation of oil palms may not be able to exploit it fully due to its initial small root system and the large leaching loss of B in Malaysian soils (Rajaratnam, 1973b).

Conclusions

The annual B requirement of oil palm increases rapidly and reaches a peak at six years after planting. B is mainly needed for the canopy development and FFB production. Majority of the absorbed B is phloem immobile but there is an indication that B in the petiole and rachis of old leaf is phloem mobile. By the time of replanting, B immobilized in the stem is sufficient to meet B requirement of oil palm in its first four years of growth and production. A management strategy is therefore required to recycle this substantial amount of B to the next generation of oil palm.

Acknowledgements

We wish to thank our Principals, Messrs. Boustead Plantations Bhd. and KL-Kepong Bhd., and Advanced Agriecological Research Sdn. Bhd. for permission to present this paper. The first author would like to thank Mr. Yoshio Nagai, Borax Asia Pte Ltd. and U.S. Borax for partial sponsorship to the symposium.

References

Bell R.W., Dell B., and Huang L., 2002 Boron requirements of plants. In Proc. Int. Workshop on All Aspects of Animal and Plant Boron Nutrition (Goldbach H.E., Brown P.H., Rerkasem B., Thellier M., Wimmer M.A. and Bell R.W., eds) New York. pp. 63–85. Kluwer, Acad.

Brown P.H., Hu H., and Roberts W.G., 1999 Occurrence of sugar alcohols determines boron toxicity symptoms of ornamental species. J. Amer. Soc. Hort. Sci. **124**, 347–352.

Gerritsma W. and Soebagyo F.X., 1999 An analysis of the growth of leaf area of oil palm in Indonesia. Expt. Agric. **35**, 293–308.

Goh K.J. and Kee K.K., 2000 The oil palm sector in Southeast Asia: Changing requirements for fertilizers, particularly P., and K. In Workshop on Improving Soil Fertility Management in Southeast Asia. pp. 21–23. November 2000, Bogor, Indonesia. IBSRAM, Thailand, Preprint.

Goh K.J., Hardter R., and Fairhurst T.H., 2003 Fertilizing for maximum returns. In Oil Palm: Management for large and sustainable yields (T. Fairhurst and R. Hardter, eds) Potash and Phosphate Institute (PPI), Potash and Phosphate Institute Canada (PPIC) and Int. Potash Inst. (IPI), Singapore, pp. 279–306.

Hyams D., 2001 CurveExpert version 1.38 - A curve fitting system for Windows. http://curveexpert.webhop.biz/

Jeje A.A., Odetola A.O., and Zimmermann M.H., 1978 Transpiration and oil accumulation rates for developing oil palm fruits Elaeis guineensis Jacq. Acta Bot. Neerl. **27**, 213–228.

John M.K., Chuah H.H., and Neufeld J.H., 1975 Application of improved azomethine-H method to the determination of boron in soils and plants. Anal. Lett. **8**, 559–568.

Jourdan C. and Rey H., 1997 Modelling and simulation of the architecture and development of the oil palm (Elaeis guineensis Jacq.) root system. Plant Soil **190**, 235–246.

Ng S.K., Cheah T.E., and Thamboo S., 1968 Nutrient contents of oil palms in Malaya. III. Micronutrient contents in vegetative tissues. Malaysian Agric. J. **46**(4), 115–126.

Rajaratnam J.A., 1972 The distribution and mobility of boron within the oil palm (Elaeis guineensis L.) I. Natural distribution. Ann. Bot. **36**, 289–297.

Rajaratnam J.A., 1973a The effect of boron deficiency on oil palm fruit yield in Malaysia. In Proc. Int. Oil Palm Conference on Advances in Oil Palm Cultivation (R.L. Wastie and D.A. Earp, eds) Inc. Soc. Planters, Kuala Lumpur, pp. 280–288.

Rajaratnam J.A., 1973b Movement and accumulation of boron in some Malaysian soils. Expl. Agric. **9**, 233–240.

Rajaratnam J.A. and Lowry J.B., 1974 The role of boron in the oil palm (Elaeis guineensis). Ann. Bot. **38**,193–200.

Rao V., Soh A.C., Corley R.H.V., Lee C.H., Rajanaidu N., Tan Y.P., Chin C.W., Lim K.C., Tan S.T., Lee T.P., and Ngui M., 1983 A critical reexamination of the method of bunch quality analysis in oil palm breeding. PORIM Occasional Paper **9**, 1–28.

Shorrocks V.M., 1997 The occurrence and correction of boron deficiency. Plant and Soil **193**, 121–148.

StatSoft 2004 STATISTICA (data analysis software system), version 7. www.statsoft.com.

Tinker P.B. and Nye P.H., 2000 Solute movement in the rhizoshere. Oxford University Press, Oxford, United Kingdom. pp. 342–346.

Tinker P.B. and Smilde K.W., 1963 Dry matter production and nutrient content of plantation oil palms in Nigeria. II Nutrient content. Plant Soil **19**(3), 350–363.

Genotypic Differences of Boron Nutrition in Plants

Physiological and Genetic Base of Boron Efficiency in *Brassica napus*

Fangsen Xu [1, 2] and Yunhua Wang [1]

[1]Microelement Research Center, Huazhong Agricultural University, Wuhan 430070, China

[2]National Key Laboratory of Crop Genetic Improvement, Huazhong Agricultural University, Wuhan 430070, China.

Introduction

Boron (B) is an essential microelement for growth and development of higher plants, and *Brassica napus* is one of the most sensitive crops to B deficiency. In China, there are more than 6.67 million hectares of oil rapeseed distributed mainly in East, Middle and Down Stream valley of Yangtze River, and Southwest, which are deficient or seriously deficient in available B (Liu 1996). *Brassica* grown in these areas usually shows B-deficiency symptoms or at least yield depression without application of B fertilizer. However, there are significant differences between *Brassica napus* cultivars in their response to B-deficiency (James et al. 2000a, b, 2001; Hu et al. 1991, Yuan and Liu 1990, Xie and Yang 1994, Xie et al.1994). Thus, it is possible that screening for B-efficient germplasm of *Brassica napus* can serve to breed for cultivars better adapted to soils low in available B.

Below, we will summarize our research on screening for B efficiency, studies about the underlying physiological mechanism, and the genetic basis of B efficiency in *Brassica napus*.

Screening for B-efficient germplasms in *Brassica napus*

In 1991 and 1995, 210 cultivars of *Brassica napus* were selected from germplasm resource pools at the Genetic Breeding Institute of Rapeseed, Huazhong Agricultural University and Oil Crop Research Institute, Agricultural Academy of China, respectively. The screening of B-efficient germplasm was conducted according to the two-pace method (Wang and Lan 1995a, b). By the screening method, we finally selected eight cultivars as being particularly B-efficient and two as B-inefficient. The B-efficient cultivars could grow well still at 0.25 $mg{\cdot}kg^{-1}$ soil B (hot water soluble = HWS) and with high seed yields, whereas the B-inefficient cultivars usually did not produce a significant seed yield at that B level (Wang and Lan 1995a, b; Wang et al. 1999; Chu 1999).

F. Xu et al. (eds.), Advances in Plant and Animal Boron Nutrition, 205–211.

According to the result of the two-pace method, the response of *Brassic napus* to B deficiency at seedling stage was always different from the performance during the entire growth period, which the B efficiency coefficient obtained at seedling stage coincided to 90% with the performance over the entire growth period (Chu et al. 1999). The B-efficient cultivars show some typical agronomic characters with lighter leaf color, earlier flowering and a shorter growth period (Fangsen et al. 2002). One strategy for a higher B efficiency thus seems to be saving B from vegetative growth.

Physiological mechanism of B efficiency

Root system growth and nutrient uptake

Cao (1996) studied the root system growth of B-efficient cultivar "Tezao 16" and B-inefficient cultivar "Bakow" in a B deficient solution culture. It was found that the B-efficient cultivar had a better developed root system than the B-inefficient cultivar under B deficiency. The root system parameters of "Bakow", such as root dry weight, root length, total of absorption area of root system, active absorption area, and the ratio of active to total absorption area were significantly decreased by B deficiency for 15 d. In "Tezao 16", however, these parameters were hardly affected (Cao et al. 1997).

The root system is the principal plant organ for nutrient uptake. The difference of root morphological characters between B-efficient and B-inefficient cultivars under B deficiency significantly affects uptake, utilization of B, which will finally result in the observed growth difference. B absorption rates of "Bakow" were significantly lower than that of "Tezao 16" at 0.01-0.50 mg/L B, but when B concentration exceeded 0.50 mg/L, the B uptake rate of "Bakow" was higher than that of "Tezao 16" (Cao et al. 1997). When B was deficient during the growth period, the average of B accumulation in shoot of five B-efficient cultivars was significant higher than that of B-inefficient cultivars, which reached 2.0 times and 4.1 times at the flowering stage and the silique stage, respectively (Yu et al. 1999). Thus, a stronger B uptake capacitiy under limiting conditions is likely one of the main physiological mechanisms of B efficiency in *Brassica napus*.

B efficient cultivars of *Brassica napus* not only showed a better B absorption, but also higher uptake of most other mineral nutrients (Geng et al. 1998; Yu et al. 1998). B deficiency also affected root exudates and composition of bleeding sap differently between cultivars differing in their in B efficiency. Under B deficiency, secretion of soluble sugars decreased significantly in B-inefficient cultivars, whereas there was no difference in the B-efficient ones. B levels did not affect the total content of amino acids in root exudates of various B efficient cultivars. There was, however, genotypic difference in its amino acid composition. In B-inefficient cultivars, B deficiency lead to a decrease of exudation rates, and soluble sugar and amino acid contents (Cao et al. 1997).

Distribution and utilization of B

For a long time, most scientists thought of B uptake being a passive process, and that transport of B in the plant was mainly controlled by its transpiration (Bowen 1972; Pate 1975; Raven 1980). But there were some reports about a partial retranslocat in the phloem of radish and cauliflower (Shelp 1987); *Brassica napus* (Shen Z. G. and Shen K. 1994; Shen Z. G. et al.) and cotton (Xie et al. 1992).

Xu et al. (1998) studied B distribution and re-use of various B efficient cultivars and F_1 generations at seedling and bolting stage in pot culture. It was found that B distribution in deficient plants differed to that under normal B levels. Under the condition of B deficiency, B concentration in the oldest leaf of B-inefficient cultivars was higher than that of B-efficient cultivars at both stages, but it was the opposite in the youngest leaf (Xu et al. 1998). The results were identical to that of Xiong et al. (1995), where a lower B level in basal leaves in the efficient cultivars but two to four fold higher levels in buds and floral organs in B efficienct cultivars under B deficiency (Xiong et al. 1995). This points to a higher re-translocation rate in the B efficient cultivars.

Yu et al. (1999) also reported that B accumulation and accumulation ratio in different organs of *Brassica napus* differed between development stages. During vegetative growth, B was mainly accumulated in the vegetative organs, whereas with the initiation of reproductive growth, B accumulated mostly in the reproductive organs, reached the highest values at budding stage. The B accumulation ratio of reproductive to vegetable organs was significantly higher in B-efficient cultivars than in B-inefficient ones at flowering and budding stage under severe and intermediate B deficiency. This indicates again that B-efficient cultivars possess a higher capacity to transfer B from old leaf to young leaves (Yu M. et al. 1999) as one of the physiological mechanisms of B efficiency in *Brassica napus*

Du et al. studied the B form in *Brassica napus* using a sequential extraction procedure. B in *Brassica napus* was divided into three forms: free B, loosely- and firmly bound B. One B-efficient cultivar (9589), one B-inefficient cultivar (95105) and the F_1 hybrid (95105×9589) were studied under two B levels in soil culture. The results showed levels of the three B forms were different between parents and their F1 progeny. Free B and bound B in the B efficient cultivar (9589) were significantly lower than in the inefficient cultivar (95105), and the F_1 hybrid (95105×9589) ranked intermediate. The loosely-bound B concentrations, however, showed the opposite tendency (Du et al. 2000).

B and calcium relationship in various B efficient cultivars

There are indications for a more direct relationship between B and Ca in higher plants: both are mainly found in the cell wall. It has been shown that BRGII complexes are further stabilized by Ca^{2+} (see review by O'Neill et al. 2004, and Matoh 1997). Early in 1937, Naftel reported that adding lime in acid soil should accelerate B deficiency of plant (Naftel 1937). B was negatively related to calcium in absorption and transportation (Zhan and liu 1994; Bennett and Mathias 1973). But it was also reported that B and Ca in plants also showed positive relationship (Peng et al. 1995; Xiong et al. 1995).

Wang and Lan (1995) reported that calcium contents in leaf, stem, bud and flower of *Brassica napus* were negatively correlated with HWS B concentrations in the soil, whereas a positive correlation was obtained for pollen walls. Wang (1999) studied the relationship between B and Ca in various B efficient cultivars with nutrition solution and cell suspension culture He found that elevation of B or Ca concentrations from the deficiency level to normal level in the nutrition solution enhanced growth of plant, alleviated B or calcium deficiency symptoms, and increased the absorption of B and Ca in two B efficient cultivars and the proportion of B and Ca contained in top leaves. When B level exceeds 0.5 mg/L and Ca level is over 100 mg/L further increase in B or Ca seemed to exert inhibitory effects. However, evident difference in responding to the application levels of B and Ca between the B-efficient and B-inefficient cultivars was found that there was a mutual enhancement in B-inefficient cultivars at lower Ca and higher B levels, by contraries, B and Ca showed positive relationship in B-efficient cultivars at lower B and higher Ca concentrations (Wang et al. 2003).

Effect of B deficiency on enzyme activity in various B efficient cultivars

Peroxidase (POD), catalase (CAT) and superoxide dismutase (SOD) are three key enzymes in plants for protection against reactive oxygen species (ROS). Malonyl dialdehyde (MDA) is the main product of superoxide reaction in the plasmamembrane due to the attack of active oxygen to membrane. So the activities of POD, CAT, SOD plus the MDA content can be used as indices for the ability of a plant to deal with stresses, raising ROS levels.

Geng et al. (1999) reported that activities of SOD and CAT decreased significantly in B-inefficient cultivars under B deficiency, resulting in a drastic increase of MDA and membrane permeability. B-efficient cultivars, however, were hardly affected (Geng et al.1999, Wang 1999). When B was deficient, there were significant differences in POD activities of differently B efficient cultivars (Yang et al. 1999). This suggests that the resistance of *Brassica napus* to B deficiency was tightly related to the ROS protection system of plants.

At the same time, B deficiency decreased the activities of peptidase, proteinase and acid phosphatase in old leaves of B-inefficient cultivars, which finally resulted in a decreased re-use of N, P, and K (Geng et al. 1999).

Genetic base of B-efficiency

There exists a wide genetic polymorphism between B-inefficient and B-efficient cultivars. The studies on esterase and peroxidase isoenzyme activities of various organs at seedling, budding and flowering stages showed that there are two kinds of B efficient cultivars with respect to their specificity of isoenzyme bands at the different growth stages. The difference in the esterase isoenzyme bands of seedling leaves are the most evident. There was an isoenzyme which was clearly distinguishable for B-efficient cultivars, but which was absent in the B-inefficient ones in two consecutive years (Xu et al. 1998).

Xu et al. (2001) selected two B-inefficient and three B-efficient cultivars as parents to get twelve-reciprocal first generation crosses. We found that the response of the F_1 to B-deficiency was identical to that of their respective B-efficient parents, which indicates that the B-efficiency trait in *Brassica napus* is dominant. Results of the segregating F_2 indicated that the B-efficiency trait is controlled by one major gene according to the ratio of B efficient to B inefficient individuals, fitting well into the expected 3:1 ratio. The F_2 population was derived from a cross between B-efficient cv. “Qingyou 10”, and the B-inefficient “Bakow”. Using a molecular marker technique, the major gene was mapped on the ninth linkage group, and genome-wide QTL analyses detected one major locus near the major gene, which explained 64.0% of the phenotypic variance. At the same time, three minor loci on three linkage groups were also detected (Xu et al. 2001).

The B efficiency coefficient was tightly linked with the cv.s growth period. The average of the B efficiency coefficient of four B-efficient cultivars was 0.917, and time from sowing to bolting and maturity were 91.0 d and 181.8 d, respectively, whereas, these parameters for B-inefficient cultivars were 0.229, 139.5 d and 201.0 d, respectively. The B efficiency coefficient of individuals in two F_2 generations was negatively correlated to time to bolting and maturity at the 1% significance level (Xu et al. 1998, 1999). Thus, taking duration of the growth period as selection parameter for B-efficient oilseed-rape germplasm seems to be easy and effective.

Although we were able to obtain B-efficient germplasm in *Brassica napus* on the basis of cv.s with rapid development, and considering the wide range of B effects on growth, development, physiological and biochemical reactions in higher plants, the true mechanism(s) of B-efficiency in *Brassica napus* is(are) still not well understood. As pointed out above, further B efficiency mechanisms in *Brassica napus* may involve: (1) higher uptake ability of B by a more developed root system; (2) reutilization by an increased mobility of B in plants;

(3) adaptation to B-deficiency stress by more efficient enzyme systems protecting from ROS. Taking into account results from QTL analysis, B-efficiency in rape seems be controlled by one major and several minor genes at the same time.

Acknowledgement

The research work involved in the review was supported by National Nature Science Foundation of China (39770439, 30170549, 30471041) and National 863 High Technology Program in China (No. 2004AA222190).

Reference

Bennett O.L. and Mathias E.L., 1973 Growth and chemical composition of crown vetch as affected by lime, B., soil source, and temperature regime. Argon J., **65**: 587–591.

Bowen J.E., 1972 Effect of environmental factors on water utilization and B. accumulation and translocation in sugarcane. Plant cell physiol., **12**: 703–712.

Cao X.Y., Liu W.D., and Pi M.M., 1997 The research on B. absorption characteristics of two rape genotypes responding differently to B. deficiency. J. of Huazhong Agricultural University, **16**(5):336–339.

Chu H.Y., Yu M., Wang Y.H. et al., 1999 Study on differences of B use efficiency of rape (Brassica napus) cultivars. J. Of Huazhong Agricultural University, **18**(2): 134–138.

Du, C.W., Wang Y.H., Xu, F.S., Yang Y.H., and Wang H.Y., 2002 Study on the Genetic Mechanism of B. Efficiency in Rape Cultivars (Brassica napus L.). J. of Plant Nutrition, **25**(2): 231–244.

Xu F.S., Yang Y.H., Wang Y.H., and Wu L.S., 2002 Boron uptake and retranslocation in cultivars of Brassica napus differing in boron efficiency. In Boron in Plant and Animal Nutrition, Eds Heiner E. Goldbach et al., P127–135, Kluwer Academic/Plenum Publisher, New York.

Geng M.J., Cao Y.Y., and Zhu D.W., 1999 Effect of B deficiency on physiological characteristic of different rape cultivars at seedling stage. Plant nutrition and fertilizer science, **5**(1): 81–84.

Geng M.J., Cao Y.Y., and Zhu D.W. et al., 1998 Effect of B deficiency on physiological characteristic of different rape cultivars in flowering stage. Chinese Journal of Oil Crop Science, **20**(1): 70–73.

James C.R. Stangoulis, Harsharn S. Grewal, Richard W. Bell and Robin D. Graham. 2000a Boron efficiency in oilseed rape: I. Genotypic variation demonstrated in field and pot grown Brassica napus L., and Brassica juncea L. Plant and Soil, **225**: 243–251.

James C.R. Stangoulis, Patrick H. Brown, Nacer Bellaloui, Robert J. Reid and Robin D. Graham. 2001 The efficiency of boron utilisation in canola. Aust. J. Plant Physiol., 28, 1109–1114.

James C.R. Stangoulis, Michael J. Webb and Robin D. Graham 2000b Boron efficiency in oilseed rape: II. Development of a rapid lab-based screening technique. Plant and Soil **225**: 253–261.

Liu Z., 1996 Microelements in Soils of China. Jiangsu Science and Technology Publishing House, Nanjing, China, pp. 63–88.

Matoh, T., 1997 Boron in plant cell walls. Plant and Soil **193**: 59–70.

O'Neill M.A., Ishii T., Albersheim P., and Darvill A.G., 2004 Rhamnogalacturonan II: Structure and function of a borate cross-linked cell wall pectic polysaccharide. Annual Review of Plant Biology **55**: 109–139.

Raven J.A., 1980 Short-and long-distance transport of boric acid in plants. New phytologist, 84：231–249.

Shelp B.J., Shattuck V.I., and Proctor J.T.A., 1987 B nutrition and mobility, and its relation to the elemental composition of greenhouse grown root ropes II. Radish. Commun In Soil Sci Plant Anal, **18**(2): 203–219.

Shen Z.G., Shen K., and Zhang X.S., 1992 B absorption and mobility in rape seedlings as affected by calcium. Chinese Bulletin of Botany, **9**(4): 33–37.

Wang H.Y., Wang Y.H., Du C.W., Xu F.S., and Yang Y.H., 2003 Effects of Boron and Calcium Supply on Calcium Fractionation in Plants and Suspension Cells of Rape Cultivars with Different Boron Efficiency Journal of Plant Nutrition, **26**(4): 789–806.

Wang Y.H. and Lan L.F., 1995 A study on the B efficiency of rape (Brassica napus L). Journal of Huazhong Agricultural University, Sup. Sum. **21**: 71–84.

Cao X.Y., Liu W.D., and Pi M.M., 1997 Composition of root exudates and root bleeding sap of a B-efficient and inefficient oil seed rape genotype. In B is Soils and plants, Eds Bell R.W., Rerkasem B. pp. 171–174, Kluwer Academic Publishers, Netherlands

Xiong S.L., Wu L.S., Wang Y.H. et al., 1995 Absorption and distribution of B in different varieties of rape (Brassica napus). J. of Huazhong Agricultural University, Sup. Sum. **21**: 85–91.

Xu F.S., Wang Y.H., and Li J.C., 1998 Inheritance of rapeseed (Brassica napus) B efficiency in F_1 generation. Plant nutrition and fertilization, **4**(3): 305–301.

Xu F.S., Wang Y.H., Ying W.H., and Meng J.L., 2002 Inheritance of boron nutrition efficiency in Brassica napus. Journal of Plant Nutrition **25:** 901–912.

Xu, F.S., Wang Y.H., and Meng J.M., 2001 Mapping the B nutrition efficiency gene(s) in Brassica napus using molecular markers featured in Brassica and Arabidopsis. Plant Breeding, **120**(4): 319–324.

Xue J.M., Yang Y.A., and Ye Z.Q., 1995 Effect of B on the growth, yield and quality of different rape cultivars. J. of Zhejiang Agricultural University, **21**(1): 66–70.

Xue J.M. and Yang Y.A., 1994 Genetypic variation in response of rape (Brassica napus) to B deficiency. J. Of Zhejiang Agricultural University, **20**(4): 422–426.

Yamauchi T., Hara T., and Sonoda Y., 1986a Distribution of calcium and B in the pectin fraction of tomato leaf cell wall. Plant Cell Physiol., **27**: 729–732.

Yang Y.H., Wu L.S., Wang Y.H. et al., 1999 Effect on B on cell-wall enzyme activity of different rape cultivars. Plant nutrition and fertilizer science, **5**(4): 341–346.

Yu M., Chu H.Y., Wu L.S. et al., 1999 Study on B uptake, distribution in different B efficiency rape (Brassica napus) genotypes. Chinese Journal of Crop Sciences, **21**(1): 49–52.

Yu M., Chu H.Y., Wu L.S. et al., 1999 Study on calcium uptake and distribution in rape (Brassica napus) cultivars with different B use efficiencies. J. Of Huazhong Agricultural University, **18**(5): 440–443.

Genotypic Differences in Tolerance to B Deficiency in Mango in India

M. Edward Raja
Indian Institute of Horticultural Research, Bangalore, India
E-Mail: medward@iihr.ernet.in

Introduction

India has largest area under mango in the world at 1.1 m ha and it is also the fruit crop cultivated in largest area in Inda but its average productivity is lowat 6-8t ha^{-1} $year^{-1}$. It is considered as a hardy crop and hence is grown under rainfed situation with low input management and is also less profitable compared to many other fruit crops. Among the reason for its low productivity and profitability, altenrnate year bearing or orregular bearing is the major one (Yadav1997). Out of 15-20 important commercial cultivars only Neelum, Mallika and Totapuri are considered regular bearers (RB) but their market acceptibility is limited. Premium and profitable cultivars like Alphonso, banganapalli prefered in lopcal and export markets are alternate bearers (AB) and farmers are at a disadvantage. Since B deficiency (Fig.1) is considered as one of the important causes of low productivity and quality problems (Edward M. Raja, 2005) it was decided to investigate whether inability to tolerate B deficiency could be a cause for alternate bearing. Scott (1952) observed cultivar differences in B deficiency tolerance in grapes grown in same field. Out of 44 cultivars evaluated only 14 were severely affected 12 were not affected. B susceptible cultivars when grafted on various rootstocks did not show any B deficiency symptomes. Fe efficiency in Mngo rootstock has been used in Israel (Kadman and Gazit 1976). The mechanisms of tolerance are as follows: Low metabolic requirement, root Geometry and their distribution in different horizon, and translocation of B fromroot to shoot through xylem and movement from old leaves or shoot to young roots through phloem (Graham 1984). Though B is generally phloem immobile, some members of the Apple have ability to translocation by carbohydrate derivatives like sorbital (Brown and Hu 1996). Wall and Andus (1962) indicated that the T3238 tomato mutant is susceptible to B deficiency because it lacks the ability to transport b to the

F. Xu et al. (eds.), Advances in Plant and Animal Boron Nutrition, 213–224.

top of the plant. According to Kelly and Gavelman (1962) Ca and K levels and Ca/B ratio decides the B nutrition and not the total Bconcentration. Hence this study was initiated to study thye factors governing the defferential tolerance of regular and alternate bearing mango cultivars to B deficiency.

Materials and methods

First a survey was undertaken in mango orchards to study the B status of the important mango cultivars in varying climates: Inhumid konkan region known for Alphonso mango (AB) aswell as the semi arid region of vijayawada region of AndhraPradesh southern karnataka and Nortern Tamilnadu. The young leaves from non-fruiting terminals were collected. Soil samples at 100m radial deference from the 0-30cm layer were collected. The leaves were analyzed for ,B, (Gaines and Mitchell 1979) and N by Kjeldal and K by flame photometer method. The soil samples were examined for physico-chemical properties (Black 1965) and the available B by midified Azomethine H method (Parker and Gardner 1981).

The study on the mechanism of genotypic tolerance to B deficiency was carried out in the IIHR farm bangalore in 2004-2005. the four year old RB cultivars Neelum, Mallika and Totapuri and three AB cultivars Alphonso, banganpalli and Arka Punith were identified, and the bark of the primary, secondary and tertiary branches (shoots bearing inflorescence) and the leaves in the retiary branches where inflorescence is borne were collected in the pre flowering stage (Nov) and were analyzed for B,Ca and K. There were six trees as replication for the six cultivars chosen as treatments. The oservation on change in leaf B Ca for 3years(1st 3rd and 4th year) and yield data (3-4th year) for comparing the RB and AB were recorded in at harvest stage. The differential response of RB and Abcultivars Neelum and Alphonso to B application was studied in Konkan, maharashtra with three treatments no boron, solubor 0.1% spray three times ans soil application of borax at 250g/tree with RBD statical design. Observation on date of flowering, yield, present spongy tissue and fruit weight were recorded. The results were statically analyzed by ANOVA and are presented below.

Results

B *status of mango cultivars*

The soil and climate characteristic (Table 1) also indicate a wide range fromk a humid (Region I) to semi arid climate (Region II to IV) with a rainfall as high as 2500 mm yr^{-1} to as low as 600mm yr^{-1} but the available B (hws B) is high in the low rainfall Dharamapuri region with a mean of 0.62 mg $kg^{-1}$to a low of 0.32 mg kg^{-1} in the Konkan region. These soils are taxonomically alfosols in RegionII to IV and incept soils in the Region I. The soils are low in organic matter(0.47%OC) characteristic of soils of and tropical climate. Base saturation is low in Konkan soils (55%) whereas in the semi arid Dharmapuri soil it is the highest(87%). The soils do not have a very good physical property in the semi arid region (II to IV) since the bulk density is high at mean value of 1.34 g c^{-1} which is due to the hard setting nature of these soils, due to poor structure.

The result on survey of the mango orchards in different states on RB and AB cultivars are presented Table 2 for leaf nutrient status. There is no major difference in the leaf B status of the different cultivars but the Ca/B ratio varied with cultivars. The Ca/b ratio, which influences the active b in the lamina, favours the RB since the men Ca/B ratio is 395 whereas in the AB it is higher at 436, which distinctly gives an unfavourable metabolic active B status for the AB. The leaf nitrogen is very high in high rainfall Konkan region (1.59-1.75%) and low in the low rainfall Dharamapuri region (0.73-0.92%) with bangalore and Vijayawada regions occupying intermediate position.

The studies on B distribution in different types of branches (primary, secondary and tertairy) and leaves on tertiary branches in the AB and RB are presented in Table 3. It indicated in AB, the B was high in bark of primary branches (29-30 mg kg^{-1}) moderate in secondary (23 mg kg^{-1}) but low in tertiary barks and the leaf B is also low at 21 mg kg^{-1} indicating a poor mobil;ity and consequently low B in leaves. The cultivar Banganpalli has maintained a highest mobility and Alphonso the least among AB.

In the RB, the primary bark had low B (19 mg kg^{-1}) and the tertiary the highest (32.0 mg kg^{-1}) indicating a better mobility towards the sink, resulting in s higher B status of 40.3 mg kg^{-1} in the leaves. This indicated a distinct difference in phloem mobility of B between alternate and regular beares. The field studies on the effect of age on Ca status on B nutrition of AB and RB (Table4) indicated in the first year the alternate and regular bearers did not differ significantly in the leaf B as well as Ca/B ratio whereas in the 3rd year the men Ca/B ratio of AB is very high (420) but the men fruit yield was low 1.97 kg $tree^{-1}$ whereas the regulars had a lower Ca/B ratio of 329 but the men yield was high at 6.3 kg. Studies on differential reswponse of AB and RB to b (Table 5) indicated the cv Alphonso one of the most important AB responded significantly by a increased yield of 63% over control for foliar spray of solubor 0.1 three times in july, November and January.

The soil application of B at 250 g Borax/tree also reswulted in significant yield response of 44 % whereas the RB Neelum recorded increased yield of 28% and 13% over no B control for 0.1% solubor spray and soil application of Borax. In addition B application resulted in better quality in terms of reduction in spongy tissue by 11-13% early flowering by 2-4 weeks and better fruit weight in the AB Alphonso wfereas RB did not record very significant improvement the quality factors due to B application.

Table 1. Soil characteristics of mango orchards in different regions

Location and climate	Cultivar	Soil characteristics pH	Available B mg kg^{-1}	OC %	CEC $Cmol^{(+)} kg^{-1}$	Base % saturation	Texture	Bulk density g cc^{-1}
RegionI. Konkan Humid	Alphonso	4.9	0.31	0.61	6.4	58	S.loam	1.24
(2500 mm rainfall)	Totapuri	4.8	0.33	0.54	8.4	52	loam	1.14
RegionII. Bangalore	Alphonso	6.2	0.46	0.43	11.4	74	loam	1.33
Semi arid	A.Punith	6.3	0.53	0.53	12.4	82	loam	1.38
(750 mm raifall)	Mallika	6.2	0.46	0.41	10.4	72	S.loam	1.30
“	Neelum	6.1	0.50	0.51	12.7	82	loam	1.28
“	Totapuri	6.7	0.41	0.47	14.2	75	S.loam	1.38
“ III. Vijayawada AP	Banganpalli	6.2	0.46	0.46	12.3	78	loam	1.28
900 mm rainfall	Totapuri	6.3	0.68	0.41	16.3	72	S.loam	1.32
“	Neelum	6 4	0.50	0.61	24.1	74	loam	1.40
“ IV. Dharamapuri	Alphonso	7 8	0.71	0.35	24.2	85	loam	1.38
600 mm rainfall	Neelum	7.4	0.64	0.43	26.2	86	loam	1.32
(semiarid)	Totapuri	7.6	0.52	0.41	20.4	92	loam	1.42

Table 2. Growth and leaf nutrient status of mango cultivars in different climate regions

				Leaf nutrient status					
	Location	Cultivar	Visible B de	B def B(mg kg^{-1})	Ca %	K %	N %	Ca/B	Tree canopy m^3
Region I	Konkan Humid	Alphonso(AB)	Severe	20	1.43	1.37	1.75	510	6.3
	(2500 mm rainfall)	Totapuri(RB)	Mild	28	1.00	1.17	1.59	454	3.8
Region II	Bangalore	Alphonso(AB)	Mod	22	1.35	0.86	1.02	422	4.8
	Semi arid	Arka Punith(AB)	Mod	34	1.42	0.40	1.17	417	4.9
	750 mm rainfall	Mallika(RB)	Healthy	23	0.95	1.22	1.40	413	3.2
		Neelum(RB)	Healthy	24	0.90	1.24	1.30	375	3.1
Region III	Vijayawada	Totapuri(RB)	Mild	28	1.13	1.14	1.10	403	34
	900 mm	Banganpalli(AB)	Mod	40	1.68	0.70	0.801	420	6.3
	"	Totapuri(RB)	Healthy	28	1.19	1.02	0.94	373	4.3
	"	Neelum(RB)	Mild	28	1.01	1.04	1.24	361	3.8
Region IV	Dharamapuri	Alphonso(AB)	Mod	42	1.73	0.82	0.92	412	5.2
	00 mm	Neelum(RB)	Healthy	36	1.31	1.14	0.82	364	2.9
	(rainfall)	Totapuri(RB)	Mild	34	1.44	1.12	0.73	423	2.2

Table 3. The B content in different Mango cultivars (mg kg^{-1})

	Variety of Mango	Bark			
Sl.No	Alternate bearers (AB)	Primary Branch	Secondary Branch	Tertiary Branch	Physiologically Mature leaves
1	Alternate Alphonso	20	16	14	19
2	Banganapalli	37	33	31	21
3	Arka Punith	31	20	16	23
	Mean	29.3	23	20.3	21
4	Mallika	20	26	27	32
5	Neelum	20	20	34	42
6	Totapuri	17	25	36	47
	Mean	19	23.6	32.3	40.3

Table 4. Effect of age and calcium status on B nutrition in Mango cultivars

Cultivar	I Year			III Year				IV Year			
	B Mg kg^{-1}	Ca %	Ca/B Ratio	B Mg kg^{-1}	Ca %	Ca/B Ratio	Fruit Yield kg $tree^{-1}$	B Mg kg^{-1}	Ca %	Ca/B Ratio	Fruit Yield kg $tree^{-1}$
Alternate - Bearers(AB)											
1. Alphonso	22	0.72	327	24	0.94	391	2.5	22	1.25	568	-
2. Banganapalli	21	0.73	347	26	0.96	369	2.4	27	1.35	500	-
3. Arka Punith	20	0.68	340	24	1.02	501	1	23	1.25	543	-
Mean	21	0.72	338	23.3	0.97	420	2.0	24	1.28	537	-
Regular Bearers (RB)											
1. Amarapali	22	0.73	327	24	0.82	342	12.6	29	0.9	310	9.4
2. Mallika	23	0.74	322	27	0.8	296	3.8	36	0.97	269	14.3
3. Sindhu	21	0.66	314	24	0.78	325	1.3				
4. Totapuri	23	0.73	317	26	0.98	376	2.2	24	1.15	479	3.3
5. Neelum	24	0.65	270	27	0.83	307	4.6	27	0.94	348	4.9
Mean	22.3	0.7	310	25.6	0.84	329	6.5	29.2	0.98	343	10.8

Table 5. Differential response of alternate and regular bearing mango cultivars to Boron

	Alphonso (alternate bearer)					Neelum regular bearer			
Treatments	Date of Flowering	Mean yield Kg $tree^{-1}$	spongy tissue	Leaf B mg kg^{-1}	Mean fruit wt g $fruit^{-1}$	Date of Flowering	Mean yield Kg $tree^{-1}$	Fruit wt g $fruit^{-1}$	Leaf B g kg^{-1}
Control (No Boron)	Jan I Week	40.5	35	22	195	Jan III week	62.3	170	26
Solubor 0.1% (July, Nov & Jan)	Dec I Week	65.5	22	78	240	Jan I week	79.7	220	70
250g borax/tree (July)	Dec III Week	58.4	24	54	200	Jan II week	70.3	208	60
CD%	-	9.2	-	14	26.2	-	8.2	22	9

Discussion

Though mango occupies the largest area among fruit crops in India its productivity is low due to many important reasons among which its alternate year bearing or irregular bearing habit is the most important reason (Yadav 1997). Though cultivars differ in this trait, some cultivars like Neelum, Totapuri, Mallika and Amarapalli are comparatively regular bearers but they are not premium cultivars with a desirable color, shape and other market acceptable qualities. B deficiency is one of the most important disorders affecting both yield and quality of mango and field observation indicated that cultivars differ in the susceptibility to B deficiency. Due to the low organic matter status and high pH of the soils in India (Govindaraja 1965) available B is low and B deficiency is common in India. Besides mango is a B loving crop too (Agarwala et al. 1988) and its deficiency is observed wide spread both in humid tropics with acid alfisols and also in semi arid regions of India. The survey of the mango orchards indicated that Konkan where the premium cultivar Alphonso AB is cultivated has the most acidic soil (Table-1) with a lowest abailable B (0.32 mg kg^{-1}) and the high rainfall, coarse soil texture, and poor root growth due to Al toxicity ad high humidity aggravate the B deficiency problem. In the other three regions where the semi arid climate prevails the high soil pH and base status and poor structure indicated by high bulk density affects B uptake and utilization.

The study on the growth and leaf nutrition status (Table-2) of AB and RB cultivar indicated a wide variability in Ca/B ratio. It is observed in the humid region the mango growth is vigorous indicated by a big canopy whereas in the semi arid area the tree size is small indicated by a smaller canopy resulting in a challenging situation since in the humid Konkan region the need for B is more but the supply is less resulting in severe B deficiency. In spite of the adverse situation in Konkan the cultivar Totapuri was not affected very much by B deficiency establishing a cultivar difference. Similarly in the semi arid region (Vijayawada, Dharmapuri and Bangalore) also there were difference in tolerance to B deficiency indicated by the low leaf B status and visible symptoms. It was observed the AB had a wider Ca/B ratio compared to the RB. It is also a fact that out of the 15-20 commercial cultivars of mango in India excepting the three under the present study almost all are AB but with better consumer acceptability hence there is a need to study the reasons for cultivar difference, since according to Fangsen Xu et al. (2002) crops adopt different strategies to tide over B deficiency so that the crop and B management practices can be tuned to the crop's genetic makeup to handle B deficiency stress.

The study on mobility of B through phloem is also an index of tolerance mechanism of B deficiency and the analysis of bark of primary, secondary and tertiary branches (Table – 3) indicated in AB, the B content of primary bark was more than tertiary bark whereas in the regular bearers the bark of the tertiary branches had more B than the primary ones. This indicated better phloem mobility in RB compared to the alternate ones and it has been observed by

Brown and Hu (1996) that in species where sorbitol is the one of the carbohydrates produced in the leaves there was better mobility. In the study on the effect of age on B and Ca/B ratio indicated that the poor B nutrition in AB is also due to not only poor mobility in the tertiary branches but also the high Ca/B ratio in the leaves.

The studies on the effect of age and Ca status on B nutrition (Table-4) has indicated in the early stage (I year) both RB and AB do not differ significantly in the Ca and B nutrition but in the fourth year, the AB have recorded a mean Ca of 1.28% and a mean Ca/B ratio of 537 compared to 0.98% of 343 of RB indicating a wider Ca/ B ratio as the plant grows, aggravating the B deficiency. Boron is one of the nutrient having important roles in hormone and nucleic acid metabolism and hence has a very important role in yield.

In the evaluation of mango cultivars as recorded in Brazil a range in B response from no increase in yield in cv Winter to double yield in Tommy Atkins to five times increase in Haden 2H and Vandyke was recorded by Rosetto et al. (2001). In the present study also (Table-5) vast cultivar difference were exhibited. The AB Alphonso responded more significantly to B than the regular bearing Neelum. The fruit quality also improved by B application (Shorrocks and Nicholson, 1980) resulting in reduced Spongy tissue incidence. Early flowering and increase in fruit size were also recorded for B application in AB cv Alphonso.The wide Ca/B ratio and consequent poor B nutrition (Table-4) may also be because of the favorable situation for continuous Ca uptake in AB.

The AB tree is of bigger size compared to the regular bearers since, in the off year when there is no or less yield the tree uses the organic and inorganic nutrients for build up of tree frame work and the tree roots grow radially and depth wise increasing the root volume. This consequently results in foraging of more soil volume and resulting in more Ca uptake in soil where the base saturation is high (above 75%) in the present study.

The on set of fruitset results in cessation of root activity in perennial crops due to poor supply of photosynthates to the roots. Since Ca is taken up in the root tips, in AB the root number and root tips are more due to large root system with a better supply of photosynthates (fruiting once in 2 years only) to the roots and hence Ca uptake is more, resulting in a wider Ca/B ratio (Faust 1980). Wide Ca/B ratio according to Kelly ad Gavelman (!962) has resulted in B deficiency and poor yield in Red beet. Besides restricting Ca at root levels by RB also may be one of the reasons for narrow Ca/B ratio.

The study on the response of AB and RB to B (Table 5) has indicated very significant response to B by AB Alphonso confirming the earlier finding that AB suffer due to poor B mobility as well as wider Ca/B ratio and they need exogenous B application for better yield and higher quality.

Conclusion

The alternate bearing habit is one of the important cause for low productivity and profitability of mango in India and poor nutrition especially widespread B deficiency is one of the causes since it affects yield by reducing fruitset and reduces post harvest life. But to the disadvantage of the farmer, the AB like Alphonso and Banganapalli are the ones having consumer acceptibility and export market and high profit to farmers.

This study indicated the RB cultivars are able to translocate B to the sink effectively and also keep it metabolically active by a narrow Ca/B ratio than the alternate bearers. Though in the early stage (1 Year) the B nutrition pattern is the same for the both types as the tree ages (4th Year) the difference in Ca/B ratio increases resulting wider CA/B ratio in alternate bearers and less and irregular yield. Due to the poor B nutrition status as the tree ages, the AB need external B supply and result in significant yield response to foliar spray of 0.1% solubor than soil application of 250 g borax /tree. The response by regular bearer to B was on a lower scale. The climate and soil also play a major role in expression pf genotypic differences in B nutrition. The RB can also be evaluated as rootstock for efficient B nutrition.

Acknowledgement

The author acknowledges the facilities and support givern by Director,IIHR, for this work. The analytical help given by Mr.Mazhar Jameel and help given by K.Usha in manuscript preparation is also gratefully acknoledged.

References

Agrawala S.C., Nauriyal B.D., Chittralekhs C., and Shrma C.D., 1988 Manganese, zinc and Boron deficiency in mango Sci Hort. **35**: 99-107.

Black C.A., 1965 Methods of soil analysis, Part II Agronomy monograph, **9:** American Society of Agronomy, Madison, Wisconsis, USA.

Brown P.H and Hu H.N., 1996 Phloem mobility of boron is species dependent, evidence for phloem mobility in sorbitol-rich species. Ann. Bot. **77**:497-505.

Xu F.S., Wang Y.H., Ying W.H., and Meng J.L., 2002 Inheritance of Boron nutrition. Efficiency in Brasica, napus J. Plant nutrition. 25:901-912.

Faust M., 1980 Interaction between nutrition uptake on photosynthesis. Acta Horticulturae. **92**: 193-199.

Gaines T.P and Mitchell G.A., 1079 Boron determination in plant tissue by Azomethine H method Commun. Soil Science Plant Anal. 10:1099-1108.

Graham R.D., 1984 Breeding for nutrition characteristics in cereals Adv. In Plant nutrition **1**: 57-102.

Graham R.D and Nambiar 1981 Advances in research on copper deficiency incereals. Aust. J. Agric. Res. **32**: 1009-1037.

Govindarajan 1965 Soils of India ICAR publication, Krishi Bhavan New Delhi.

Kelley J.F. and Gavelman W.H., 1960 Variability in red beet (Beta vulgaris) to boron nutrition Proc. Amer Soc. Hort Sci. **76**: 409-415.

Kelley J.F. and Gavelman W.H., 1960 Variability in mineral composition of Red beet (Beta vulgaris) to varieties in relation to boron nutrition. Proc. Amer. Soc Hor. Sci. **76**:416-424.

Kadmand A. and Gazit S., 1976 Selection of mango rootstocks for adverse soil and water areas. Acta Hort. **57**: 8-87.

Scott L.E., 1952 An instance of Boron deficiency in the grape under field condition Proc. Amer. Soc. Hort. Sci. **38**:373-388.

Rorsetto C.J., Furlani P.R., Bortolettor J.A., and Igue T., 2001 Differential response of Mango varieties to Boron. Acta Hort **320**: 120-128.

Shorrocks V.M. and Nicholson D.D., 1980 Influence of Boron deficiency on fruit quality. Acta. Hort. **92**: 103-108.

Yadav I.S., 1992 mango research in India. The past 50 years Ind. Hort. **42**: 10-17.

Differences of Cell Wall Constituents between B-Efficient and B-inefficient Rape Cultivars

Yuhua Yang [1, 2], Lijun Wang [1], Min Yu [1, 3], Changwen Du [1], Yunhua Wang [1] and Lishu Wu [1]

[1]*Microelement Research Center*, Huazhong Agricultural University, Wuhan, 430070, China;

[2]Agricultural College, Yangtze University, Jinzhou, 434103, China;

[3]Department of Horticulture, Foshan University, Foshan, 528000, China; yumin0820@tom.com

Introduction

Previous research showed that the amount of boron (B) required for root growth in a B-inefficient rape cultivar was higher than that in a B-efficient cultivar, which caused a rapid inhibition of root growth in the B-efficient cultivar when it suffered B deficiency (Hu et al. 1994; Cao et al. 1996; Xiong et al. 1995). The typical symptom of B deficiency is inhibition of root tip growth caused by the disruptions of structure and elasticity of cell walls (Hu and Brown 1994). The composition and properties of pectin may play a critical role in determining the physical characteristics of the cell wall (Jarvis 1984). Boron is bound to pectin in cell walls by the formation of borate-ester cross-links with pectin which affects colloidal state and properties of pectin and the elasticity of cell wall (Hu et al. 1996). While differences in B requirements among species were known to correlate with differences in pectin content of cell walls, it is not know if differences in cell wall composition can explain known cultivar differences in B efficiency. The objective of this study was to elucidate mechanism of B efficiency in B-efficient and B-inefficient rape cultivars by examining wall composition, with particular emphasis on pectin.

Materials and methods

Materials and culture methods

Rape cultivars (*Brassica napus*) for the experiment (a B-inefficient cultivar, 95105 and a B-efficient cultivar, 95090) were used reported in the previous screening study on B efficiency of seed rape [7]. Rape seeds were soaked, and then planted onto double-layer yarn until the formation of the first enphyllum and then transferred to 7 L black plastic pots with 25 seedlings per pot and 3 replicates. According to the seedling sizes, the strength of culture solution was increased from 1/4 to ½ and then to complete Hoagland and Arnon nutrient

F. Xu et al. (eds.), Advances in Plant and Animal Boron Nutrition, 225–230.

solution mixture. Culture solution was changed per 7 d in seedling stage and every 4-5 d in bud and flowering stages. Roots, upper leaves (seedling: the top 2 leaves; flowering: the top 3 leaves) and lower leaves (seedling: the bottom 2 leaves; flowering: the bottom 3 leaves) in seedling (4 true leaves) and flowering stages were sampled, respectively. The samples were rinsed with distilled water, dried using absorbent paper, weighed, cooled in liquid nitrogen and stored at -20°C.

Extraction of cell wall and determination of cell wall constitutions

According to Huber method (Huber and Nevins 1979), fresh samples (80.0 g) were homogenized in ice-cold phosphate buffer (10 mmol/L pH7.0) using a porcelain mortar. The homogenate was filtered by a Bruce tundish, and then washed with 2-3 L ice cold homogenizing buffer. The cell walls were then washed with acetone (1 L -20℃) followed by an additional cold buffer wash. During the washing process the cell wall was maintained in suspension by continuous stirring.

The extracted cell walls were dried at 70℃ and ground in a carnelian mortar. 0.3 g of the cell wall sample was weighed into a plastic centrifugation tube with 100-fold volume of imidazole solution (0.5 mol/L pH7.0), and the mix solution was continuously stirred for 24 h at 25℃ and then was centrifuged. The supernatant was removed. The residue was washed by extraction solution and centrifuged twice again. The supernatant were dialyzed using 1000 MWT dialysis bag against distilled water and was dried at 55℃, which was analyzed for chelated pectin (pectin 1). Alkaline-soluble pectin (pectin 2) and hemicellulose were extracted using 50 mmol/L Na_2CO_3 containing 20 mmol CDTA /L and 4 mol/L KOH, respectively. The residue was dried at 55℃ and weighed for cellulose analysis.

Determination methods for total sugar, uronic acid, and 3-Deoxy-D-manno -2-Octulosonic acid (KDO) were anthrone colorimetry, hydroxydiphenyl colorimetry (Blumenkrantz and Asboe-Hansen 1973) and thiobarbituric acid colorimetry (Karkhanis et al. 1987), respectively.

Results

Difference of cell wall constitutions of different rape cultivars

The content of cellulose in root cell wall was higher than that in other organs of plant which accounted for 38-44% of cell cell wall dry weight; the proportions of hemicellulose and pectin in cell wall were 34-38% and 22-29%, respectively (Table 1). With aging of roots, cellulose gradually increased, while the pectin and hemicellulose decreased. There was no significant difference in composition of root cell wall between B-efficient and B-inefficient cultivars at either seedling or flowering stages.

Table 1. Composition of root Cell wall in rape cultivars (% dry weight)

Stage	Cultivar	Pectin 1	Pectin 2	Total pectins	Hemicellulose	Cellulose
Seedling	95090	7.85±0.23	18.9±1.93	26.7	35.8±2.83	38.9±0.71
	95105	10.5±1.35	18.3±2.32	28.7	37.0±1.63	37.2±0.78
		ns	ns	ns	ns	ns
Flowering	95090	10.5±0.28	11.7±0.38	22.3	33.6±1.88	43.9±1.64
	95105	11.7±0.19	12.0±0.39	23.7	35.9±1.99	41.1±2.48
		ns	ns	ns	ns	ns

Show T-test in Table to indicate significant differences

Percentage of cellulose and hemicellulose in the upper leaves was lower than that in roots and accounted for 24-30% and 22-28% of cell wall dry weight, respectively (Table 2). However, the amount of pectin in upper leaves was significant higher than that of roots.

Unlike the root, the percent of pectin and hemicellulose (with exception of 95105) of the upper leaves in the seedlings was lower than that in the flowering stage, whereas the percent of cellulose in seedling stage was higher.

The percent of cellulose, hemicellulose and total pectin of upper leaf was not significantly different between the two cultivars, whereas pectin fractions were significant different. The chelated soluble pectin of B-inefficient cultivar was significant lower than that of the B-efficient cultivar. However, the levels of alkaline pectin of B-inefficient cultivar were higher than that of the B-efficient cultivar. Differences in pectin fractions at the flowering stage were greater than that in the stage of seedling.

Table 2. Composition of upper leaf cell wall in rape cultivars (% dry weight)

Stages	Cultivars	Pectin1	Pectin2	Total pectins	Hemicellulose	Cellulose
Seedling	95090	24.7±0.80	19.5±0.91	44.2	25.6±1.38	28.9±0.78
	95105	19.4±1.44	24.2±2.75	43.5	28.0±1.67	29.5±0.32
		*	*	ns	ns	ns
Flowering	95090	37.6±0.45	10.8±0.79	48.4	28.0±1.00	23.8±1.00
	95105	28.5±1.24	22.6±1.32	51.1	22.9±0.86	26.1±0.51
		*	*	ns	ns	ns

The percent of cellulose, hemicellulose and pectin in cell walls of lower leaves were similar to those of upper leaves (Table 3). However, there was no difference in pectin levels between cultivars. The chelated soluble pectin was significant lower than alkaline pectin, which accounted for 15% and 35% of cell cell wall dry weight, respectively.

Table 3. Constitution of lower leaf cell wall in rape cultivars（% dry weight）

Stages	Cultivars	Pectin1	Pectin2	Total pectins	Hemicellulose	Cellulose
Seedling	95090	15.9±1.23	32.1±1.32	48.0	26.6±1.25	26.0±1.26
	95105	15.1±1.03	34.7±2.15	49.8	24.9±0.95	27.1±2.25
Flowering	95090	15.3±0.23	33.9±0.24	49.2	24.7±0.35	26.9±0.61
	95105	14.4±0.98	36.0±0.64	50.4	21.4±0.54	28.1±0.13

Differences in pectin constituents of upper leaf cell walls between rape cultivar

Concentrations of sugars associated with pectin 1 (chelated soluble pectin) and urone were lower than that of pectin 2, this suggested that the extracts of pectin 1 included cell wall proteins and cell residues besides pectin not sure what the evidence is to draw this conclusion. The concentration of total sugar in pectin of cell walls of the B-inefficient cultivar was lower than that of B-efficient cultivar, but no significant differences existed between the levels of urone of both cultivars. The concentration of KDO of the B-inefficient cultivar was much higher than that of the B-efficient cultivar. KDO (3-Deoxy-D-manno-2-Octulosonic acid) is a unique component of a pectin polysaccharide---RG-II, which contains two molecules of KDO (York et al. 1985). So the amount of KDO represents the amount of RG-II of pectin. RG-II is a specific B binding site in cell walls and there is significant positive correlation between the amount of KDO and B amounts of cell walls (r^2=0.778) (Matoh et al. 1996).

Total sugar concentrations in extracts of pectin 2 (alkaline pectin) were higher than those in pectin 1 (Table 4). However, the concentrations of total sugar and urone from pectin 2 in cell walls of the B-efficient cultivar were higher than in the B-inefficient cultivar. The concentration of KDO of pectin 2 was lower than that of pectin 1, suggesting that B-binding of pectin 2 is lower than that of pectin 1.

Table 4. Difference in pectic fractions of upper leaf cell wall in rape cultivars

Cultivars	Fraction of pectin 1			Fraction of pectin 2		
	Total sugar	Urone	KDO	Total sugar	Urone	KDO
	-mg/g dry pectin—		A548	-mg/g dry pectin-		A548
95090	189±5.3	217±8.8	0.045	295±10.0	282±8.8	0.024
95105	148±6.0	223±7.4	0.084	255±11.5	240.7±10.9	0.024

Discussions

Hu et al. (1996) analyzed 14 species including mostly vegetables and Gramineae species, for pectin contents in cell walls and their B requirement for growth. There was a significant positive correlation between boron required and the contents of cell wall pectin. However, in oilseed rape, there is no significant difference in the amounts of total cell wall pectin between the two cultivars which differed in B efficiency, except for the pectin constituents in upper leaves. Alkaline pectin was the main component of pectin in the B-inefficient cultivar, while chelated soluble pectin was the main component in the B-efficient cultivar. Chelated soluble pectin is one kind of oligomer bound loosely to cell walls, and it gradually decreases with maturity of tissues and the level in cell walls is sensitive to environmental factors. Alkaline pectin has higher polymerization and is more stable. B-inefficient cultivar contains more alkaline soluble pectin suggesting that the pectin in its new emerging leaves has stable polymerization in cell walls. For the B-efficient cultivar, polymerization of the pectin in new emerging leaves appears to be lower. From above preliminary results, it is suggested that chelated soluble pectin in B-inefficient cultivar presents more boron-binding sites in younger leaves, and implies that formation of cell walls in emerging leaves of the B-inefficient cultivar needs more boron. However, our understanding of the functions of pectin fractions in B binding, and their relationship to formation of cell walls and B efficiency remains unclear.

Acknowledgement

We are grateful to the finance support by the National natural Science Foundation of China (NSFC No. 39970421).

Reference

Blumenkrantz N. and Asboe-Hansen G., 1973 New method for quantitative determination of uronic acids. Analytical Biochemistry **54**, 484–489.

Cao X., 1996 Study on physiological properties of roots under B deficiency among rape genotypes. Ph D Dissertation. Wuhan, Huazhong Agri. Univ.

Hu H.H. and Brown P.H., 1994 Localization of boron in cell cell walls of squash and tobacco and its association with pectin: Evidence for a structural role of boron in the cell cell wall. Plant Physiology **105**, 681–689.

Hu H., Brown P.H., and Labavitch J.M., 1996 Species variability in boron requirement is correlated with cell cell wall pectin. J. Exp. Bot. **47**, 295, 227–232.

Hu Q., Xu G., and Shi R., 1990 Studies on mechanism for genetic differences of boron nutrient of different rape cultivars. J. Nanjing Agri. Univ. **13** (1), 80–86 (in Chinese).

Huber D.J. and Nevins D.J., 1979 Autolysis of the cell cell wall β–D-glucan in corn coleptiles. Plant Cell Physiology **20** (1), 201–212.

Jarvis M.C., 1984 Structure and properties of pectin gels in plant cell cell wall. Plant Cell Environ. **7**, 153–164.

Karkhanis Y.D., Zeltner J.Y., Jackson J.J., and Carlo D.J., 1978 A new and improved microassay to determine 2-keto-3-deoxy octonate in lipopolysaccharide of Gram-Negative bacteria. Analytical Biochemistry. **85**, 595–601.

Matoh T., Kawaguchi S., and Kobayashi M., 1996 Ubiquity of a borate- rhamnogala acuronan --II complex in the cell cell walls of higher plants. Plant Cell Physiol **37**, 636–640.

Wang Y., Wu L., Yu M., and Zhu H., 1999 Study on differences of utilization efficiency of boron among rape cultivars. Proceedings of International Symposium on Fertilizer and Agricultural Developments. Beijing: China Agriculture Publishing House, 533–535.

Xiong S., Wu L., Wang Y., and Lan L., 1995 Absorption and distribution of boron of different rape cultivars. J. Huazhong Agri. Univ. (Supplement), **21**, 85–91.

York W.S., Darvin A.G., and Mcnhi M., 1985 3-Deoxy-D-manno-2-Octulosonic acid (KDO) is a component of rhamnogalacturonan---II. A pectic polysaccharide in the primary cell cell walls of plants. Carbohydrate Research **138**, 109–126.

Determination of Suitable Maize (*Zea mays* L.) Genotypes to be Cultivated in Boron-Rich Central Anatolian Soil

E.E. Hakki[1,*], E. Atalay[1], M. Harmankaya[2], M. Babaoglu[1], M. Hamurcu[2] and S. Gezgin[2]

[1]Department of Field Crops, Faculty of Agriculture, Selcuk University, 42079 Kampus, Konya – TURKEY
e-mail: eehakki@selcuk.edu.tr

[2]Department of Soil Science & Plant Nutrition, Faculty of Agriculture, Selcuk University, 42079, Kampus, Konya - TURKEY

Introduction

The maize (*Zea mays* L.) (2n=20) is one of the most widely grown staple plants of the world, ranking first followed by the rice and the wheat. Considerable increases in production and acreage were achieved in all over the world during the last quarter of the 20th century. The production of maize in Turkey has also increased steadily during the last decade (FAOSTAT). Central Anatolian Region has been traditionally considered as the cool season cereals (mainly wheat and barley) storehouse of Turkey with a production of about 4 million tones from 3.5 million ha acreages. Although maize has only recently been introduced to the region, it has been well accepted by the farmers as a valuable alternative crop to sugar beet which requires a compulsory 4 years crop rotation. It is expected that maize cultivation may expand to 100 000 ha in the coming years due to increasing demand for food and animal feed. Now, hybrid maize is the leading type of maize grown on most farms.

Both deficiencies and toxicities of microelements can suppress plant growth. When present at increased levels of bio-availability, both essential micronutrients and non-essential metals are toxic (Baker and Brooks, 1989). Boron is one of the most important micronutrients that presents both deficiency and toxicity symptoms in a quite narrow range. B occurs in many rocksand soils at total concentrations of 5-50 mg kg^{-1}, and is normally present in plant leaf tissue at concentrations of 10-50 mg kg^{-1}. However, many species, including important cereals such as wheat and maize, are quite sensitive to elevated B levels in their tissues, showing severe toxicity symptoms at tissue levels of about 50 mg kg^{-1}. Such levels can be found in tissues when the available soil B exceeds 3 mg kg^{-1}. Previous studies denote South Australian dry lands (Cartwright et al. 1984, 1986), the valleys along the southern coast of Peru (Mason, 1967), soils of Israel (Ravikovitch et al. 1961), the Andes foothills in northern Chile (Caceres et al. 1992) among other B rich soils in different continents.

F. Xu et al. (eds.), Advances in Plant and Animal Boron Nutrition, 231–247.

B contents of 898 soil samples from 7 provinces in Turkey; Konya, Afyon, Karaman, Aksaray, Nigde, Nevsehir and Kayseri were surveyed by Gezgin et al. (2002). These regions encompass 3.5 million ha of cultivated land in Central Southern Anatolia. According to the survey, nearly 27% of soils in these provinces contained low levels of plant available B which can be corrected by external B applications. Maize was found to be the crop with the best positive response towards B application in deficient soil (Sakal and Singh, 1995). However, another 18% of soils in this region contain B at more than the critical upper level for available soil B, which is considered to be 3 mg kg^{-1} (Keren & Bingham, 1985) for most crops. Thus, the size of soil with B problem is summing up a total of almost half of the arable soil. Accordingly, strategies should be developed by breeding B-tolerant genotypes in high-B soils and B-efficient genotypes in B-defficient soils. This approach could offer enormous advantages at such sites by helping to widen the areas in which cereals could be cultivated without suffering yield reductions. Soil amendments by conventional techniques such as leaching or increasing pH by liming (Nable et al. 1997) for increased B adsorption on soil seem not to suit Central Anatolian conditions due to its low annual rainfall and water shortages, and the high lime content of the soils. For this reason, B-tolerant genotypes should be determined. In addition, identification of B toxicity tolerant maize cultivars supported by farmer extension services will prevent yield losses due to B toxicity.

The aim of this study was not only to determine the best hybrid maize cultivars that can be grown efficiently on the problematic soils, but also to adress the genotypical differences among extensively planted hybrid maize cultivars in terms of their efficiency to tolerate to the high boron levels.

Materials and methods

Plant material

Fifteen commercial hybrid field (dent) maize (*Zea mays indentata Sturt.)* cultivars (Table 1) were grown in a microprocessor-controlled (temperature, humidity, radiation) glasshouse. Experiments were performed in pots each containing 1826 g of basic experimental soil in a completely randomized design with 3 replicates. Basic soil contained 0.13 mg kg^{-1} available B (low) and 3.6% $CaCO_3$ with a pH of 7.1. Pots were added 0, 0.625, 1.25, 2.5, 5, 10 and 40 mg B kg^{-1} (B_0, B_1, B_2, B_3, B_4, B_5 and B_6, respectively) derived from boric acid (H_3BO_3). All pots were applied a basic fertilization procedure with a 200 mg N kg^{-1} (in the form of urea (NH_2CONH_2) solution), 100 mg P kg^{-1} (triple super phosphate), 80 mg K kg^{-1} (K_2SO_4), 5.5 mg Fe kg^{-1} ($FeSO_47H_2O$), 0.8 mg Zn kg^{-1} ($ZnSO_4\,7H_2O$), 0.3 mg Cu kg^{-1} ($CuSO_45H_2O$) and 4.5 mg Mn kg^{-1} ($MnSO_4$) to correct macro and micronutrient deficiencies. Eight seeds were planted in each pot. After seedling emergence, the number of seedlings was reduced to an even number (4 seedlings per pot). The pots were irrigated to the soil moisture capacity. Plants were grown for 61 days before they were harvested from their stems at soil level.

Table 1. Maize cultivars tested in the experiments

No	Cultivar	Registred in Turkey by	Year
1	TTM-81-19	Antalya Agric. Res. Inst., TR	1987
2	MAT-97	Antalya Agric. Res. Inst., TR	1997
3	RX-770	May-Agro, TR	2001
4	PIAVE	May-Agro, TR	Prod. permit
5	DK-585	Monsanto	2003
6	DK-647	Monsanto	2000
7	LUCE	Pan Tohum, TR	1997
8	ADA-95.10	Sakarya Agric. Res. Inst., TR	2000
9	TTM-815	Sakarya Agric. Res. Inst., TR	1985
10	LG-55	Sapeksa, TR	1992
11	LG-60	Sapeksa, TR	1992
12	T-1595	Sapeksa, TR	1992
13	T-1915	Sapeksa, TR	1992
14	BC-566	Sapeksa, TR	1992
15	P-3397	Pioneer	1996

Toxicity observations

Plants were visually screened for their responses to B. Seedling emergence rates and typical B toxicty symtoms (red, black spots on the leaves) were taken as indicators of B toxicity.

ICP-AES analyses

Extractable B concentrations in the basic soil were determined according to the method of Cartwright et al. (1983) by extraction with 0.01 M mannitol plus 0.01 M $CaCl_2$ using a soil:solution ratio of 1:5 and a shaking time of 16 h. The B extracted was determined by ICP-AES.

Plant samples were carefully washed with deionised water followed by immersion in a 0.2 N HCl solution and rinsing again with deionized water to remove any traces of soil and were then oven-dried at 65°C for 48 h before dry weights were measured. Samples (0.5 g) of finely ground plant material were digested with concentrated HNO_3 in a microwave system (CEM, Mars5 model). The B in the extracts was analysed by ICP-AES (Varian-Vista model) (Nyomora et al. 1997) in at least 4 plant samples with 3 replicates. The B standard used was from Merck, Germany.

DNA isolation and ISSR assay

Genomic DNA was extracted using a standard protocol (Hakki et al. 2001) with minor modifications (Instead of half seed extraction, whole seed was utilized and 70% ethanol washes were twice). DNA concentrations were determined by an Eppendorf BioPhotometer. DNA samples were also run on 1% agarose (Promega) gels with 1 μg of *Hind* III-digested λ DNA (Fermentas Life Sciences).

Out of 25 primers, ten most suitable (in terms of repeatability, scorability and the ability to distinguish between varieties) were selected for identification. Each reaction contained 2.0 mM $MgCl_2$; 10 mM Tris-HCl (pH 8.8); 50 mM KCl; 0.8% Nonidet P40; 200 μM of each of dNTPs; 0.5 μM primer; 20 ng DNA template and 0.4 units of Taq DNA Polymerase (Fermentas Life Sciences) in a final reaction volume of 25 μl. After a pre-denaturation step of 3 minutes at 94°C, amplification reactions were cycled 35 times at 94°C for 1 minute, at annealing temperature for 50 seconds and 72°C for 1 min in Eppendorf Mastercycler gradient thermocycler. A final extension was allowed for 10 min at 72 °C. Upon completion of the reaction, amplified products were loaded onto a 1.5% agarose/1 × Tris-Borate EDTA gel and electrophoresed at 4 V/cm.

Data analyses

All elemental data were analysed using computerised statistical packages (MINITAB and MSTATC). ISSR assays were repeated at least twise for each primer and only the reproducible fragments were scored, with special emphasis on the repeatability of the bands that present polymorphism. Each DNA fragment generated was treated as a separate character and scored accordingly (1 for the presence and 0 for the absence). A rectangular binary data matrix of 15 × 64 was prepared and its statistical analysis was performed using the NTSYS-pc (Ver. 1.7, Rohlf 1992). In cluster analysis of the hybrids the unweighted pairgroup method with the arithmetic mean (UPGMA) procedure was followed.

Results

Phenotypical evaluation

Toxicity symptoms were evaluated visibly and distinctive differences were presented among cultivars to elevated B levels (Figure 1). While B_0 contained basic soil (B-deficient), B_1 was the corrected dosage by the addition of 0.625 mg B kg^{-1} soil (B-sufficient) and B_6 was supplied with toxic levels of B (B-toxic). From B_2 to B_5 pots were contained sequentially elevated B levels (B-high). Two varieties namely TTM.81-19 and DK 647 were not even able to survive during seedling stage at B6 although their seeds were able to germinate. Hence, both were found to be the most sensitive cultivars to high B levels. Varying degrees of toxicity symtoms was also observed at B_5 and B_4 almost in every plant but the severity was genotype dependent. Higher B levels resulted in substantial growth retardations in MAT-97, PIAVE, DK-585, LG-55, LG-60 and T-1595. No visible symptoms were observed in roots since the B concentration in the roots remained relatively low compared to those in leaves, even at 40 mg kg^{-1} B supply (data not given).

Elemental analyses

Effects of boron on plant dry weights

The highest plant dry weights (g. pot^{-1}) were obtained from TTM.815, LG-55 and BC-566 at B_1 (0.625 mg B kg^{-1}) (Table 2). However, MAT-97, DK-585, DK-647, ADA-95.10, LG-60 and T-1595 yielded higher dry weights at B_2 (1.25 mg B kg^{-1}) whereas RX-770, PIAVE, LUCE, T-1915 and P-3397 were the genotypes with the highest dry wt (mean 14.53) at B_3 (2.5 mg B kg^{-1}). In all, the highest dry wt was obtained from TTM.81-19 (19.09 g from 2.5 mgBkg^{-1}) followed by RX-770, LG-55, P-3397, T-1595 (17.88, 16.62, 16.60 and 16.17 respectively). However, since TTM.81-19 was unable to grow in 40 mg B kg^{-1} its mean dry wt in six different B dossages has become 12.59 while P-3397 was found to attain 14.70 g.pot^{-1} followed by RX-770 with 14.01 rank. Substantial losses in dry weights were observed in all genotypes in accordance with increased B concentrations. However this was most obvious in genotypes TTM.81-19 and DK-647 which were unable to survive at B_6 after germination. MAT-97, PIAVE, DK-585, LG-55, LG-60 and T-1595 were also highly effected genotypes. While LUCE, ADA-95.10, TTM.815, T-1915 and BC-566 were moderately effected in B_6, the genotypes with minor influences were RX-770 and P-3397. Dry weights of TTM.815 and T-1595 were also among the highest when compared in B_5 but were dramatically reduced at B_6 (Table 2).

Effects of Boron on Plant Tissue Concentration Levels

Boron concentrations in the tissues of all genotypes were increased steadily in accordance with the increasing B levels (Table 3). The highest B content in B_6 was found in MAT-97 (1364.5 mg Bkg^{-1}), followed by TTM.815 and T-1595 which were also above 1000 mg B kg^{-1} dry matter. Mean B concentration value at B_6 was 764.20 mg kg^{-1}, and 11 out of 15 genotypes had exceeded this value (Table 3). Mean tissue boron content was highly correlated to B application level

Table 2. Plant dry weights (g pot^{-1}) as affected by various boron levels

Cultivars	Bor levels (mg kg^{-1}) B_0 (0.0)	B_1 (0.625)	B_2 (1.25)	B_3 (2.5)	B_4 (5)	B_5 (10)	B_6 (40)	Mean
TTM.81-19	14.39Cab	15.30BCabc	16.68Ba	19.09Aa	12.02Dcd	10.64De	0.00Ef	12.59bcd
MAT-97	13.52BCabcd	14.74ABabc	16.03Aab	14.27ABd	12.02Cb	11.87Ccde	1.58Df	12.01de
RX770	11.67CDcde	13.61BCbc	15.62Bab	17.88Aab	14.63Bab	14.32Ba	10.40Da	14.01a
PIAVE	11.40Cde	13.27BCcd	14.73ABab	15.47Acd	12.80BCbcd	11.49Cde	0.57Df	11.39e
DK 585	10.17ABe	11.36Ad	11.72Ac	8.66Be	8.42Be	8.28Bf	0.37Cf	8.42g
DK 647	12.89Aabcd	13.99Abc	14.17Ab	9.16Be	8.80Be	8.03Bf	0.00Cf	9.58f
LUCE	12.47Bbcd	13.47ABbc	14.39ABab	14.86Acd	14.26AabcB	13.84AabcB	4.80Cbc	12.58bcd
ADA95.10	12.15Bbcde	13.86ABbc	15.39Aab	14.58Acd	14.07ABabcd	13.59ABabcd	5.89Cb	12.79bcd
TTM.815	15.11Aa	15.61Aab	14.81Aab	14.81Acd	14.79Aab	14.16Aab	3.16Bcde	13.21b
LG 55	14.44Bab	16.62Aa	14.49Bab	14.09Bd	13.54Babcd	12.88Babcd	1.15Cef	12.46bcd
LG 60	12.90BCabcd	14.64ABabc	15.61Aab	14.73ABcd	14.44ABab	11.97Cbcde	1.51Def	12.26d
T1595	13.86Babc	15.48ABabc	16.17Aab	15.09ABcd	14.75ABab	14.40Aba	2.15Cdef	13.13bc
T1915	12.56Abcd	13.67Abc	13.99Ab	14.49Acd	13.73Aabcd	12.84Aabcd	4.18Bbcd	12.30cd
BC 566	13.62ABabcd	15.18Aabc	14.74Aab	14.17ABd	13.63ABabcd	12.27Babcde	4.92Cbc	12.65bcd
P3394	14.14Bab	15.65ABab	16.14ABab	16.60Abc	15.32ABa	14.48ABa	10.58Ca	14.70a
Mean	13.02C	14.43B	14.98A	14.53AB	13.15C	12.34D	3.46E	

LSD values= 1.986 (overall); 0.5129 (between B levels) (row); 0.7508 (between cultivars) (column), $p< 0.01$

Capital letters are used to indicate variation between the dry matter contents of the cultivars in different B levels (rows).

Small letters are used to indicate variation between the genotypes in each level of B with respect to shoot dry matter contents (column)

Table 3. The effect of B application on the shoot B concentration (mg B kg^{-1}) of the maize cultivars

	B levels (mg kg^{-1})							
Cultivars	B_0(0.0)	B_1(0.625)	B_2(1.25)	B_3(2.5)	B_4(5)	B_5(10)	B_6(40)	Mean
TTM.81-19	14.0Da	16.4Da	22.8Dabc	55.0Ca	82.0Bde	229.0Aa	0.0Ej	59.88k
MAT-97	16.9Fa	20.0Fa	30.3Ea	54.5Da	155.9Ca	225.0Bab	1364.5Aa	266.72a
RX770	12.4Ea	19.0Ea	19.7Ebc	36.5Dcde	74.7Cef	146.2Bfg	795.1Ag	157.65h
PIAVE	12.6Ea	17.3Ea	19.4Ebc	34.6Dde	77.9Cef	217.6Bb	559.4AI	134.10j
DK 585	12.7Fa	17.7EFa	22.7Eabc	44.1Dbcd	106.8Cb	162.7Bd	843.1Af	172.83f
DK 647	10.8Ea	13.4DEa	20.7Dabc	45.9Cabc	75.5Bef	157.3Ade	0.0Fj	46.23l
LUCE	11.8Ea	16.4Ea	20.8Eabc	35.3Dde	69.9Cfg	138.0Bgh	792.9Ag	155.00hi
ADA95.10	11.9Fa	16.9EFa	21.8Eabc	44.6Dbcd	91.5Cc	150.2Bef	937.4Ad	182.04d
TTM.815	8.7Fa	15.6Fa	26.2Eabc	51.7Dab	73.8Cef	193.7Bc	1104.1Ab	210.55b
LG 55	11.6Fa	14.7Fa	29.1Eab	43.4Dbcd	88.4Ccd	130.5Bh	839.6Af	165.31g
LG 60	7.0Fa	12.3Fa	21.7Eabc	43.8Dbcd	80.8Cde	227.1Ba	843.7Af	176.64e
T1595	7.5Fa	15.2EFa	18.4Ec	35.9Dcde	73.8Cef	161.4Bd	1019.1Ac	190.18c
T1915	8.4Fa	13.4EFa	17.8Ec	32.0De	63.5Cgh	137.9Bgh	797.1Ag	152.87i
BC 566	9.4Fa	12.7Fa	22.7Eabc	43.1Dbcd	63.3Cgh	134.2Bh	900.7Ae	169.46f
P3394	12.5Fa	14.6EFa	21.9DEabc	30.5De	56.5Ch	133.0Bh	666.5Ah	133.64j
Mean	11.2G	15.7F	22.4E	42.1D	82.3C	169.6B	764.20A	

LSD values= 9.005 (overall); 2.325 (between B levels) (row); *3.403* (between cultivars) (column), $p< 0.01$
Capital letters are used to indicate variation between the B concentration of the cultivars in different B levels (rows)
Small letters are used to indicate variation between the genotypes in each level of B with respect to shoot B concentrations (column)

with the exception of B_1 and B_2. Considering mean B contents of all levels, MAT-97 contained about 5 times more B in its tissues when compared with the lowest mean B content genotype, namely DK-647. Hence large genotypic variations existed that predominantly determined the responses of the plants to boron.

Figure 1.Maize cultivars grown in pots containing 40 mg kg^{-1} B showing large genotypic variations with respect to B response

Table 4. Scorable fragments generated by ISSR primers utilised and their polymorphisms

Primer used	Primer Sequence	Number of Bands	Number of Polymorphic bands	Polymorphism (%)
M 1	$(AGC)_6G$	7	6	85.71
M 2	$(ACC)_6G$	3	3	100
M 5	$(GA)_9C$	5	5	100
M 7	$(AG)_9C$	7	6	85.71
M 8	$(AC)_9G$	5	5	100
M 9	$(AC)_8CG$	6	6	100
M 15	$(CA)_8AG$	3	3	100
M 16	$(CA)_8GC$	9	8	88.88
M 17	$CAG(CA)_8$	13	12	92.3
M 18	$CGT(CA)_8$	6	5	83.33
Total		64	59	92.18

Molecular analysis results

Genomic DNA isolated from 15 maize hybrids was successfully amplified following optimization of the amount of template DNA, $MgCl_2$ concentration and *Taq* DNA polymerase, and also the temperatures of the amplification reactions. Out of 25 primers screened in a preliminary experiment, ten primers showing consistently reproducible and highly scorable fragments were selected as the most suitable primers for varietal identification (Table 4). Assaying ISSR variation with these primers yielded 64 bands, 59 of which were polymorhic. Three of these ISSR patterns obtained using primers M1, M7 and M18 were given in Fig. 2.

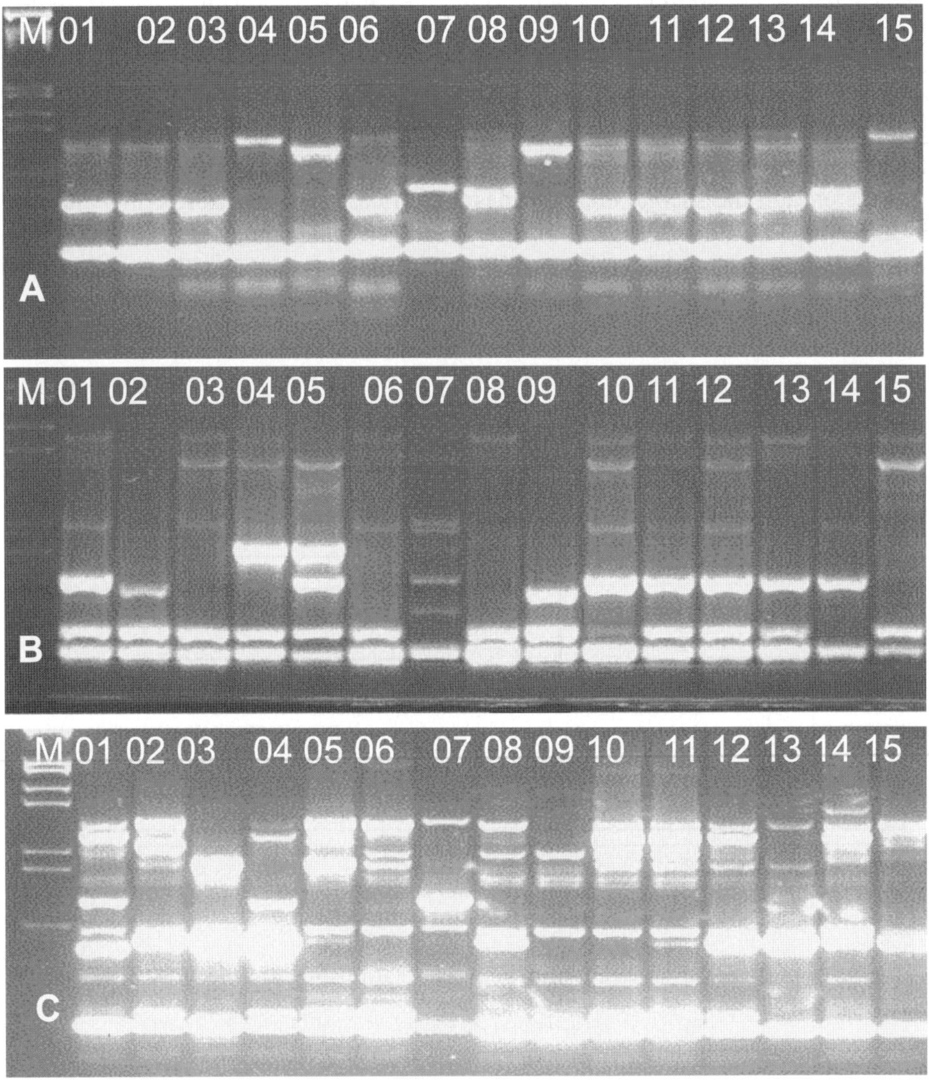

Figure 2. Amplification profiles of maize cultivars by A. M1, B. M18 and C. M7 ISSR primers. Lane numbers from 1-15 are as per the cultivar list in Table 1. M; the molecular weight marker

The primer sequences, the number of fragments they amplified, and the level of polymorphism was also given in Table 4. The number of bands varied from 3 to 13, and the number of polymorphic bands per primer ranged between 3 and 12 with a mean of 5.36 with the primer employed. The primer M17 amplified the highest number of ISSR bands (13 bands 12 of which were polymorphic). The ISSR data matrix was used to compute pairwise genetic distances of the varieties according to Nei and Li's (1979) coefficient. These coefficients were employed to generate a dendogram of maize genotypes using UPGMA (Fig. 3). Hybrid maize varieties could be successfully discriminated by the utilization of all the ISSR profiles. ISSR variation was relatively high with an overall polymorphism of 92.18 % among the genotypes used. The varieties PIAVE, DK-585 and LUCE were clearly seperated from the remaining genotypes which formed two major clusters: the first one includes most of the highly boron sensitive varieties as a subcluster; and the second one includes the efficient genotypes RX-770 and also P-3397 (Fig. 3). Thus, in general, the ISSR generated molecular data conform with the morphological and the ICP-AES analyses data with respect to B contents of the cultivars and their relatedness.

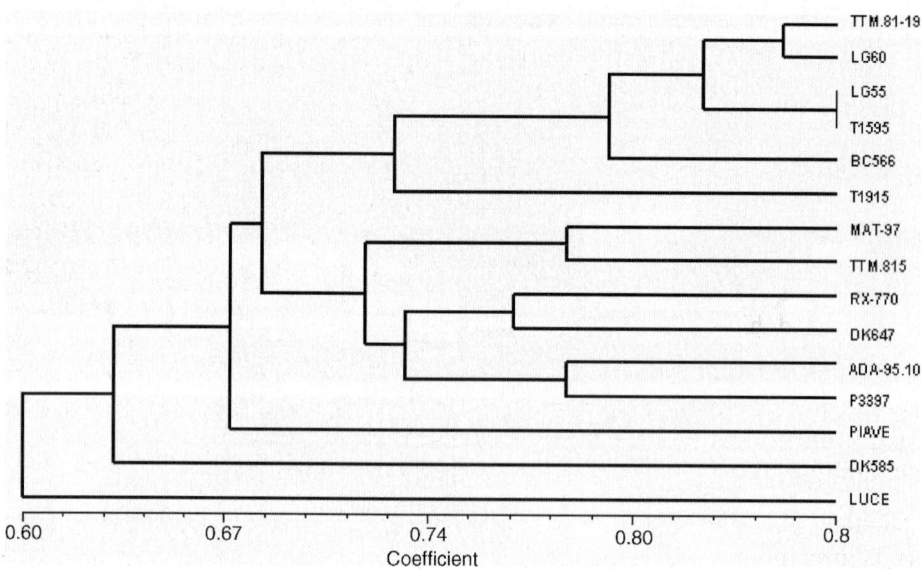

Figure 3. Dendogram showing the genetic relationship among 15 hybrid maize cultivars

Discussion

Maize requires an efficient micronutrient supply throughout its growth. Growth will retard both in excessiveness and deficiency of any nutrient in the growth medium. In the case of boron supply, maize is one of the most sensitive crops. Deficiency and/or toxicity of B can be seen in different parts of the world, including USA, Brasil, Bangladesh, Bulgaria, China, Finland, India, Pakistan, South Africa, Sweeden, Thailand, Russia, Southern Australia, Middle East, Peru, Israel (Cartwright et al. 1984, 1986; Nable et al. 1997). Our research group demonstrated that, Central Anatolian region of Turkey encompassing 3.5 M ha agricultural land has both deficiency (about 27%) and toxicity (18%) problems with respect to B in the soil (Gezgin et al. 2002).

While deficiency problem can be corrected by B efficient fertilization, toxicity problem remains to be solved using much complicated measures. In order to use the soil containing high levels of B for agriculture, either the soluble B contents must be reduced to non-phytotoxic levels for the proposed crop or a tolerant genotype should instead be used. Tolerant genotypes are usually evidenced where high bioavailable boron is present in the soil or in ground water (Jamjod, 1996, Paul et al. 1992a, Bagheri et al. 1994). Frequently used method of amelioration of boron laden soil is leaching with excess water. However, this should be done very carefully since there is the risk of lossing the other essential plant nutrients from the soil too. Soil amendments or removing B from the soil by means of B-hyperaccumulator plants are other creative alternatives. However, these plants are generally not good candidates (e.g. *Gypsophila sp.*) for such large scale applications since they are mostly perennial herbaceous species with long top and thick roots penetrating into the soil which may also make the soil polluted by their roots. Seed multiplication and their cultivation in large acreages are also among other problems for such species. Viability of the only B-hyperaccumulator plant available to our knowledge, *Gypsophila*, first reported/discovered by our research group from a boron mining area of Eskisehir, Turkey (Babaoglu et al. 2004), was only maintained through laboratory micropropagation.

Genetic engineering applications remain to be evaluated whenever suitable genes become available and studies mature after the isolation of the first B-transporter gene (*bor-1*) from Arabidopsis (Rohlf, 1992).

Genetic variation in response to high boron concentrations is well documented at both inter- and intra-specific levels (Nable et al. 1997). Studies for long identified a wide range in response to B within genus or species (Oertli and Kohl, 1961; Oertli and Roth, 1969; Francois and Clark, 1979). A number of plants including wheat, barley, rice, peas, citrus, peach, pear, almond have demonstrated a wide range of intraspecific variation in response to boron (Paul et al. 1988a, 1991a, 1991b, 1992b; Bagheri et al. 1992, 1994, 1996; Picchioni and Miyamoto, 1991; Paliwal and Mehta, 1973; Nable, 1988; Mahboobi et al. 2001; Karabal et al. 2003).

In preliminary studies conducted by Gunes and Alpaslan (2000), it was found that B application reduces the dry weight in some maize genotypes tested. They used eight genotypes (Furio, Riogrande, Sele, DK743, Helix, Missouri, Betor and Poker) that were not included in the present study which used fifteen maize hybrids (Table 1) and 6 B levels, up to 40 mg B kg^{-1}, under glasshouse conditions to the soil with 0.13 mg kg^{-1} bioavailable B content. In addition, hybrids selected were the varieties the most widely planted in Central Anatolian region of Turkey.

Assuming 0.625 mg kg^{-1} boron application results in optimum B dosage for maize planted to a boron defficient soil, also evidenced in the region (Gezgin et al. 2002), response of the varieties were tested to arithmetic increments. Considering dry wt as a valuable parameter to determine the most efficient boron concentration, it is interesting to denote that different varieties response best to differing bioavailable B contents in the soil. For instance, TTM.815, LG-55 and BC-566 responded best to the B_1 while most of the varieties (namely MAT-97, DK-585, DK-647, ADA-95.10, LG-60, T-1595, TTM.81-19, RX-770, PIAVE, LUCE, T-1915 and P-3397) achieved their highest dry wt contents in B_2 and B_3. Not surprisingly, B_2 and B_3 had the highest mean dry wt in turn, followed by B_1. The varieties found to be the most efficient to high boron levels (namely P-3397 and RX-770) have been within the genotypes with their highest dry wt contents in B_3. However, this was found not to be the criteria to determine the efficiency of the varieties since TTM.81-19 had also its pick value in B_3 which was dramatically decreased in successive high dossages. TTM.81-19 was one of the two genotypes unsuccessfull to grow in B_6, the other being DK-647. Considerable growth retardations were determined in all the genotypes tested at the B_4 and B_5 boron levels, RX-770 and P-3397 seemingly less effected and their dry weight contents were about the same as the genotypes moderately effected by high boron concentrations. Dramatical reductions in dry wt of all the plants were evidenced excluding the two efficient genotypes. When all the B concentrations applied are considered the mean dry wt contents of the two efficient genotypes RX-770 and P-3397 again ranked the highest values. When the ratios of B concentrations and dry weights on B_5 are estimated it clearly denotes that sensitive genotypes accumulate twice as much as B per unit dry wt. For instance, sensitive genotype TTM.81-19 had a value of 21.52 (the highest one) followed by DK-585 (19.65) and DK-647 (19.59). On the other hand the B tolerant genotype P-3394 had the lowest score (9.19) which was followed by LUCE (9.97) and RX-770 (10.21). These values alone are good indicators of tolerance capacity of the relevant genotypes. More profound results were generated when these ratios are taken from the B_6 selective dossage. P-3394 and RX-770 had 63.00 and 76.45 values (B content per unit dry wt) respectively while these values become zero on TTM.81-19 and DK647 due to their inability to survive on 40 mg kg^{-1} B applied soil. The ratio had become 2,278.65 on DK-585 which indicates that it is the next highly sensitive genotype which was followed by PIAVE (981.40), MAT-97 (863.61) and LG-55 (730.09). These values are well-correlated with the literature data indicating an exclusion mechanism for the physiology of tolerance to boron toxicity (Nable et al. 1997). Root B contents did not reveal reasonable data as cleaning the roots and their healthy analyses were extremely difficult. Evidence support the idea that closely related species or genotypes susceptible to B toxicity have higher concentrations of B in upperground parts of plants than do tolerant genotypes. A number of studies including the ones with sunflower, lemon, cultivated tomato, cereals and legumes (Eaton, 1944; Francois and Clark, 1979; Toledo and Spurr, 1984; Bagheri et al. 1992, Chhipa and Lal, 1990; Paul et al. 1988a, 1992a, 1992b; El-Motaium et al. 1994) support this hypothesis. Thus, B tolerant genotypes P-3397 and RX-770 were able to maintain lower B concentrations in their leaves and shoots than the susceptible genotypes TTM.81-19 and DK-647 which were followed by the genotypes with increasing levels of susceptibily. This is in line with Nable et al. (1997) who reported that boron tolerant genotypes are able to maintain lower B concentrations in their shoots than susceptible genotypes. However, it is in contrast with the mechanism of B hyperaccumulator *Gypsophila sphaerocephala* discovered by our group (Babaoglu et al. 2004), and some previous reviews about the general characteristics of hyperaccumulator plants (Baker and Brooks, 1989) that accumulate high levels of elements preferably in their aerial tissues. *Gypsophila* plants are not using an exclusion mechanism but instead, they accumulate the B in their leaves, seeds/spikes and stems (about 3350, 2100 and 230 mg B kg^{-1} dry matter). Yet, no information is available on the detoxification mechanism or safe cellular compartmentation for the accumulation of B within these plants. Recently, B adsorption capacity of plant cell walls was thought to play a role in controlling B tolerance of plant species as the cell wall was the main binding site for B in plants (Goldberg and Grieve, 2003). Further studies are required in order to fully cover the mechanisms whether it is genotype specific or any common mechanism is available in the plant kingdom.

Molecular markers continue to acquire novel utilities in terms of genetic analysis of crop plants. Research on complex physiological traits such as salt tolerance are one of the suggested application areas (Flowers et al. 2000). The primary objective of the present study was to evaluate the massively grown hybrid maize genotypes in terms of their response to boron and to observe the genotypical relationships of these plants using molecular marker technologies. Any correlation between phenotypical data and that of the genotype would help breeders in designing crossing experiments where response to B will be considered as a trait. In this study, the genotypical analyses were conducted using the 10 ISSR primers found to be reproducible from the 25 primers used in prescreening. Selected primers yielded a high level of polymorphism (92.18% among the genotypes) and a complete varietal analysis of the samples were possible. ISSR DNA markers produce more consistent results when compared with RAPDs. They are more reliable and informative which is the reason why they are recently preferred dominant markers. Based on the information generated by the dendogram PIAVE, DK-585 and LUCE are the genotypes seperated from the remaining ones which formed two major clusters. One of the groups include most of the highly boron sensitive varieties as a subcluster; and the other group includes the efficient genotypes RX-770 and also P-3397. The most interesting observation was that the efficient genotype RX-770 was the most closely related to DK-647, one of the two genotypes found to be the most sentitive ones, the other being TTM.81-19. This observation could aid breeders to transfer this trait, possibly controlled by a dominant nuclear gene, to their improved genetic material in breeding programs. This study may also help to accelarate the decisions of the breeders to start with the best available genotype when B is considered. Beyond the phenotypical observations and element accumulation and/or dry wt contents of plants based on the effects of B concentrations applied that could be successfully utilized as screening strategies, genotypical relations are also have the high potential to reduce the time spend in targetted breeding programs. In this regard, while distribution of majority of the polymorphic bands (electromorphs) amplified by ISSR primers in our study were more or less in the expected frequencies in B tolerant and B sensitive genotypes, some of the bands did show skewed profile. The later included the fragments generated using primers M17 and M18, which were present in the selected B-tolerant samples and absent in B-sensitive ones. Such polymorphic fragments stand greater chances of having a linkage with the genomic regions, which may have significant contributions towards the B-tolerance as well as other relevant traits like salt tolerance (Kaushik et al. 2003). Identification of such polymorphic loci (possibly by applying further ISSR primers or better yet by using AFLP technique to generate more fragments) and confirming further their relationships with the B tolerance will obviously improve our understanding of the mechanism and improving the efficacy of breeding programs.

References

Babaoğlu M., Gezgin S., Topal A., Sade B., and Dural H., 2004 Gypsophila sphaerocephalafenzl ex Tchihat.: A boron hyperaccumulator plant species that may phytoremediate soils with toxic B levels. Turk J Bot. **28**: 273–278.

Bagheri A., Paull J.G., Rathjen A.J., Ali S.M., and Moody, D.B., 1992 Genetic variation in the response of pea (Pisum sativum L.) to high soil concentrations of boron. Plant and Soil **146**: 261–269.

Bagheri A., Paull J.G., and Rathjen A.J., 1994 The response of Pisum sativum L. germplasm to high concentrations of soil boron. Euphytica **75**: 9–17.

Bagheri A., Paull J.G., and Rathjen A.J., 1996 Genetics of tolerance to high concentrations of soil boron in peas (Pisum sativum L.). Euphytica **87**: 69–75.

Baker, A.J.M. and Brooks R.R., 1989 Terrestrial higher plants which hyperaccumulate metallic elements: a review of their distribution, ecology and phytochemistry. Biorecovery **1**: 81–126.

Caceres L., Gruttner D., and Contreras N., 1992 Water recycling in arid regions: Chilean case. Ambio **21**: 138–144.

Cartwright B., Zarcinas B.A., and Mayfield A.H., 1984 Toxic concentrations of B in a red-brown earth at Gladstone, South Australia. Aust. J. Soil **22**: 261–272.

Cartwright B., Zarcinas B.A., and Spounce L.A., 1986 Boron toxicity in South Australian barley crops. Aust. J. Agric. **37**: 351–359.

Chhipa B.R. and Lal P., 1990 A comparative study on the effect of soil B on yield, yield attributes and nutrient uptake by susceptible and tolerant varieties of wheat. An. Edafol. Agrobiol. **48**: 489–498.

Eaton F.M., 1944 Deficiency, toxicity and accumulation of boron in plants. J. Agric. **69**: 237–277.

El-Motaium R., Hu H., and Brown P.H., 1994 The relative tolerance of six Prunus rootstocks to boron and salinity. J. Am. Soc. Hort. Sci. **119**: 1169–1175.

FAOSTAT: http://faostat.fao.org

Flowers T.J., Koyama M.L., Flowers S.A., Sudhakar C., Singh K.P., and Yeo A.R., 2000 QTL: their place in engineering tolerance of rice to salinity. J Exp Bot **51**: 99–106.

Francois L.E. and Clark R.A., 1979 Boron tolerance of twenty-five ornamental shrub species. J. Amer. Soc. Hort. Sci. **104**: 319–322.

Gezgin S., Dursun N., Hamurcu M., Harmankaya M., Onder M., Sade B., Topal A., Soylu S., Akgun N., Yorgancilar M., Ceyhan E., Ciftci N., Acar B., Gultekin I., Isık Y., Seker C., and Babaoglu M., 2002 Determination of B contents of soils in Central Anatolian cultivated lands and its relations between soil and water characteristics. Boron in Plant and Animal Nutrition. Edited by Goldbach et al., Kluwer Academic / Plenum Publishers, New York.

Goldberg S. and Grieve C.M., 2003 Boron adsorption by maize cell walls Plant and Soil **251**: 137–142.

Gunes A., Alpaslan M., Ozcan H., and Cıkılı Y., 2000 Tolerance to boron toxicity of maize (Zea mays L.) cultivars widely cultivated in Turkey. Turk. J. Agric. For. **24**: 277–282.

Hakki E.E., Savaksan C., and Akkaya M.S., 2001 Genotyping of Anatolian doubled-haploid durum lines with SSR markers. Euphytica **122**: 257–262.

Jamjod S., 1996 Genetics of boron tolerance in durum wheat Ph D Thesis, The University of Adelaide, South Australia.

Karabal E., Yucel M., and Oktem H.A., 2003 Antioxidant responses of tolerant and sensitive barley cultivars to boron toxicity. Plant Science **164**: 925–933.

Kaushik A., Saini N., Jain S., Rana P., Singh R.K., and Jain R.K., 2003 Genetic analysis of a CSR10 (indica) × Taraori Basmati F3 population segregating for salt tolerance using ISSR markers. Euphytica **134**: 231–238.

Keren R. and Bingham F.T., 1985 Boron in Water, Soil and Plants. In: Stewart, B.A. (ed.), Adv. In Soil Sci. **1**: 229–276.

Mahboobi H., Yucel M., and Oktem H.A., 2001 Cell wall uronic acid concentrations of resistant and sensitive cultivars of wheat and barley under boron toxicity. J. Plant Nutr. **24** (12): 1965–1975.

Masson L., 1967 Alaunos problemas relacionados con la calinidad in los vales de la costa sur. Oficeing Nacional de Evaluacion di Recursos Nationals. Lima, Peru.

Nable R.O., Banuelos G.S., Paull and J.G., 1997 Boron Toxicity. Boron in Soils and Plants: Plant and Soil **198**: 181–198.

Nable R.O., 1988 Resistance to boron toxicity amongst several barley and wheat cultivars: a preliminary examination of the resistance mechanism. Plant and Soil **112**: 45–57.

Nei M. and Li W.-H., 1979 Mathematical model for studying genetic variation in terms of restriction endonucleases. Proceedings of the National Academy of Sciences, USA **76**: 5269–5273.

Nyomora A.M.S., Sah R.N., and Brown P.H., 1997 Boron determination in biological materials by inductively coupled plasma atomic emission and mass spectrometry: effects of sample dissolution methods. Fresenius J. Anal. Chem. **357**: 1185–1191.

Oertli J.J. and Kohl H.C., 1961 Some considerations about the tolerance of various plant species to excessive supplies of boron. Soil Sci. **92**: 243–247.

Oertli J.J. and Roth J.A., 1969 Boron nutrition of sugar beet, cotton and soybean. Agron. J. **61**: 191–195.

Paliwal K.V. and Mehta K.K., 1973 Interactive effect of salinity, SAR and boron on the germination and growth of seedlings of some paddy (Oryza sativa) varieties. Plant and Soil **39**: 603–609.

Paull J.G., Cartwright B., and Rathjen A.J., 1988a Responses of wheat and barley genotypes to toxic concentrations of soil boron. Euphytica **39**: 137–144.

Paull J.G., Rathjen A.J., and Cartwright B., 1991a Major gene control of tolerance of bread wheat (Triticum aestivum L.) to high concentrations of soil boron. Euphytica **55**: 217–288.

Paull J.G., Rathjen A.J., and Cartwright B., 1991b Tolerance to high concentrations of boron for the amphiploid of Triticum aestivum-Agropyron elongatum. Plant and Soil **133**: 297–299.

Paull J.G., Nable R.O., and Rathjen A.J., 1992a Physiological and genetic control of the tolerance of wheat to high concentrations of boron and implications for plant breeding. Plant and Soil **146**: 251–260.

Paull J.G., Nable R.O., Lake A.W.H., Materne M.A., and Rathjen A.J., 1992b Response of annual medics (Medicago spp.) and field peas (Pisum sativum) to high concentrations of boron: genetic variation and the mechanism of tolerance. Aust. J. Agric. **43**: 203–213.

Picchioni G.A. and Miyamoto S., 1991 Growth and boron uptake of five pecan cultivar seedlings. Hort Science **26**: 386–388.

Ravikovitch S., Margolin M., and Navroth J., 1961 Microelements in soils of Israel. Soil Sci. **92**: 85–89.

Rohlf F.J., 1992 NTSYS-pc: Numerical taxonomy and multivariate analysis system, version 2.0. State Univ. of New York, Stony Brook, NY.

Sakal R. and Singh A.P., 1995 Boron research and agricultural production. In micronutrients res. Agric. Prod. (Ed., Tandon, Hls) P:1–31 Fert. Dev. And Cons. Org. New Delhi, India.

Takano J., Noguchi K., Yasumori M., Kobayashi M., Gajdos Z., Miwa K., Hayashi H., Yoneyama T. and Fujiwara T., 2002 Arabidopsis boron transporter for xylem loading. Nature **420**(6913):337–40.

Toledo J. and Spurr J., 1984 Plant growth and boron uptake by Lycopersicon esculentum and L. cheesmanii f.minor. Turrialba **34**: 111–115.

Part II

Boron in Animals and Humans

Dietary Boron: Evidence for Essentiality and Homeostatic Control in Humans and Animals

Curtiss D. Hunt

U.S. Dept. Agriculture, Agricultural Research Service, Grand Forks Human Nutrition Research Center, PO Box 9034, Grand Forks, ND 58202-9034, USA; Email: chunt@gfhnrc.ars.usda.gov

Introduction

Boron has served essential function since the early evolution of life as evidenced by the essentiality of the element for the heterocystous Cyanobacteria, predominant organisms during the Middle Pre-Cambrian Period (Bonilla et al. 1990). The element boron is essential for at least some organisms in all phylogenetic kingdoms. The first natural biomolecules found to contain boron were antibiotics produced by species in Eubacteria (Bonilla et al. 1990; Chen et al. 1981; Hutter et al. 1967; Sato et al. 1978; Schummer et al. 1994). Boron is required for specific brown algae and diatoms in Stramenopila and for all higher plants in Viridiplantae (Lovatt and Dugger 1984). Specific fungi have a demonstrated physiological response to boron (Bennett et al. 1999), an important finding because species in the kingdom Fungi are thought to share a common ancestor with animals exclusive of plants (Carney and Bowen 2004). This review summarizes the evidence that boron satisfies the criteria for essentiality in humans and higher animals (Expert Consultation WHO/FAO/IAEA 1996; Frieden 1984; Mertz 1970; Underwood and Mertz 1987): 1) it reacts with biological material or forms chelates; 2) it is present in healthy tissues of different animals at comparable concentrations; 3) toxicity results only at relatively high intakes; 4) tissue concentrations during short term variations in intake are maintained by homeostatic mechanisms; 5) depletion prevents growth and completion of the life cycle; 6) depletion consistently results in reduction of a physiologically important function; and 7) when an integral part of an organic structure, depletion causes reduction in performance of a vital function.

Boron and biological compounds

Boron Chemistry

Boron satisfies the essentiality criterion that the element reacts with biological material or forms chelates. The chemical characteristics of boron favor formation of complexes between the element and many biomolecules. Organoboron compounds are those organic compounds that contain B-O bonds,

F. Xu et al. (ed.), Advances in Plant and Animal Boron Nutrition, 251–267.

i.e., the orthoborates ($B(OR)_3$, $(RO)B(OR')_2$ and $(RO)B(OR')(OR'')$, and orthoborates of polyhydric alcohols (Greenwood and Earnshaw 1984). Organoboron compounds include B-N compounds, because B-N is isoelectronic with C-C (Greenwood 1973). Only organoboron compounds are apparently important in biological systems and they are the result of interaction with OH or amine groups. As described below, organoboron complexes occur in plants and are produced in vitro with biomolecules isolated from animal tissues.

The most probable form of dietary boron after ingestion and subsequent hydrolysis (Greenwood and Earnshaw 1984) is orthoboric acid (common name: boric acid) $B(OH)_3$. Boric acid accepts a hydroxyl ion (a Lewis acid) to form the tetrahedral anion $B(OH)_4^-$ (Reaction 1) (Greenwood 1973):

$$B(OH)_3 + 2H_2O \leftrightarrow H_3O^+ + B(OH)_4^- \qquad pK_a = 9.25\ (25^\circ C) \qquad (1)$$

At typical physiological boron concentrations (0.006 - ~9.0 µmol/L) in plants, animals, or humans, inorganic boron is essentially present only as the mononuclear species boric acid $B(OH)_3$ and as borate $B(OH)_4^-$ (Weser 1967). Within the normal pH range of the gut and kidney, $B(OH)_3$ prevails as the dominant species (pH 1: ~100% $B(OH)_3$; pH 9.3: 50%; pH 11: ~0%) (Spivack and Edmond 1987). Undissociated (and uncharged) boric acid is very soluble in water [$B(OH)_3$-saturate solution at 20°C = 0.75 mol/L] and its permeability coefficient for transport across the lipid bilayer is several orders of magnitude higher than that of ions (Takano et al. 2005), on the same order as urea (Raven 1980). The similarity between the molecular radii of boron (2.57 Å) and water (2.82 Å) suggests that boron can replace the water molecules that hydrate the polar head groups of lipids (Verstraeten et al. 2005).

Boron Biochemistry

Boron esters

There are a vast number of mono or polyhydroxy compounds that contain one or more hydroxy groups with suitable positions for interaction with boron to form boroesters, specific organoboron complexes. For example, boric acid reacts with a suitable dihydroxy compound to form the corresponding boric acid monoester ("partial" esterification) (e.g., Structure 1) that retains the trigonal-planar configuration and no charge. Borate may react with a suitable dihydroxy compound to form the corresponding borate monoester ("partial" esterification; monocyclic) (Structure 2) with a tetrahedral configuration and a negative charge. A compound of similar configuration and charge is also formed when a boric acid monoester forms a complex with an available hydroxyl group. These two types of boromonoesters can react with another dihydroxy compound to give a corresponding spiro-cyclic borodiester ("complete" esterification) that is a chelate complex with a tetrahedral configuration and negative charge (Structure 3) (Van Duin et al. 1984).

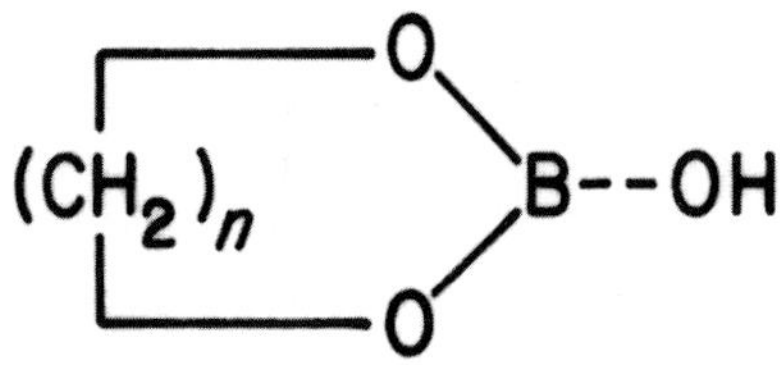

(Structure 1)

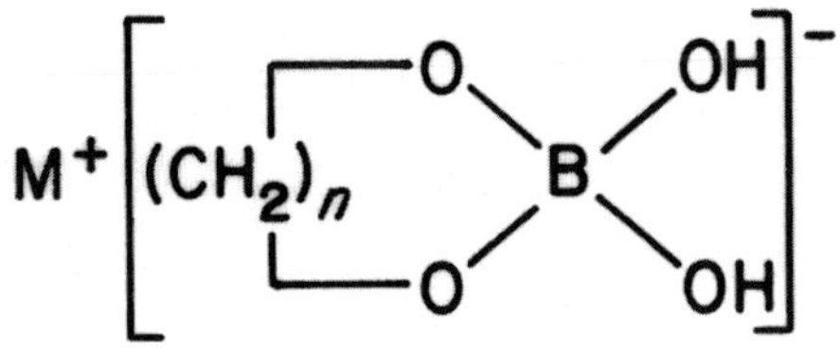

(Structure 2)

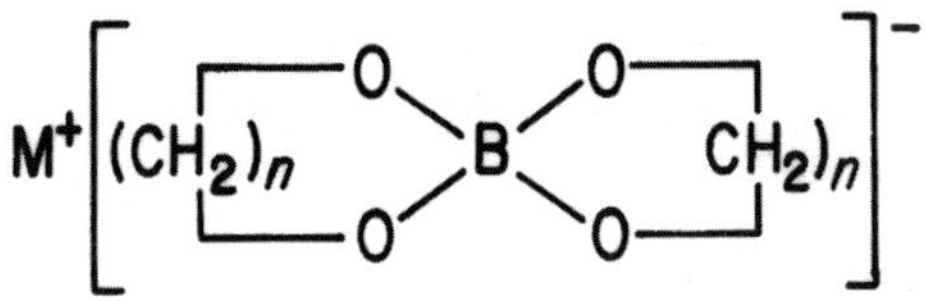

(Structure 3)

Boroester formation is facilitated typically by the presence of adjacent and *cis* hydroxyl groups on the ligand (Zittle 1951). The first natural biomolecules found to contain boron were antibiotics including tartrolon B (Structure 4) characterized by a boron atom bound to four oxy groups (Schummer et al. 1994). The discovery of the currently recognized boron dependent biomolecules was achieved because the bound boron formed four coordinate covalent bonds with the ligand, creating a thermodynamically stable complex that is almost undissociable in water (Gerrard 1961; Thellier et al. 1979). However, many boron interactions with biological ligands are rapidly reversible (Berezin et al. 1967; Hausdorf et al. 1987) and difficult to detect with current techniques. Several biologically important sugars and their derivatives (sugar alcohols, onic and uronic acids) contain the relevant *cis*-diol conformations that promote boron complexation (Raven 1980). In fact, the high affinity of boric acid for adjacent and *cis*-hydroxyl groups probably exerted considerable evolutionary pressure on the selection of sugars as structural components instead of energy substrates as described below.

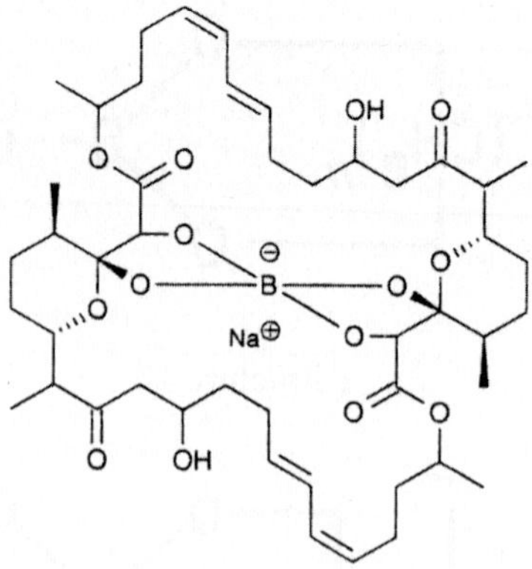

(Structure 4)

Boron-Nitrogen Compounds

The known ability of boron to form reversible covalent bonds with the nitrogen atom of amine groups (Bachovchin et al. 1988; Bone et al. 1989; Ivanov et al. 2002) suggests the possibility of a large array of biochemicals other than polyols that can react with boron to form complexes. The serine proteases are major proteolytic enzymes with critical roles including control of the coagulation system (Kettner et al. 1988). Nanomolar concentrations of peptide boronic acids (Ivanov et al. 2002; Kettner and Shenvi 1984) and millimolar concentrations of boric acid (Berezin et al. 1967) form reversible covalent adducts with the active site serine of the enzymes (Structure 5).

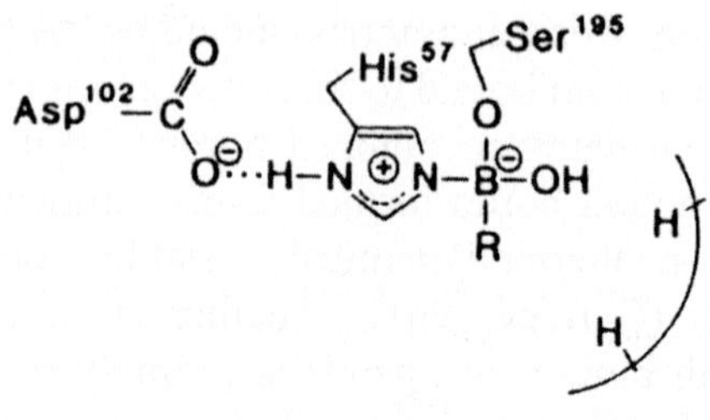

(Structure 5)

Boron concentrations in healthy tissues

Boron satisfies the essentiality criterion that it is present at comparable concentrations in healthy tissues of different animals. Similar concentrations of boron (µg/mL) were reported in the plasma of humans (0.017-0.191) (Ferrando et al. 1993; Hunt et al. 1997; Iyengar et al. 1990; Mauras et al. 1986; Usuda et al. 1997; Wallace et al. 2002; Ward 1993), rats (0.038-0.039) (Seaborn and Nielsen 1994; Vaziri et al. 2001), chicks (0.047-0.152) (Hunt 1989; Hunt 1996), cows (0.052-0.153) (Hunt 1996; Small et al. 1997), lambs (0.163) (Hunt 1996), pigs (0.126) (Hunt 1996), and horses (0.227) (Hunt 1996). Liver boron concentrations (µg/g; dry weight) are similar in humans (1.1-5.4) (Shuler et al. 1990; Ward 1987), rats (0.51) (Bai and Hunt 1996), chicks (1.01-4.4) (Rossi et al. 1993), and cows (3.3) (Ward 1987); for brain tissue (µg/g; dry weight), similar in humans (0.87) (Shuler et al. 1990), rats (0.64) (Bai and Hunt 1996), and chicks (1.01-1.05); for bone tissue (µg/g; dry weight), similar in humans (1.6) (Ward 1993), rats (1.3) (Bai and Hunt 1996), chicks (0.59-0.64) (Bai and Hunt 1996; Hunt et al. 1994; Wilson and Ruszler 1996), and mule deer (1.7) (Stelter 1980).

Boron toxicity

Boron satisfies the essentiality criterion that an essential mineral has a range of safe exposures and it has a toxic range when its safe exposure is exceeded. Boron produces toxicity in all tested biological organisms when excessive amounts are absorbed. Boron dose response experiments to determine embryo-larval malformations in the frog *X. laevis* have demonstrated the pattern of areas of survival, deficiency, optimization, toxicity, and lethality that are characteristic of an essential element (Fort et al. 2002). When expressed on a molar basis, the average dietary intake of boron for, as example, American postmenopausal women [0.10 mmol (1.11 mg)/d] is substantially greater than that of manganese [0.04 mmol (2.42 mg)/d], copper [0.01 mmol (0.76 mg)/d], or molybdenum [0.0008 mmol (0.076 mg)/d]. Even so, boron has a low order of toxicity; the Dietary Reference Intake upper limit for boron (mg/d) is 20 for adults, 17 for adolescents aged 14-18 years, and 11, 6, and 3 for children aged 9-13, 4-8, and 1-3 years respectively (Food and Nutrition Board: Institute of Medicine 2001).

A literature survey conducted in 1949 of boric acid or borax poisoning found 86 cases and a mortality of 48.8% with many fatalities associated with accidental massive doses given to infants (Pfeiffer 1949). In 1988, an independent retrospective examination of 784 cases of acute single boric acid ingestions in Maryland and Washington, DC found no fatalities or severe manifestations of toxicity in older children and adults after boric acid ingestions of up to 88.8 g (15.54 g boron) (Litovitz et al. 1988). At present, death from boron poisoning is exceptionally rare probably because of the emphasis placed on maintaining electrolytic balance and supporting kidney function during the worst part of the illness (Litovitz et al. 1988).

Boron homeostatic mechanisms

Dietary Intakes of Boron

Boron is ubiquitous in the environment with an average concentration of 4600 µg/L (430 µmol/L) in seawater, a much higher concentration compared to seawater content of molybdenum [9.6 µg (0.1 µmol)/L], iron [5.6 µg (0.1 µmol)/L], zinc [4.9 (0.075 µmol)/L], copper [3.2 µg (0.05 µmol)/L], or manganese [2.0 µg (0.036 µmol)/L] (McClendon 1976). Boron consumption varies considerably among individuals and by sex-age group. For example, boron intake for infants aged 0 to 6 months is 0.75±0.14 mg/d (mean±SE) (1st percentile, 0.03; 99th percentile, 6.40 mg/d); for males aged 51 to 70 years, 1.34±0.02 mg/d (1st percentile, 0.39; 99th percentile, 3.34 mg/d); for lactating females, 1.39±0.16 mg/d (1st percentile, 0.38; 99th percentile, 3.49 mg/d) (Food and Nutrition Board: Institute of Medicine 2001). The range of dietary boron intakes within a sex-age group arises from a variety of factors. For example, compared with animal-based food products, plant-based products are much richer sources of dietary boron (Hunt and Meacham 2001). Furthermore, most plant species within the subclass *Dicotyledoneae*, which includes fruits [i.e., raw pears: 2.27 µg (0.21 µmol) B/g], vegetables, tubers and legumes have much higher concentrations of boron than do species from the subclass *Monocotyledoneae*, especially gramineaceous species (the grasses) including rice [0.09 µg (0.008 *µmol*) B/g], corn, barley, and wheat (Hunt and Meacham 2001). For this reason, diets that provide only 0.36 mg B/2000 Kcal (and otherwise nutritionally adequate) are prepared easily by excluding vegetables, tubers, nuts, and legumes (Hunt et al. 1997).

Evidence for Boron Homeostatic Control

Boron satisfies the criterion of essentiality that tissue concentrations during short term variations in intake are maintained by homeostatic mechanisms. Several new lines of evidence support the concept of boron homeostasis. For example, human milk boron concentrations are under apparent homeostatic control. Concentrations of boron in milk from mothers of exclusively breast-fed healthy full-term infants in St. John's, Newfoundland, Canada (Hunt et al. 2004), and Houston, TX, USA (Hunt et al. 2005) did not vary widely and were stable over the first four months of lactation (Figure 1). Mean concentrations of milk boron were similar between the test populations (Houston: 39 µg [3.61 µmol]/L; St. John's: 29 µg [2.68 µmol]/L) despite probable differences in dietary boron intake. The consistent range in milk boron concentration coupled with a distinct pattern of no change over time (i.e., stable) in two separate populations suggest that lactating mothers who self-select diets may maintain homeostatic control over milk boron concentrations.

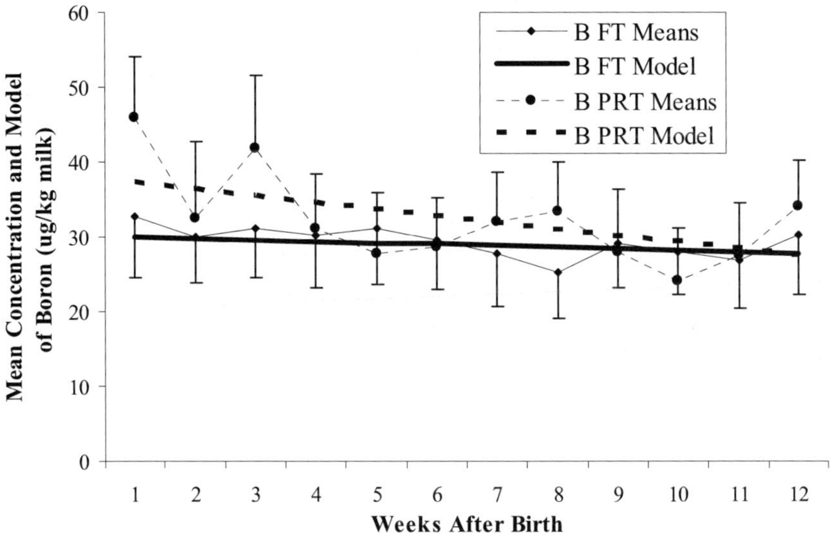

Figure 1. Model and mean (±SE) concentrations of boron in breast milk from mothers of full-term (FT) and premature (PRT) infants; n = 9 per group over the 12 wk after birth. During the first 12 wk of lactation, prematurity affected the rate of change in concentrations (P = 0.01) (Hunt et al. 2004)

There is considerable evidence for the presence of boron concentration against a gradient across mammalian cell membranes. In a metabolic study with postmenopausal women, gastrointestinal absorption of inorganic boron and subsequent urinary excretion was near 100% (Hunt et al. 1997). However, a 9.0-fold increase in dietary boron (0.36 mg to 3.0 mg B/day; an increase from the 5th to 95th percentile of usual boron intake) during the study yielded only a 1.5-fold increase in plasma boron concentrations (5.92 ± 4.16 μg to 8.79 ± 5.18 μmol B/L). The relatively small change in blood boron values coupled with large increases in urinary boron values indicates a strong boron gradient across the plasma membrane in the kidney. Other investigators have also reported a remarkably narrow range of boron concentrations in whole blood from subjects with unknown dietary histories (Clarke et al. 1987). In female rats, supplementation with high amounts of boron (9.25 mmol/L water) for 21 days, caused an increase in plasma boron concentrations but an undefined homeostatic mechanism concurrently eliminated any excess of boron from the liver and brain against their own concentration gradients (Magour et al. 1982). In yearling beef heifers, the percent of filtered boron reabsorbed by the kidneys decreased significantly with increased boron intake (Green and Weeth 1977).

The recent report (Ralston and Hunt 2004) that cultures of either RAW 264.7 cells or HL60 cells retain intracellular boron against a concentration gradient indicated the presence of intracellular boron binding species or the existence of boron specific transporters on the plasma membrane. Most likely, the repeated demonstration of the concentration of boron against a gradient indicates the existence of boron specific transporters. This line of evidence for the homeostatic control of boron is enhanced further by the discovery of a specific mammalian borate transporter, NaBC1, expressed in the basolateral membranes of epithelial cells (Park et al. 2004), a topic that will be described in detail elsewhere in this symposium. The recent identification of the boron transporter, BOR1 (AtBor1), in the flowering plant Arabidopsis thaliana (Takano et al. 2005; Takano et al. 2002) and its mammalian homolog, BTR1, a newly discovered bicarbonate transporter superfamily member (Parker et al. 2001), provides further evidence for the homeostatic control of boron in humans and mammals.

Boron depletion and the life cycle

Boron satisfies the criterion of essentiality that depletion prevents growth and completion of the life cycle in animal models. The original finding (Hunt and Nielsen 1981) that boron deprivation can impair growth has been the basis for further research in several independent laboratories. Boron deprivation reduced feed efficiency in weanling pigs (Armstrong et al. 2000) and growth in chicks (Bai and Hunt 1996) and rats (Nielsen et al. In press). In experiments with zebrafish, sperm from low-boron males successfully fertilized eggs from boron females (Rowe and Eckhert 1999), but 92% of the embryos (compared to 37% of controls) died within 10 days. However, the low-boron embryos could be rescued from death if repleted with boron during the first hour after fertilization. Studies with the South African clawed frog, *Xenopus laevis* (Fort et al. 2002), indicate that specimens fed a low-boron diet in a low-boron culture media produced a substantially higher number of necrotic eggs and fertilized embryos than frogs fed a boron-sufficient diet. By 96 hours of development, none of the larvae from boron-deficient adults and maintained in low-boron culture media developed normally. In rat dams fed a low-boron diet (0.04 μg/g), the number of implantation sites was reduced significantly compared to dams fed a boron-adequate diet (2.00 μg/g) (Lanoue et al. 1998).

Boron depletion and physiological/structural abnormalities

Boron satisfies the criterion of essentiality as defined by who (Expert Consultation WHO/FAO/IAEA 1996) that “an element is considered essential to an organism when reduction of its exposure below a certain limit results consistently in a reduction in a physiologically important function, or when the element is an integral part of an organic structure performing a vital function in the organism.” Because boron is essential for at least some species in each of the phylogenetic kingdoms, it is not surprising that boron deprivation appears to perturb physiological processes in frogs, zebrafish, chicks, rats, pigs, and humans.

There is new evidence that dietary boron may be important when biological responses to insulin are impaired, a condition referred to as insulin resistance. As a dietary ingredient, boron decreased peak pancreatic in situ insulin release in chicks (Figure 2). In rats, dietary boron decreased plasma insulin concentrations but did not change glucose concentrations (Bakken and Hunt 2003). Earlier, dietary boron deprivation was reported to induce a modest but significant increase in fasting serum glucose concentrations in older volunteers (men and women) fed a low-magnesium, marginal copper diet, (Nielsen 1989). Together, these findings may be relevant in understanding diabetes mellitus because of the reasonable argument that β-cell "exhaustion" might explain the β-cell deterioration that occurs during excessive insulin demand (Reaven 1999; Sprietsma and Schuitemaker 1993). That is, it is possible that β-cells too easily induced to secrete mass quantities of insulin are more readily damaged, which eventually can cause them to stop functioning and result in diabetes mellitus. The findings to date suggest that physiological amounts of boron may help reduce the amount of insulin required to maintain plasma glucose.

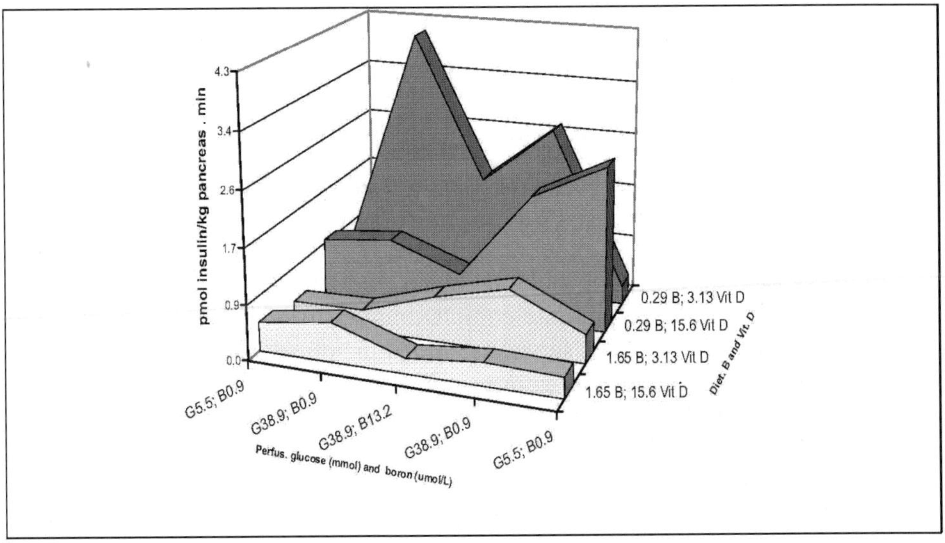

Figure 2. Alteration in peak insulin secretion from isolated, perfused pancreata from 1-d-old cockerels fed a diet containing 0.29 (boron-low) or 1.65 mg B/kg and supplemented with cholecalciferol (Vit D) at 3.13 (inadequate) or 15.60 (adequate) μg/kg for 26-37 d. Perfusion phases varied in the total amount of glucose (G: 5.5 or 38.0 mmol/L) and boron (B; 0.9 or 13.2 μmol/L) in the perfusate (Bakken and Hunt 2003)

In one or more animal species, boron deprivation induces signs of increased risk of prostate cancer, altered mineral and steroid metabolism, defective bone structure, or an abnormal inflammatory response to antigens. New advances in these areas are described in detail elsewhere in this symposium.

Identification of boron –containing biomolecules

Evolutionary pressure from boron on selection of energy substrates

Robust proof of boron essentiality in humans and animals will include identification of boron-containing biomolecules in these species. There are several guideposts in the search for these biomolecules. The high affinity of boric acid for adjacent and *cis*-hydroxyl groups probably exerted considerable evolutionary pressure against the selection of certain sugars as energy substrates. Sugars with five-membered rings are called furanoses, and those with six-membered rings are called pyranoses (Zubay 1988). Compounds in a configuration where there are *cis*-diols on a pyranoid ring (e.g., the pyranoid form of alpha-D-glucose) form weaker complexes with boron than do compounds configured to have *cis*-diols predominately on a furanoid ring (e.g., apiose and ribose). Thus, in early plants, boron may have driven the selection of sucrose as the mobile storage carbohydrate because other algae sugars form a tight complex with boron (Bonilla et al. 1990; Pfeiffer and Braverman 1982). The near absence of an alpha-furanose form of D-glucose in aqueous solutions (Zubay 1988) suggests that glucose was selected as the aldose for general energy metabolism because of its lower reactivity with boric acid.

Evolutionary pressure from boron on selection of structural sugars

Apiose

Apiose is one of only two natural sugars that occurs in its physiological derivatives in the strongly-borate-complexing erythrofuranose configuration (Loomis and Durst 1992). This sugar, not found in animals or humans, may have been selected for, not against, its borate-complexing furanose capability (Loomis and Durst 1992) and hence, structural roles. Therefore, it is an example of a boron-containing biomolecule with a likely undiscovered analog in the animal kingdom. Apiose residues in monomers of rhamnogalacturonan II (RG-II), a pectic polysaccharide in the primary cell walls of plants, complex only with a boron atom to form a borodiester (Structure 6). The cross-linked RG-II dimers are thought to be “load-bearing,” acid-labile linkages that are hydrolyzed by a decrease in wall pH during auxin-induced cell expansion (Loomis and Durst 1992; O'Neill et al. 1996).

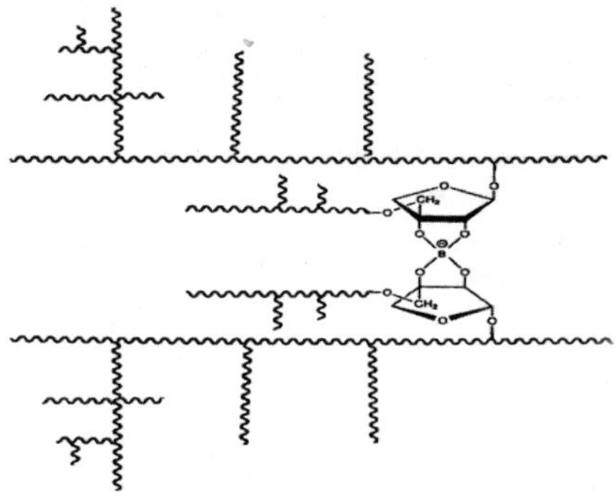

(Structure 6)

Ribose

Ribose is the other of only two natural sugars that occurs in its physiological derivatives in the strongly-borate-complexing erythrofuranose configuration (Loomis and Durst 1992). Ribose derivatives have a strong affinity for boron and therefore that sugar may have been selected for structural functions especially in nucleotides (Loomis and Durst 1992). Nucleotide-boron adduction has been demonstrated for a number of ribose-containing nucleotides and cofactors all with different affinities for boron (Structure 7) (Ralston and Hunt 2001). S-adenosylmethionine (SAM), the predominant methyl donor in biological methylations and a versatile cofactor in a variety of physiological processes, has the highest known affinity for boron of all animal and human biocompounds examined (Ralston and Hunt 2001). Next in rank are members of the diadenosine phosphate (Ap_nA) family of biomolecules Ap_3A, Ap_4A, Ap_5A, and Ap_6A, signal nucleotides present in all cells with active protein synthesis (McLennan 1992).

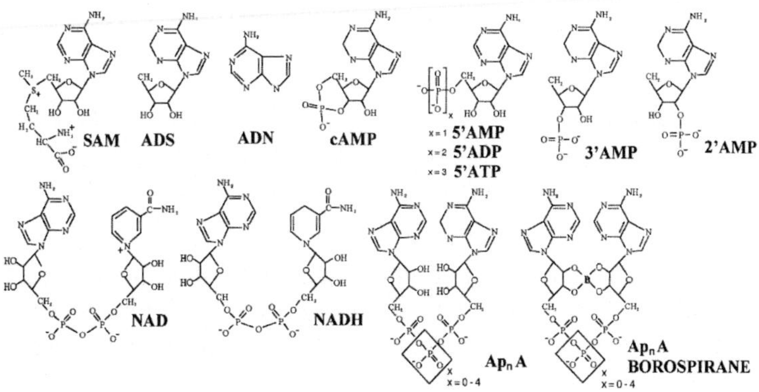

(Structure 7)

The physiological consequences of boron binding with SAM or the Ap_nA molecular species remain unexplored but probably are not trivial. Because the Ap_nA molecules function in cell signalling, it is not surprising that a cell-to-cell communication signal that requires boron was reported recently (Chen et al. 2002). The signal, autoinducer-II (AI-II), is produced by a bacterium and is derived from a ribose moiety, S-ribosylhomocysteine. Furthermore, the signal binds to the primary receptor to form a furanosyl borate diester complex (Structure 8).

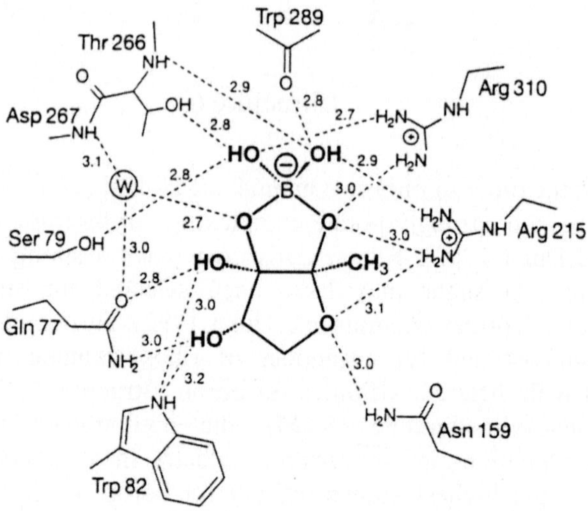

(Structure 8)

NAD[+] (Structure 7) follows Ap_3A in rank for boron affinity (Ralston and Hunt 2001) and oxidoreductase enzymes that require pyridine (e.g., NAD[+] or NADP) or flavin (e.g., FAD) nucleotides are well known to be competitively inhibited by borate or its derivatives (Kim et al. 2003; Smith and Johnson 1976). Reversible enzymatic inhibition as an essential role for an element is unusual. However, there is irrefutable evidence that boron serves to inhibit or dampen several metabolic pathways in plants (Lovatt and Dugger 1984; Shkol'nik and Il'inskaya 1975).

Based on the structure of known boron-containing biomolecules and the regulatory nature of those enzymes that are strongly inhibited by boron, it is predicted that several biomolecules waiting discovery are derived from ribose and serve as signaling molecules that interact with the cell surface. It is predicted that they are probably comprised of mirror or near-mirror halves stabilized by a single boron atom.

References

Armstrong T.A., Spears J.W., Crenshaw T.D., and Nielsen F.H., 2000. Boron supplementation of a diet for weanling pigs improves feed efficiency and bone strength characteristics and alters plasma lipid metabolites. J. Nutr. **130**: 2575–2581.

Bachovchin W.W., Wong W.Y.L., Farr-Jones S., Shenvi A.B., and Kettner C.A., 1988 Nitrogen-15 NMR spectroscopy of the catalytic-triad histidine of a serine protease in peptide boronic acid inhibitor complexes. Biochem. **27**: 7689–7697.

Bai Y. and Hunt C.D., 1996 Absorption and distribution of boron in rats following a single oral administration of boron. Proc. N. D. Acad. Sci. **50**:53.

Bai Y. and Hunt C.D., 1996 Dietary boron enhances efficacy of cholecalciferol in broiler chicks. J. Trace Elem. Exp. Med **9**: 117–132.

Bakken N.A. and Hunt C.D., 2003 Dietary boron decreases peak pancreatic in situ insulin release in chicks and plasma insulin concentrations in rats regardless of vitamin D or magnesium status. J. Nutr. **133**: 3516–3522.

Bennett A., Rowe R.I., Soch N., and Eckhert C.D., 1999 Boron stimulates yeast (Saccharomyces cerevisiae) growth. J Nutr **129**: 2236–2238.

Berezin I.V., Vill K.H., Martinek K., and Yatsimirshii A.K., 1967 Reversible inactivation of a-chymotrypsin resulting from interaction of Cu^{++} ions with the imidazole group of a histidine residue. Molek. Biol. **1**: 719–728.

Bone R., Frank D., Kettner C.A., and Agard D.A., 1989 Structural analysis of specificity: alpha-lytic protease complexes with analogues of reaction intermediates. Biochem. **28**: 7600–7609.

Bonilla I., Garcia-González M., and Mateo P., 1990 Boron requirement in cyanobacteria. Its possible role in the early evolution of photosynthetic organisms. Plant Physiol. **94**: 1554–1560.

Carney G.E. and Bowen N.J., 2004 p24 proteins, intracellular trafficking, and behavior: Drosophila melanogaster provides insights and opportunities. Biol Cell **96**:271–278.

Chen T.S.S., Chang C.-J., and Floss H.G., 1981 On the biosynthesis of boromycin. J. Org. Chem. **46**: 2661–2665.

Chen X., Schauder S., Potier N., Van Dorsselaer A., Pelczer I., Bassler B.L., and Hughson F.M., 2002 Structural identification of a bacterial quorum-sensing signal containing boron. Nature **415**: 545–549.

Clarke W.B., Koekebakker M., Barr R.D., Downing R.G., and Fleming R.F., 1987 Analysis of ultratrace lithium and boron by neutron activation and mass-spectrometric measurement of ^{3}He and ^{4}He. Int. J. Rad. Appl. Instrum. **38**: 735–743.

Expert Consultation WHO/FAO/IAEA. 1996 Trace elements in human nutrition and health Geneva: WHO. p. 343.

Ferrando A.A., Green N.R., Barnes K.W., and Woodward B., 1993 Microwave digestion preparation and ICP determination of boron in human plasma. Biol. Trace Ele. Res. **37**: 17–25.

Food and Nutrition Board: Institute of Medicine. 2001 Dietary reference intakes for vitamin A, vitamin K, arsenic, boron, chromium, copper, iodine, iron, manganese, molybdenum, nickel, silicon, vanadium, and zinc Washington, D.C.: National Academy Press.

Fort D.J., Rogers R.L., McLaughlin D.W., Sellers C.M., and Schlekat C.L., 2002 Impact of boron deficiency on Xenopus laevis: a summary of biological effects and potential biochemical roles. Biol Trace Elem Res **90**: 117–142.

Frieden E., 1984 A survey of the essential biochemical elements. In: Frieden E editor. Biochemistry of the essential ultratrace elements New York: Plenum Press. pp. 1–15.

Gerrard W., 1961 The Organic Chemistry of Boron London: Academic Press. pp. 1–308.

Green G.H. and Weeth H.J., 1977 Responses of heifers ingesting boron in water. J. Anim. Sci. **46**: 812–818.

Greenwood N.N., 1973 Boron. In: Bailar J.J., Emeléus H., Nyholm R., Trotman-Dickenson A editors. Comprehensive Inorganic Chemistry. First ed. Oxford: Pergamon Press Ltd. pp. 665–990.

Greenwood N.N. and Earnshaw A., 1984 Chemistry of the Elements Oxford, Great Britain: Pergamon Press. pp. 155–242.

Hausdorf G., Krüger K., Küttner G., Holzhütter H.-G., Frömmel C., and Höhne W.E., 1987 Oxidation of a methionine residue in subtilisin-type proteinases by the hydrogen peroxide/borate system--an active site-directed reaction. Biochim. Biophys. Acta **952**: 20–26.

Hunt C.D., 1989 Dietary boron modified the effects of magnesium and molybdenum on mineral metabolism in the cholecalciferol-deficient chick. Biol. Trace Elem. Res. **22**: 201–220.

Hunt C.D., 1996 Dietary boron deficiency and supplementation. In: Watson RR editor. Trace Elements in Laboratory Rodents Boca Raton, FL: CRC Press. pp. 229–253.

Hunt C.D., Butte N.F., and Johnson L.K., 2005 Boron concentrations in milk from mothers of exclusively breast-fed healthy full-term infants are stable during the first four months of lactation. J. Nutr. **135**: 2383–2386

Hunt C.D., Friel J.K., and Johnson L.K., 2004 Boron concentrations in milk from mothers of full-term and premature infants. Am J Clin Nutr **80**: 1327–1333.

Hunt C.D., Herbel J.L., and Idso J.P., 1994 Dietary boron modifies the effects of vitamin D_3 nutriture on indices of energy substrate utilization and mineral metabolism in the chick. J. Bone Miner. Res. **9**: 171–181.

Hunt C.D., Herbel J.L., and Nielsen F.H., 1997 Metabolic response of postmenopausal women to supplemental dietary boron and aluminum during usual and low magnesium intake: boron, calcium, and magnesium absorption and retention and blood mineral concentrations. Am. J. Clin. Nutr. **65**: 803–813.

Hunt C.D. and Meacham S.L,. 2001 Aluminum, boron, calcium, copper, iron, magnesium, manganese, molybdenum, phosphorus, potassium, sodium, and zinc: concentrations in common Western foods and estimated daily intakes by infants, toddlers, and male and female adolescents, adults, and seniors in the United States. J. Am. Diet. Assoc. **101**: 1058–1060; Table, http://www.eatright. com/images/journal/ 0901/hunt_tables.pdf.

Hunt C.D. and Nielsen F.H., 1981 Interaction between boron and cholecalciferol in the chick. In: Gawthorne J., White C., editors. Trace Element Metabolism in Man and Animals-4. Canberra: Australian Academy of Science. pp. 597–600.

Hutter R., Keller-Schierlein W., Knusel F., Prelog V., Rodgers G.C. Jr., Suter P., Vogel G., Voser W., and Zahner H., 1967 Stoffwechselprodukte von Mikroorganismen. Boromycin (The metabolic products of microorganisms. Boromycin). Helv. Chim. Acta **50**: 1533–1539.

Ivanov D., Bachovchin W.W., and Redfield A.G., 2002 Boron-11 pure quadrupole resonance investigation of peptide boronic acid inhibitors bound to alpha-lytic protease. Biochemistry **41**: 1587–1590.

Iyengar G.V., Clarke W.B., and Downing R.G., 1990 Determination of boron and lithium in diverse biological matrices using neutron activation-mass spectrometry (NA-MS). Fres. J. Anal. Chem. **338**: 562–566.

Kettner C.A., Bone R., Agard D.A., and Bachovchin W.W., 1988 Kinetic properties of the binding of alpha-lytic protease to peptide boronic acids. Biochem. **27**: 7682–7688.

Kettner C.A. and Shenvi A.B., 1984 Inhibition of the serine proteases leukocyte elastase, pancreatic elastase, cathepsin G, and chymotrypsin by peptide boronic acids. J. Biol. Chem. **259**: 15106–15114.

Kim D.H., Marbois B.N., Faull K.F., and Eckhert C.D., 2003 Esterification of borate with NAD+ and NADH as studied by electrospray ionization mass spectrometry and 11B NMR spectroscopy. J Mass Spectrom **38**: 632–640.

Lanoue L., Taubeneck M.W., Muniz J., Hanna L.A., Strong P.L., Murray F.J., Nielsen F.H., Hunt C.D., and Keen C.L., 1998 Assessing the effects of low boron diets on embryonic and fetal development in rodents using in vitro and in vivo model systems. Biol. Trace Elem. Res. **66**: 271–298.

Litovitz T.L., Klein-Schwartz W., Oderda G.M., and Schmitz B.F., 1988 Clinical manifestations of toxicity in a series of 784 boric acid ingestions. Am. J. Emerg. Med. **6**:209–213.

Loomis W.D. and Durst R.W., 1992 Chemistry and biology of boron. BioFactors **3**: 229–239.

Lovatt C.J. and Dugger W.M., 1984 Boron. In: Frieden E editor. Biochemistry of the Essential Ultratrace Elements New York: Plenum Press. pp. 389–421.

Magour S., Schramel P., Ovcar J., and Mäser H., 1982 Uptake and distribution of boron in rats: interaction with ethanol and hexobarbital in the brain. Arch. Environm. Contam. Toxicol. **11**: 521–525.

Mauras Y., ANg K.S., Simon P., Tessier B., Cartier F., and Allain P., 1986 Increase in blood plasma levels of boron and strontium in hemodialyzed patients. Clin. Chim. Acta **156**: 315–320.

McClendon J.H., 1976 Elemental abundance as a factor in the origins of mineral nutrient requirements. J. Mo. Evol. **8**:175–195.

McLennan A.G., 1992 Dinucleoside phosphates--an introduction. In: McLennan AG editor. Ap_4A and other dinucleoside polyphosphates Boca Raton: CRC Press. pp. 1–9.

Mertz W., 1970 Some aspects of nutritional trace element research. Fed. Proc. **29**: 1482–1488.

Nielsen F.H., 1989 Dietary boron affects variables associated with copper metabolism in humans. In: Anke M, Baumann W, Bräunlich H, Brückner C, Groppel B, Grün M editors. 6th International Trace Element Symposium 1989 Jena, DDR: Karl-Marx-Universitat, Leipzig and Friedrich-Schiller-Universitat. pp. 1106–1111.

Nielsen F.H., Stoecker B.J., and Penland J.O., In press. Boron is a dietary factor for bone microarchitecture and central nervous system function. In: Xu F. editor. The third international symposium onall aspects of plant and animal boron nutrition Wuhan, China: Springer.

O'Neill M.A., Warrenfeltz D., Kates K., Pellerin P., Doco T., Darvill A.G., and Albersheim P., 1996 Rhamnogalacturonan-II, a pectic polysaccharide in the walls of growing plant cells, forms a dimer that is covalently cross-linked by a borate ester. J. Biol. Chem. **271**: 22923–22930.

Park M., Li Q., Shcheynikov N., Zeng W., and Muallem S., 2004 NaBC1 is a ubiquitous electrogenic Na(+)-coupled borate transporter essential for cellular boron homeostasis and cell growth and proliferation. Mol Cell **16**: 331–341.

Parker M.D., Ourmozdi E.P., and Tanner M.J., 2001 Human BTR1, a new bicarbonate transporter superfamily member and human AE4 from kidney. Biochem Biophys Res Commun **282**: 1103–1109.

Pfeiffer C.C., 1949 Is boric acid harmless? Mod. Hosp. **74**: 106–110.

Pfeiffer C.C. and Braverman E.R., 1982 Epochal trace elements and evolution. Agents Action **12**: 412–415.

Ralston N.V.C. and Hunt C.D., 2001 Diadenosine phosphates and S-adenosylmethionine: novel boron binding biomolecules detected by capillary electrophoresis. Biochim. Biophys. Acta **1527**: 20–30.

Ralston N.V.C. and Hunt C.D., 2004 Transmembrane partitioning of boron and other elements in RAW 264.7 and HL60 cell cultures. Biol. Trace Elem. Res. **98**: 181–192.

Raven J.A., 1980 Short- and long-distance transport of boric acid in plants. New Phytol. **84**: 231–249.

Reaven G.M., 1999 Insulin resistance: a chicken that has come to roost. Ann N Y Acad Sci **892**:45–57.

Rossi A.F., Miles R.D., Damron B.L., and Flunker L.K., 1993 Effects of dietary boron supplementation on broilers. Poul. Sci. **72**: 2124–2130.

Rowe R.I. and Eckhert C.D., 1999 Boron is required for zebrafish embryogenesis. J. Exp. Biol. **202**: 1649–1654.

Sato K., Okazaki T., Maeda K., Maeda K., and Okami Y., 1978 New antibiotics, aplasmomycins B and C. J Antibiot (Tokyo) **31**: 632–635.

Schummer D., Irschik H., Reichenbach H., and Höfle G., 1994 Antibiotics from gliding bacteria, LVII. Tartrolons: new boron-containing macrodiolides from Sorangium cellulosum. Liebigs Ann. Chem. **1994**: 283–289.

Seaborn C.D. and Nielsen F.H., 1994 Boron and silicon: effects on growth, plasma lipids, urinary cyclic AMP and bone and brain mineral composition of male rats. Environ. Tox. Chem. **13**: 941–947.

Shkol'nik M.Y. and Il'inskaya N.L., 1975 Effect of a boron deficiency on activity of glucose-6-phosphate dehydrogenase in plants with different boron requirements. Fiziol. Rastenii **22**: 801–805.

Shuler T.R., Pootrakul P., Yarnsukon P., and Nielsen F.H., 1990 Effect of thalassaemia/immunoglobin E disease on macro, trace and ultratrace element concentrations in human tissues. J. Trace Elem. Exp. Med. **3**: 31–43.

Small J.A., Charmley E., Rodd A.V., and Fredeen A.H., 1997 Serum mineral concentrations in relation to estrus and conception in beef heifers and cows fed conserved forage. Can. J. Anim. Sci. **77**: 55–62.

Smith K.W. and Johnson S.L., 1976 Borate inhibition of yeast alcohol dehydrogenase. Biochem. **15**: 560–565.

Spivack A.J. and Edmond J.M., 1987 Boron isotope exchange between seawater and the oceanic crust. Geochim. Cosmochim. Acta **51**: 1033–1043.

Sprietsma J. and Schuitemaker G., 1993 Diabetes can be prevented by reducing insulin production. Med. Hypotheses **42**: 15–23.

Stelter L.H., 1980 Baseline levels of selected trace elements in Colorado oil shale region animals. J. Wildlife Dis. **16**: 175–182.

Takano J., Miwa K., Yuan L., von Wiren N., and Fujiwara T., 2005 Endocytosis and degradation of BOR1, a boron transporter of Arabidopsis thaliana, regulated by boron availability. Proc Natl Acad Sci U S A **102**: 12276–12281.

Takano J., Noguchi K., Yasumori M., Kobayashi M., Gajdos Z., Miwa K., Hayashi H., Yoneyama T., and Fujiwara T., 2002 Arabidopsis boron transporter for xylem loading. Nature **420**: 337–340.

Thellier M., Duval Y., and Demarty M., 1979 Borate exchanges of Lemna minor L. as studied with the help of the enriched stable isotopes and of a (n, alpha) nuclear reaction. Plant Physiol. **63**: 283–288.

Underwood E.J. and Mertz W., 1987 Introduction. In: Mertz W editor. Trace Elements in Human and Animal Nutrition--Fifth Edition San Diego: Academic Press. pp. 1–19.

Usuda K., Kono K., and Yoshida Y., 1997 Serum boron concentration from inhabitants of an urban area in Japan. Biol. Trace Ele. Res. **56**:167–178.

Van Duin M., Peters J.A., Kieboom A.P.G., and Van Bekkum H., 1984 Studies on borate esters I. The pH dependence of the stability of esters of boric acid and borate in aqueous medium as studied by ^{11}B NMR. Tetrahedron **40**: 2901–2911.

Vaziri N.D., Oveisi F., Culver B.D., Pahl M.V., Andersen M.E., Strong P.L., and Murray F.J., 2001 The effect of pregnancy on renal clearance of boron in rats given boric acid orally. Toxicol Sci **60**: 257–263.

Verstraeten S.V., Lanoue L., Keen C.L., and Oteiza P.I., 2005 Relevance of lipid polar headgroups on boron-mediated changes in membrane physical properties. Arch Biochem Biophys **438**:103–110.

Wallace J.M., Hannon-Fletcher M.P., Robson P.J., Gilmore W.S., Hubbard S.A., and Strain J.J., 2002 Boron supplementation and activated factor VII in healthy men. Eur J Clin Nutr **56**: 1102–1107.

Ward N.I., 1987 The determination of boron in biological materials by neutron irradiation and prompt gamma-ray spectrometry. J. Rad. Nuc. Chem. **110**: 633–639.

Ward N.I., 1993 Boron levels in human tissues and fluids. In: Anke M., Meissner D., Mills C., editors. Trace Element Metabolism in Man and Animals-8 Gersdorf, Germany: Verlag Media Touristik. pp. 724–728.

Weser U., 1967 Chemistry and structure of some borate polyol compounds of biochemical interest. In: Jorgensen C., Neilands J., Nyholm R., Reinen D., Williams R., editors. Structure and Bonding New York: Springer-Verlag. pp. 160–180.

Wilson J.H. and Ruszler P.L., 1996 Effects of dietary boron supplementation on laying hens. Br Poult Sci **37**: 723–729.

Zittle C.A.z., 1951 Reaction of borate with substances of biological interest. In: Ford F. editor. Advances in Enzymology New York: Interscience Publishers. pp. 493–527.

Zubay G., 1988 Biochemistry New York: Macmillan.

Dietary Boron: Evidence for a Role in Immune Function

Jerry W. Spears * and Todd A. Armstrong

Department of Animal Science, Box 7621, North Carolina State University, Raleigh, NC 27695-7621.

* Corresponding author: e-mail: Jerry_Spears@ncsu.edu

Introduction

Research in a number of animal species indicates that boron is of nutritional importance and findings continue to support the concept that boron is an essential trace element (Nielsen, 2002). Recent studies indicate that dietary boron affects various immune processes (Hunt and Idso, 1999; Armstrong et al. 2001). The immune system consists of an array of interrelated components that function to protect the host animal against foreign materials, including pathogenic organisms. The immune system can be divided into innate and adaptive or acquired immunity. Innate immunity is non-specific in regard to foreign organisms that it will attack, and consists of physical barriers to organisms such as skin and internal mucous membranes, as well as components that are induced by exposure to foreign material such as phagocytic cells and complement. The innate immune system responds rapidly to invasion of the host by foreign materials; however, its effect is of relatively short duration. The adaptive immune system develops slower in response to attack by various invaders, but is more substained and leads to proliferation of lymphocytes and synthesis of antibodies specifically directed at the organism invading the host. In addition, the adaptive immune system also results in immunological memory that provides long-term immunity against future attacks by the same organism. This paper will review research indicating a role for boron in immune function.

Boron and inflammation

Inflammation is a protective response of tissues to microbial invasion or injury and is an important component of the innate immune system. The inflammatory response is vital for functioning of the immune system because it focuses immune cells and other defense molecules, such as antibodies and complement components, on the local site of infection. Events occurring early during the inflammatory response are centered around dilation of local blood vessels that leads to increased blood flow and permeability of blood vessels. These vascular changes allow phagocytic cells, such as neutrophils and monocytes, and other serum components to enter the affected tissue and limit the spread of infection and tissue damage.

F. Xu et al. (eds.), Advances in Plant and Animal Boron Nutrition, 269–276.

Although inflammation is an essential component of the host defense system, excessive inflammation can result in a number of inflammatory diseases. Osteoarthritis is a major inflammatory disease in humans and also in some animals. Newnham (2002) reported at the last Boron Symposium that boron supplementation can alleviate arthritic signs in some humans. In a double-blind study, twenty patients with confirmed osteoarthritis were given 6 mg of boron per day or a placebo for 8 weeks (Fracp et al. 1990). Five of the seven patients in the boron-supplemented group that completed the study reported improvements in subjective measures of their arthritic condition. In the placebo group only one of eight patients that completed the study reported improvements during the 8 week period.

Studies with animals also indicate that dietary boron reduces local inflammatory response (Hunt and Idso, 1999; Armstrong et al. 2001; Armstrong and Spears, 2003). Increasing dietary boron from 0.1 to 2.1 mg/kg diet reduced paw swelling in rats following intradermal injection of M. butyricum to induce arthritis (Hunt and Idso, 1999). Boron supplementation (5 mg/kg diet) to a diet containing approximately 2 mg boron/kg decreased localized swelling response following intradermal administration of phytohemagglutinin (PHA) in pigs (Armstrong et al. 2001; Armstrong and Spears, 2003). Research in chicks has showed that PHA causes an inflammatory response shortly after intradermal injection due to infiltration of neutrophils and macrophages (McCorkle et al. 1980). The reduced inflammatory response in boron-supplemented pigs was evident by 6 hours and the swelling response continued to be diminished by boron through 48 hours post PHA injection (Figure 1).

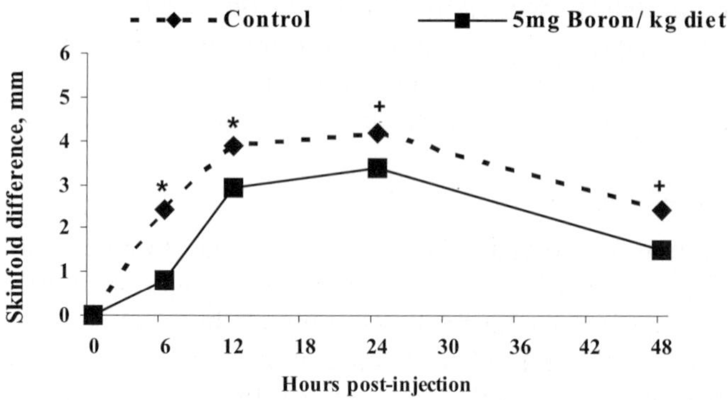

Figure 1. Effect of dietary boron on the inflammatory response in pigs following intradermal injection of phytohemagglutinin

A number of organic synthetic boron compounds including amine cyanoboranes, amine carboxyboranes, and boron analogues of peptides have also been shown to exhibit anti-inflammatory activity when given intraperitoneal in mice (Hall et al. 1980; Sood et al. 1990). Inflammation was induced in these studies by injection of carrageenin in to the hind foot.

The inflammatory process is also critical for normal repair and healing of damaged tissue. Monocytes and macrophages activated by the inflammatory response will not only phagocytose invading microorganisms, but also destroy damaged cells (including neutrophils) and tissues. Macrophages release collagenases and elastases that destroy damaged connective tissue. However, macrophages are also responsible for releasing angiogenic and growth factors that activate and promote proliferation of fibroblast that synthesize and secrete matrix constituents of connective tissues such as collagen and proteoglycans.

Treatment with boric acid has been reported to improve wound healing (Dousset et al. 2000). In nude mice surgically implanted with sponges on their back, boric acid treatment increased release of angiogenic factors, including heat shock protein 70, tumor necrosis factor-alpha (TNF-α) and vascular endothelium growth factor (Dousset et al. 2000). They suggested that boron improves wound healing by promoting synthesis and release of angiogenic factors. Human fibroblasts grown in culture increased release of TNF-α and TNF-α mRNA in response to boric acid addition (Benderdour et al. 1998). Boric acid addition to fibroblasts decreased synthesis of extracellular matrix micromolecules and increased release of collagen and proteoglycans into the culture medium (Benderdour et al. 1998).

The mechanism(s) whereby boron modifies the inflammatory response has not been defined. A number of possible modes in which boron may alter inflammatory processes have been suggested based largely on indirect research findings (Hunt and Idso, 1999).

Early in the inflammatory response phagocytic cells, primarily neutrophils, are recruited to the site of infection where they attempt to destroy the invading organisms. Tissue macrophages are later activated and they serve to destroy not only microorganisms, but also damaged tissue and dying neutrophils. The process of destroying invading organisms or other foreign materials involves the generation of reactive oxygen (superoxide radical, hydrogen peroxide, hydroxyl radical) and nitrogen species (nitric oxide) that are toxic to the invading organism. However, reactive species generated during phagocytosis can contribute to tissue damage during chronic inflammation (Kaur and Halliwell, 2000). It has been suggested that boron may reduce the production of NADPH, the major electron donor for production of reactive oxygen species during phagocytosis (Hunt and Idso, 1999). This hypothesis was based on studies in plants, indicating that boron inhibits 6-phosphogluconate dehydrogenase, a key enzyme involved in the production of NADPH via the pentose-phosphate pathway.

Boron may also reduce tissue damage from inflammation by hastening the destruction of reactive oxygen species by increasing activities of key antioxidant enzymes (Hunt and Idso, 1999). A recent study (Pawa and Ali, 2004) supports this hypothesis. In this study, rats were administered orally 0.07 or 0.45 mg boron/kg body weight for 3 days prior to injection of thioacetamide. Thioacetamide induces liver necrosis that is at least partly due to increased production of reactive oxygen species. Boron appeared to reduce liver damage in rats given thioacetamide, based on serum enzymes and liver lipid peroxidation measurements used to assess liver injury. Activities of the antioxidant enzymes, glutathione peroxidase and catalase, were higher in rats dosed with boron following administration of thioacetamide. Interestingly, boron did not affect antioxidant enzymes in control rats not given thioacetamide (Pawa and Ali, 2004). These results suggest that boron increases activity of certain antioxidant enzymes during periods of elevated production of reactive oxygen species. Chronic inflammation is a situation where enhanced production of reactive oxygen radicals could limit the ability of the antioxidant defense system to rapidly destroy these species.

Boron and cytokines

Cytokines are proteins produced and secreted primarily by immune cells that regulate immune responses by promoting communication among various cell types. Over 50 different cytokines have been isolated and many cytokines interact by either inducing or suppressing the expression of other cytokines (Aggarwal and Puri, 1995). Cytokines are also responsible for many of the clinical signs of infectious diseases, including fever and anorexia.

Recently the effect of dietary boron on production of certain cytokines was studied in pigs following intramuscular injection of E. coli lipopolysaccharide (LPS; Armstrong and Spears, 2003). Lipopolysaccharide, also known as endotoxin, is produced by gram-negative bacteria and stimulates immune processes, and production and release of various cytokines. Within 2 hours after LPS administration, serum TNF-α concentrations were greatly elevated (Table 1). Supplementation of 5 mg boron/kg to the control diet (contained 2 mg boron/kg) increased serum concentrations of TNF-α following LPS injection. Pigs were given 100 mg LPS/kg body weight in experiment 1 and 25 mg LPS/kg body weight in experiment 2. It is well documented that LPS causes large quantities of TNF-α to be released, primarily from macrophages due to upregulated TNF gene transcription and translation (Wang and Tracey, 1999). Although TNF-α is considered a pro-inflammatory cytokine, it produces a wide array of biological effects, including induction of some cytokines that are anti-inflammatory. It also stimulates the production of certain acute phase proteins that can scavenge oxygen radicals, inactivate serine proteases and enhance wound healing. Cell culture studies with human fibroblasts (Benderdour et al. 1998) and chick embryo cartilage (Benderdour et al. 1997) also support a role for boron in TNF-α release. In both of these cell lines boric acid addition increased TNF-α release by cells.

Table 1. Effect of dietary boron and lipopolysaccharide (LPS) on serum concentrations of tumor necrosis factor-α in pigs[a]

Time	+ LPS		- LPS	
	Control	5 mg B/kg	Control	5 mg B/kg
	pg/mL			
Experiment 1[b]				
0 h	32	87	30	44
2 h[c,d,e]	2,190	6,834	29	45
6 h	39	115	28	44
24 h	32	113	30	46
Experiment 2[f]				
0 h[c]	35	40	39	54
2 h[e]	1,962	2,948	39	57
6 h[e]	40	48	37	52
24 h[e]	35	36	36	44

[a]From Armstrong and Spears, 2003; [b]Pigs in LPS treatments were given 100 μg LPS/kg BW; [c]Boron effect ($P < 0.01$); [d]Boron x LPS ($P < 0.01$); [e]LPS effect ($P < 0.01$).; [f]Pigs in LPS treatments were given 25 μg LPS/kg BW.

Dietary boron also appeared to increase production of interferon-gamma (IFN-γ) in pigs after LPS administration (Table 2; Armstrong and Spears, 2003). This cytokine is important in both innate and adaptive immunity. It is important in activation of macrophages, cytotoxic T-lymphocytes, and natural killer cells. In macrophages, IFN-γ induces nitric oxide synthase which produces nitric oxide (Bach et al. 1999). Nitric oxide generated by this enzyme has antiviral activity.

Organic boron compounds have also been shown to alter production of cytokines. A number of amine carboxyboranes reduced in vitro release of IL-1 and TNF-α from macrophages (Hall et al. 1994). However, intraperitoneal administration of amine carboxyboranes in mice increased plasma concentrations of TNF-α and reduced IL-2 following LPS injection (Hall et al. 1994).

Born and adaptive immunity

Limited research has evaluated the effect of dietary boron on adaptive immune responses. Bai et al. (1997) evaluated antibody production following vaccination with human typhoid vaccine in rats fed a low boron diet (0.15 mg/kg) or 1.75 mg supplemental boron/kg diet. Anti-typhoid IgM and IgG concentrations were lower in rats receiving the low boron diet at 30 days post vaccination. However, addition of 5 mg boron/kg to a control diet containing 2.0 mg boron/kg did not affect antibody production in pigs injected with sheep red blood cells (Armstrong et al. 2001). In vitro blastogenic response of lymphocytes isolated from pigs to mitogen stimulation was also not affected by dietary boron (Armstrong et al. 2001). The discrepancy between the findings of Bai et al. (1997) and Armstrong et al. (2001) may relate to differences in the boron content of the control diet (0.15 vs 2.0 mg/kg). If the control diet fed to pigs had been lower in boron, perhaps responses in adaptive immunity to boron supplementation would have been observed.

Table 2. Effect of dietary boron and lipopolysaccharide (LPS) on serum concentrations of interferon-gamma in pigs[a]

	+ LPS		- LPS	
Time	Control	5 mg B/kg	Control	5 mg B/kg
	pg/mL			
Experiment 1[b]				
0 h	8	3	3	1
2 h[c]	238	263	4	1
6 h[c,d,e]	135	298	5	1
24 h[c]	19	51	4	2
Experiment 2[f]				
0 h[d]	6	20	4	18
2 h[c]	237	451	2	5
6 h[c]	78	191	4	9
24 h[c]	13	25	2	3

[a]From Armstrong and Spears, 2003; [b]Pigs in LPS treatments were given 100 μg LPS/kg BW; [c]LPS effect ($P < 0.01$); [d]Boron effect ($P < 0.01$); [e]Boron x LPS ($P < 0.01$); [f]Pigs in LPS treatments were given 25 μg LPS/kg BW.

Conclusions

The immune system is a complex network of various mechanisms and molecules that act and interact to protect the host from foreign organisms. Evidence suggests that boron may play a role in the immune response through effects on inflammation, cytokine production, and adaptive immunity. However, the exact function and mechanism of action for boron within the immune system has not been determined, and further research is warranted to understand the role of boron within the complex immune system.

References

Aggarwal B.B., and Puri R.K., 1995 Common and uncommon features of cytokines and cytokine receptors: An overview. In Human cytokines: Their role in disease and therapy (Aggarwal B.B., and Puri R.K., eds.). pp.1–24. Blackwell Science, Cambridge, MA.

Armstrong T.A., Spears J.W., and Lloyd K.E., 2001 Inflammatory response, growth, and thyroid hormone concentrations are affected by long-term boron supplementation in gilts. J. Anim. Sci. **79**, 1549–1556.

Armstrong T.A., and Spears J.W., 2003 Effect of boron supplementation of pig diets on the production of tumor necrosis factor-α and interferon-γ. J. Anim. Sci. **81**, 2552–2561.

Bach E.A., Kaplan D.H., and Schreiber R.D., 1999 Biochemistry, mechanism of action, and biology of the interferons. In Inflammation: basic principles and clinical correlates (Gallin J I and Snyderman R eds.). Lippincott Williams & Wilkins, Philadelphia, PA. pp. 487–503.

Bai Y., Hunt C.D., and Newman S.M., 1997 Dietary boron increases serum antibody (IgG and IgM) concentrations in rats immunized with human typhoid vaccine. 51 pp.181. Proc. North Dakota Acad. Sci.

Benderdour M., Hess K., Gadet M.D., Dousset B., Nabet P., and Belleville F., 1997 Effect of boric acid solution on cartilage metabolism. Biochem. Biophys. Res. Comm. **234**, 263–268.

Benderdour M., Hess K., Gadet M.D., Nabet P., Belleville F., and Dousset B., 1998 Boron modulates extraxcellular matrix and TNFα synthesis in human fibroblasts. Biochem. Biophys. Res. Comm. **246**, 746–751.

Dousset B., Benderdour M., Hess K., Mayap-Nzietchueng R., Belleville F., and Duprez A., 2000 Effects of boron in wound healing. In Trace elements in man and animals 10 (Rousse R.M., Anderson R.A., and Tavier A.E., eds.). pp. 1061–1065. Kluwer Academic/Plenum Publishers, New York.

Fracp R.L.T., Rennie G.C., and Newnham R.D., 1990 Boron and arthritis: the results of a double-blind pilot study. J. Nutr. Med. 1, 127–132.

Hall I.H., Starnes C.O., McPhail A.T., Wisian-Neilson P., Das M.K., Harchelroad F., and Spielvogel B.F., 1980 Anti-inflammatory activity of amine cyanoboranes, amine carboxyboranes, and related compounds. J. Pharm. Sci. 69, 1025–1029.

Hall I.H., Chen S.Y., Rajendran K.G., Sood A., Spielvogel B.F., and Shih J., 1994 Hypolipidemic, anti-obesity, anti-inflammatory, anti-osteoporotic, and anti-neoplastic properties of amine carboxyboranes. Environ. Health Perspect **102** (Suppl 3), 21–30.

Hunt C.D., and Idso J.P., 1999 Dietary boron as a physiological regulator of the normal inflammatory response: a review and current research progress. J. Trace Elements Exp. Med. **12**, 221–233.

Kaur H., and Halliwell B., 2000 Evolving measurements of radical products – how much do they tell us about inflammation. In Free radicals and inflammation (Winyard P.G., Blake R. and Evans C.H., eds.). Birkhauser Verlag, Basel, Switzerland. pp. 221–239.

McCorkle F., Olah I., and Glick B., 1980 The morphology of the phytohemagglutinin-induced cell response in the chicken's wattle. Poultry Sci. **59**, 616–623.

Newnham R.E., 2002 How boron is being used in medical practice. In Boron in plant and animal nutrition (Goldbach H.E., Rerkasem B., Wimmer M.A., Brown P.H.,

Thellier M., and Bell R.W., eds.). pp. 59–62. Kluwer Academic/Plenum Publishers, New York.

Nielsen F.H., 2002 The nutritional importance and pharmacological potential of boron for higher animals and human. In Boron in plant and animal nutrition (Goldbach H.E., Rerkasem B., Wimmer M.A., Brown P.H., Thellier M., and Bell R.W., eds.). pp. 37–49. Kluwer Academic/Plenum Publishers, New York.

Pawa S. and Ali S., 2004 Borax and boric acid provide different degrees of protection in liver necrosis. Trace Elements and Electrolytes **21**, 83–88.

Sood A.l., Sood C.K., Spielvogel B.F., and Hall I.H., 1990 Boron analogues of amino acids. VI. Synthesis and characterization of di-and tripeptide analogues as antineoplatic, anti-inflammatory and hypolipidemic agents. Eur. J. Med. Chem. **25**, 301–308.

Wang H. and Tracey K.J., 1999 Tumor necrosis factor, interleukin-6, macrophage migration inhibitory factor, and macrophage inflammatory protein-1 in inflammation. In Inflammation: basic principles and clinical correlates (Gallin J I and Snyderman R, eds.). Lippincott Williams & Wilkins, Philadelphia, PA. pp. 471–486.

Boron as a Dietary Factor for Bone Microarchitecture and Central Nervous System Function*

Forrest H. Nielsen[1,3], Barbara J. Stoecker[2] and James G. Penland[1]

[1]U.S. Dept. of Agriculture, Agricultural Research Service**, Grand Forks Human Nutrition Research Center, Grand Forks, ND, 58202-9034. [2]Nutritional Sciences, Oklahoma State University, Stillwater, OK 74078-6141.

[3]Correspondence, e-mail: fnielsen@gfhnrc.ars.usda.gov

*Mention of a trademark or proprietary product does not constitute a guarantee or warranty of the product by the U.S. Department of Agriculture and does not imply its approval to the exclusion of the products that may also be suitable.
**The U.S. Department of Agriculture, Agricultural Research Service, Northern Plains Area, is an equal opportunity/affirmative action employer and all agency services are available without discrimination.

Introduction

Studies by several different research groups using different experimental animals indicate that nutritional amounts of boron beneficially affect bone histomorphological and gross physical characteristics. One of the first studies suggesting that boron is essential for higher animals found that boron improved bone calcification in chicks fed a diet deficient but not completely lacking in vitamin D (Hunt and Nielsen, 1981). At the microscopic level, boron deprivation (0.465 mg/kg diet) exacerbated the distortion of marrow sprouts (location of calcified scaffold erosion and new bone formation) and the delay in initiation of cartilage calcification in bones during marginal vitamin D deficiency (Hunt, 1996). Boron deprivation alone decreased chondrocyte density in the zone of proliferation in the bone growth plate (Hunt, 1996). Nutritional amounts of boron (5 mg/kg) supplemented to a basal AIN-76 diet containing 0.3 mg boron/kg enhanced the beneficial effects of 17β-estradiol on trabecular bone volume and plate density in tibias of ovariectomized rats (Sheng et al. 2001). Low dietary boron (0.98 mg/kg compared to 5 mg/kg) decreased the bone strength variable bending moment in femurs of pigs (Armstrong et al. 2000). In a factorial arranged experiment where the variables were dietary boron at 0.07 or 3 mg/kg, dietary oil as either canola or palm oil at 75 g/kg and sex, femur bending moment was decreased by boron deprivation in female rats, particularly when they were fed canola oil (Nielsen, 2004). Femur breaking stress also was decreased by boron deprivation (most markedly in females) when the diet contained canola oil. Boron deprivation (0.6 mcg/kg instead of the usual 310 mcg/kg diet) of the African clawed frog (Xenopus laevis) resulted in abnormal limb development in offspring (grown in water containing 0.6 mcg boron/L instead of the usual 100 mcg boron/L) (Fort et al. 2002).

F. Xu et al. (eds.), Advances in Plant and Animal Boron Nutrition, 277–290.

Supanutritional amounts of boron also may be beneficial to bone mechanical properties and microarchitecture. A boron supplement of 5 mg/kg to a diet containing 9.4 mg boron/kg increased bone strength in chicks (Rossi et al. 1993). In rats exposed to strenuous treadmill exercise, femur and vertebra bone mineral content and density were decreased and trabecular separation was increased. A supplement of 50 mg boron/kg diet increased bone mineral content and density, trabecular volume, and trabecular thickness in the exercised rats (Rico et al. 2002).

Brain function and eye histomorphology also are affected by dietary boron. In rats, boron deprivation resulted in decreased brain electrical activity similar to that observed in nonspecific malnutrition (Penland, 1998). In humans, boron supplementation after boron deprivation yielded changes in encephalograms that suggested improved behavior activation (e.g., less drowsiness) and mental alertness, improved psychomotor skills of motor speed and dexterity, and elicited improvements in the cognitive processes of attention and short-term memory (Penland, 1998). Dietary boron modified the effect of replacing dietary palm oil with canola oil on eye mitochondrial morphology of rats (Nielsen et al. 2004). In a factorial arranged experiment where the variables were dietary boron at 0.07 or 3 mg/kg and dietary oil as either canola or palm oil at 75 g/kg, rats fed the boron-supplemented diet with canola oil had the lowest number of rod inner segment mitochondria with a high abundance of crystal folds. Rats fed the boron-supplemented diet with palm oil had the lowest number of hydropic (swollen) mitochondria. In zebrafish, boron deficiency induced photoreceptor dystrophy; the photoreceptor cells were shortened because of reduction in myoid and outer segment regions (Eckhert and Rowe, 1999). These changes apparently were the reason boron-deficient zebrafish developed photophobia.

A possible explanation for the broad and varied response to changes in dietary boron is that it has a role that influences the physicochemical characteristics of cell membranes. This influence may alter the activity of membrane-bound enzymes such as oxidoreductase enzymes that have the ribose moiety (which binds boron) of NAD^+, and the response to hormones and cytokines or transmembrane signaling. The long-chain omega-3 polyunsaturated fatty acids (PUFA), eicosapentaenoic acid (EPA) and docosahexaenoic acid (DHA), that have broad and varied beneficial effects, including on bone turnover and neurological function, are also hypothesized to act through influencing the physicochemical characteristics of cell membranes. Thus, EPA and DHA may affect cell-to-cell interactions and the expression of various receptors for hormones and cytokines. Cell membrane changes may explain the findings that EPA supplementation of ovariectomized rats inhibited bone loss (Sakaguchi et al. 1994) and enhanced the ability of estrogen to inhibit bone loss (Schlemmer et al.

1999). DHA has been shown to promote visual acuity (e.g., brightness discrimination) and development in rats (Okuyama et al. 2001) and humans (Hoffman, 2000). Also, diets enriched with the long-chain omega-3 PUFA enhanced cognitive functions in rats (Becker and Kyle, 2001; Takeuchi et al. 2002; Moriguchi and Salem, 2003).

Because dietary boron and long-chain omega-3 PUFA affect a large number of the same biological processes apparently through influencing the responses to hormones and cytokines at the cell membrane level, a reasonable hypothesis is that the fatty acid composition of the diet will alter the response to boron deprivation, or vice versa, boron deprivation will affect the response to diets containing differing amounts of omega-3 PUFA. A corollary to this hypothesis is that a finding that the omega-3 PUFA content of the diet affects the response to boron deprivation would support the contention that boron is beneficially bioactive at the cell membrane level. This hypothesis and corollary were the stimulus for the study described below.

Materials and methods

Female Sprague-Dawley rats (Charles River/SASCO, Wilmington, MA) weighing about 115 grams were fed diets in a factorial arrangement with variables being supplemental boron as boric acid at 0 or 3 mg/kg and either 75 g safflower oil/kg or 65 g fish (menhaden) oil plus 10 g linoleic acid/kg of diet. The composition of the ground corn-casein based diet has been reported (Nielsen, 2004). The basal diet contained about 0.10 mg boron/kg as determined by inductively coupled plasma atomic emission spectroscopy after low-temperature, acid-digestion in Teflon tubes (Hunt and Shuler, 1990). Standard reference material (National Institute of Standards and Technology, Gaithersburg, MD) #1515 (apple leaves) was used for quality control purposes in the diet analysis. The diets were not pelleted and were stored at -16°C in tightly capped plastic containers.

The females were housed individually in double stainless steel cages in a room maintained at 23℃ and 50% relative humidity with a normal 12 h light and dark cycle. After 5 weeks on their respective diets and deionized water (Super Q, Millipore, Bedford, MA) provided in plastic food and water cups, six females in each treatment were bred by males that were fed a commercial rat chow. Dams and pups continued having free access to their respective diets and deionized water through gestation, lactation and post-weaning. At about age 21 days, 15 male pups from each dietary group were placed individually in single stainless steel cages in a room maintained at 23°C and 50% relative humidity and with a reversed 12 h white and red light cycle. Absorbent paper under the wire mesh cages were changed daily. Rats were weighed and provided clean cages weekly. Repeated humane handling of the males after weaning was performed to minimize distress caused by behavioral testing.

At 6 and 19 weeks of age, behavioral reactivity, more specifically anxiety, was evaluated with the elevated plus-maze procedure described by Pellow et al. (1985). The apparatus consisted of a plus-shaped maze constructed of white opaque Plexiglas with two open arms opposite each other, crossed by two enclosed arms; the maze was elevated 50 cm from the floor. More entries into the open arms indicate a rat is less anxious about falling from the maze. Numbers of entries and times spent in open and closed arms of the maze, and total numbers of arm entries (movements), were determined during two consecutive 5-min trials.

At 15 weeks of age, visual function was evaluated with a brightness discrimination procedure adapted from Tang and Ho (1998). The apparatus consisted of an enclosed Y-shaped maze constructed of black opaque Plexiglas with a black lid covering one arm (dark arm) and transparent lids covering the other two arms (light arms). Rats were placed in one of the light arms and entry into either the dark or other light arm was recorded. Number of entries and times spent in light and dark arms, and total number of arm entries (movements), were determined during six consecutive 1-minute trials; light and dark arms and starting arm were varied across trials. Ambient lighting was set to 10 or 100 lux to ascertain any dietary differences in visual ability to distinguish light and dark arms (brightness discrimination) as indicated by a preference for either the light or dark arm.

Eighteen weeks after weaning, the male rats were anesthetized with ether for the collection of blood from the vena cava with a heparin-coated syringe and needle. After euthanasia by decapitation, the brain was immediately removed and quickly frozen in liquid nitrogen. The left femur and vertebra with some attached flesh were collected for measuring bone physical characteristics. The right femur with all flesh removed was collected for mineral analysis. Bones, brains and plasma were stored at -70°C until analysis.

The study was approved by the Animal Care Committee of the Grand Forks Human Nutrition Research Center, and the lawfully acquired animals were maintained in accordance with NIH guidelines for the care and use of laboratory animals.

Plasma 8-iso-prostaglandin $F_{2\alpha}$ (8-iso-$PGF_{2\alpha}$) was determined by using a Correlate-EIA^{TM} Immunoassay Kit (#900-010, Assay Designs, Ann Arbor, MI). For mineral analysis, femurs were cleaned to the periosteal surface with cheese cloth. Both the femurs and brains were lyophilized then subjected to a wet-ash, low-temperature digestion in Teflon tubes (Hunt and Shuler, 1990). Boron, calcium, copper, iron, magnesium and zinc were determined by coupled argon plasma atomic emission spectroscopy. Standard reference material (National

Institute of Standards and Technology, Gaithersburg, MD) #1515 (apple leaves) was used for quality control.

The bone strength variable maximum force (force in kg needed to break the bone) was determined on the right femur after thawing and removing the remaining flesh. A custom-designed and –built apparatus that performed a three-point bending test the same as that performed by commercially available machines was used as described previously (Nielsen, 2004).Vertebra microarchitecture of seven randomLy selected rats from each treatment was examined by microcomputed tomography (μCT40, Scanco Medical AG, Zurich, Switzerland). The vertebrae were scanned from the proximal to distal growth plate. Contours were placed on a volume of interest beginning and ending 10 slices (20.67 μm thickness per slice) away from the growth plate in order to include in the volume of interest only the secondary spongiosa within the two growth plates. Bone morphometric variables including trabecular bone volume fraction (BV/TV) and trabecular thickness, in addition to structural model index (SMI) and connectivity density were determined. SMI indicates whether the trabeculae are more plate-like or rod-like.

Data were statistically compared by using two-way analysis of variance (SAS/STAT, Version 9.02, SAS Institute, Inc., Cary, NC) followed by Tukey's contrasts when appropriate. A p value of <0.05 was considered significant.

Results

Boron deprivation significantly depressed final weight and femur boron concentration (Table 1); dietary oil did not affect either of these variables. Both boron deprivation and safflower oil instead of fish oil in the diet increased the plasma concentration of 8-iso-$PGF_{2\alpha}$, an indicator of increased oxidative stress (Table 1).

Femur strength was reduced by both boron deprivation and feeding safflower oil instead of fish oil. As a result, the maximum force to break the femur was lower in the boron-deficient rats fed safflower oil than in boron-supplemented rats fed fish oil (Table 1). Neither dietary oil nor boron affected the concentration of calcium in the femur (Table 2). However, the concentrations of mineral elements associated with the organic matrix were altered by the dietary treatments (Table 2). Boron deprivation decreased the concentrations of iron and magnesium, and safflower oil compared to fish oil increased the concentration of iron in femur. An interaction between dietary oil and boron affected the femur copper concentration. The femur copper concentration was lower in boron-deficient than boron-supplemented rats with the effect most marked when the diet contained fish oil.

Table 1. Effect of dietary boron, oil, and their interaction on boron status indicators

Diet* Boron (mg/kg)	Oil	Final Weight (g)	Plasma 8-iso-$PGF_{2\alpha}$ (ng/mL)	Femur Boron (ng/g)	Femur Maximum Force (Newtons)
0	Fish	447±7**	10.6±1.1	190±28	153±4
3	Fish	476±7	8.2±0.6	978±81	165±4
0	Safflower	441±9	12.7±0.6	115±26	146±4
3	Safflower	487±9	11.8±0.6	920±96	155±4
Analysis of variance – p Values					
Boron		0.0001	0.04	<0.0001	0.01
Oil		0.72	0.0005	0.31	0.03
Boron x Oil		0.28	0.31	0.90	0.79

* The dietary treatments were boron supplements of 0 or 3 mg/kg and either safflower oil at 75 g/kg or fish oil at 65 g/kg plus linoleic acid at 10 g/kg. The basal diet contained about 0.1 mg boron/kg. **Mean ±SE.

Table 2. Effect of dietary boron, oil and their interaction on femur mineral concentrations

Diet*		Femur			
Boron (mg/kg)	Oil	Calcium (g/kg)	Copper (mg/kg)	Iron (mg/kg)	Magnesium (mg/kg)
O	Fish	188±1**	0.84±0.05[a#]	76±4	3308±62
3	Fish	189±2	1.47±0.09[b]	122±6	4292±170
0	Safflower	189±3	1.04±0.05[a]	108±4	3436±96
3	Safflower	191±2	1.35±0.09[b]	136±8	3948±215
Analysis of Variance – p Values					
Boron		0.46	<0.0001	<0.0001	<0.0001
Oil		0.52	0.63	0.0002	0.48
Boron x Oil		0.89	0.04	0.10	0.12

* The dietary treatments were boron supplements of 0 or 3 mg/kg and either safflower oil at 75 g/kg or fish oil at 65 g/kg plus linoleic acid at 10 g/kg. The basal diet contained about 0.1 mg boron/kg.

**Mean ±SE. #Values in the same column not followed by the same letter are significantly different ($p < 0.05$) according to Tukey's contrasts.

The data in Table 3 show that vertebral microarchitecture was altered by the dietary treatments. Trabecular thickness was greater in boron-supplemented than boron-deficient rats. Fish oil compared to safflower oil tended ($p < 0.06$) to increase trabecular thickness. The interaction between dietary boron and oil differently affected trabecular connectivity and SMI. Trabecular connectivity was increased by boron deficiency in rats fed the safflower oil; dietary boron had no effect on this variable in rats fed fish oil. SMI was significantly decreased (more plate-like than rod-like) by feeding fish oil instead of safflower oil to the boron-supplemented rats; dietary oil did not significantly affect SMI in the boron-deficient rats. BV/TV was highest in the boron-supplemented rats fed fish oil and lowest in boron-supplemented rats fed safflower oil, but this difference was not significant (boron x oil interaction < 0.07).

Table 3. Effect of dietary boron, oil and their interaction on vertebral trabecular microarchitecture

Diet[*]		Trabecular			
Boron (mg/kg)	Oil	Thickness (mm)	Connection Density (1/mm³)	Structural Model Index	Bone volume fraction (BV/TV)
0	Fish	0.082±0.002[**]	44.8±4.7[ab#]	1.12±0.12[ab]	0.211±0.011
3	Fish	0.085±0.002	46.7±2.4[ab]	0.93±0.05[a]	0.230±0.009
0	Safflower	0.078±0.001	57.4±3.5[a]	1.04±0.12[ab]	0.216±0.010
3	Safflower	0.083±0.001	41.5±2.6[b]	1.31±0.05[b]	0.199±0.007
Analysis of Variance – p Values					
Boron		0.04	0.05	0.65	0.94
Oil		0.06	0.29	0.12	0.16
Boron x Oil		0.54	0.02	0.02	0.07

[*] The dietary treatments were boron supplements of 0 or 3 mg/kg and either safflower oil at 75 g/kg or fish oil at 65 g/kg plus linoleic acid at 10 g/kg. The basal diet contained about 0.1 mg boron/kg.

[**]Mean ±SE; n=7/group.

[#]Values in the same column not followed by the same letter are significantly different ($p < 0.05$) according to Tukey's contrasts.

Table 4 shows that neurological function and brain composition were affected by the dietary treatments. In the brightness discrimination test, boron-deficient rats made more entries into the 100 lux light arm of the test apparatus than did the boron-supplemented rats. The boron effect occurred mainly when the diet contained fish oil. Boron had no apparent effect when the diet contained safflower oil. In the plus maze, the boron-deficient rats entered the open arms more often than the boron-supplemented rats fed fish oil. When safflower oil was fed, there was a tendency for boron to have an opposite effect. Interestingly, the highest brain boron concentration was found in rats fed the boron-deficient, fish oil diet. Unlike with femur, boron deprivation did not decrease the boron concentration in brain. The rats fed the boron-deficient fish oil diet also had the lowest brain copper and zinc concentrations.

Table 4. Effect of dietary boron, oil and their interaction on selected brain mineral concentrations and brightness discrimination and plus maze variables

Diet*		Brain			Lux 100	Maze
Boron	Oil	Boron	Copper	Zinc	Light Entries	Open Arm Entries
(mg/kg)		(ng/g)	(mcg/g)	(mcg/g)		
0	Fish	323±10**	9.0±0.2[a#]	52±1[a]	1.2±0.07	5.0±0.6[a]
3	Fish	229±14	11.9±0.2[b]	66±1[b]	0.9±0.05	3.7±0.4[b]
0	Safflower	233±23	11.2±0.4[b]	60±1[c]	1.1±0.06	3.5±0.3[b]
3	Safflower	204±26	12.1±0.3[b]	65±1[b]	1.1±0.06	4.5±0.5[ab]
Analysis of Variance – p Values						
Boron		0.002	<0.0001	<0.0001	0.004	0.68
Oil		0.004	0.0001	0.006	0.53	0.40
Boron x Oil		0.09	0.0005	0.0002	0.07	0.02

* The dietary treatments were boron supplements of 0 or 3 mg/kg and either safflower oil at 75 g/kg or fish oil at 65 g/kg plus linoleic acid at 10 g/kg. The basal diet contained about 0.1 mg boron/kg.

**Mean ±SE.

#Values in the same column not followed by the same letter are significantly different ($p < 0.05$) according to Tukey's contrasts.

Discussion

Finding markedly lower boron concentrations in the femur of rats fed the low-boron diet indicates that they were boron-deficient. Further evidence that the rats fed the low-boron diet were deficient is that they exhibited depressed final weights and increased plasma 8-iso-$PGF_{2\alpha}$ (an indication of increased oxidative stress). Fish oil compared to safflower oil did not affect these responses to boron deprivation but did decrease plasma 8-iso-$PGF_{2\alpha}$. The decrease was independent of boron deprivation. However, the lowest plasma concentration of 8-iso-$PGF_{2\alpha}$ was found in boron-supplemented rats fed fish oil, which indicates that fish oil and boron were complementary in reducing oxidative stress. This finding plus the lack of a significant interaction affecting weight and 8-iso-$PGF_{2\alpha}$ suggest that although boron and long-chain omega-3 PUFA affect many similar biological processes (see Introduction), they apparently do this through different biochemical mechanisms and not through directly affecting the metabolism of each other. Nonetheless, the numerous significant interactive findings in the present study indicate that boron has in vivo effects that can alter some responses to changes in the dietary intake of omega-3 PUFA, and vice versa.

The 8-iso-$PGF_{2\alpha}$ data suggest that some of the interaction between dietary boron and fatty acid composition may occur because both affect redox metabolism. That is, changing oxidative stress by one of these bioactive substances may change the magnitude of the oxidative metabolism response to the other. Support for this suggestion is that boron deprivation increased plasma glutathione in rats when their diet fat source was palm oil (Nielsen, 2004), fish oil or safflower oil (Nielsen, 2005), but had no effect when diet fat source was canola oil (Nielsen, 2004). Hunt and Idso (1999) have reviewed the evidence suggesting that boron hastens the destruction of reactive oxygen species that are scavenged and destroyed by defense mechanisms employing glutathione, superoxide dismutase and catalase. Among the evidence was the finding that boron deprivation decreased erythrocyte superoxide dismutase in humans (Nielsen, 1996). DHA has been suggested to be a biodevice to combat oxidative stress in the brain (Yavin et al. 2002). This finding is consistent with the 8-iso-$PGF_{2\alpha}$ results in the present study. Other reports, however, indicate that fish oil may increase oxidative stress. For example, fish oil was found to enhance lipid oxidation in hearts and livers of rats (Yuan and Kitts, 2003). Additionally, glutathione concentrations were reduced in red blood cells but increased in livers of the rats fed fish oil. Another study found that fish oil decreased hepatic α-tocopherol and retinol concentrations, and hepatic and spleen superoxide dismutase activity in rats (Miret et al. 2003). Nonheme iron concentration was decreased and the expression of iron regulatory protein-1 was increased by fish oil in the spleen and liver of rats. Miret et al. (2003) concluded that their findings indicated that fish oil increased oxidative stress in the liver and spleen of rats.

As indicated in the introduction, both boron and omega-3 PUFA have been shown to beneficially affect bone histomorphometric measurements in rats. Thus, it was not surprising that both enhanced femur strength as assessed by the maximum force needed for breaking. It was mildly surprising, however, to find that neither dietary boron nor fatty acid composition significantly affected the calcium concentration in the femur. Thus, the change in bone strength apparently was the result of changes in the physical characteristics of the bone through a modification of the organic matrix upon which calcification occurs. Moreover, because dietary boron and oil did not interact to affect maximum force, it is likely that they affected bone strength through different mechanisms. The femur copper, iron and magnesium findings support this suggestion. Copper and iron are involved in the formation of the organic matrix. The enzymes prolyl and lysyl hydroxylase need iron to catalyze the ascorbate-dependent hydroxylation of select prolyl and lysyl residues in collagen (Tuderman et al. 1977) before crosslinking by the copper-dependent enzyme lysyl oxidase (Siegel, 1978). Magnesium deficiency has been found to decrease collagen formation and sulfate incorporation in glycosaminoglycans in the organic matrix (Wallach, 1990). The concentrations of copper, iron and magnesium were higher in boron-supplemented than boron-deficient rat femurs. In contrast to the effect of boron supplementation, increasing the omega-3 PUFA content of the diet by feeding fish oil decreased the iron concentration in femur. Fish oil compared to safflower oil did not affect the femur concentration of magnesium or copper.

The vertebral microarchitecture data also indicate that boron and fatty acid composition each affect bone physical characteristics such as strength through different mechanisms, but the response induced by a change in the dietary intake of one of these dietary components influences the response to a change in the other. The increased trabecular thickness in boron-supplemented rats and a tendency for a similar effect in rats fed fish oil suggests that a change in thickness was partly responsible for these two dietary treatments increasing bone strength. However, trabecular thickness was not affected by an interaction between dietary boron and oil, but trabecular connectivity and SMI were. Greater trabecular connectivity usually is considered advantageous for bone strength. Thus, it was surprising that the boron-deficient rats fed safflower oil had the highest vertebra trabecular connectivity. Perhaps this was in response to having the lowest trabecular thickness. The SMI findings are prime examples for showing that boron influences the response to a change in dietary fatty acid composition or vice versa. Feeding fish oil instead of safflower oil resulted in the more preferred plate-like structure (low SMI) of trabecular bone in the vertebra only in the boron-supplemented rats. Based on the maximum force, trabecular thickness, SMI and BV/TV findings, boron and fish oil are complementary in enhancing bone strength and structure.

Brain mineral composition and visual and cognitive function tests also indicate that both boron and omega-3 PUFA have bioactivities that affect the response to changes in each other's intakes. Data from boron-deficient rats fed fish oil shown in Table 4 were noticeably different than the data from rats in the other three treatments. The boron-deficient rats fed fish oil apparently were less anxious based on more entries into open arms of the plus maze. This finding may be confounded by the results from the brightness discrimination test which indicated that visual acuity was altered by an interaction between dietary oil and boron. In contrast to boron-deficient zebrafish that become photophobic (Eckhert and Rowe, 1999), the boron-deficient rats, especially those fed fish oil, were more willing to enter areas of bright light (100 lux). The boron-supplemented rats fed fish oil was more normal in preferring to enter the bright light areas less times. The changes in the brightness discrimination and plus maze tests may be related to changes in the concentrations of minerals that affect brain function. Penland & Prohaska (2004) demonstrated lasting effects of early copper deprivation on brain copper and iron, and behavior of repleted adult rats. Johnson (2005) reviewed the role of copper in dopamine-ß-monooxygenase for neurotransmitter synthesis and copper-zinc superoxide dismutase as an antioxidant; mechanisms by which changes in brain copper likely affect behavior. Sandstead et al. (2000) reviewed the detrimental effects of zinc deficiency on brain biochemistry and function, and behavior in rats. Both zinc and copper concentrations were lowest in the boron-deficient rats fed fish oil. Interestingly, this group had the highest brain boron concentration, a finding difficult to explain. Apparently, boron deficiency alters the effect of fish oil on the ability of copper, zinc and boron to enter and exit the cell. Perhaps the transport of boron into the cell by a boron transporter (Park et al. 2004) is increased by boron deficiency and the transporter efficiency is promoted by increasing the long-chain omega-3 fatty acid composition of the cell membrane.

To summarize, feeding a boron-deficient diet decreased weight gain, femur strength, and femur concentrations of copper, iron and magnesium (minerals associated with the organic matrix), and increased the plasma concentration of 8-iso-$PGF_{2\alpha}$ (an indicator of oxidative stress) in rats. A change in the omega-3 PUFA content in the diet did not alter these boron deficiency responses. A diet high (fat provided by fish oil) compared to one low (fat provided as safflower oil) in long-chain omega-3 PUFA did not affect weight gain, but decreased the plasma concentration of 8-iso-$PGF_{2\alpha}$, increased femur strength, and decreased the femur concentration of iron in rats. Boron deprivation did not alter these responses to the change in dietary fatty acid composition. Thus, dietary boron and long-chain, omega-3 PUFA affect similar processes (bone growth and composition, oxidative metabolism), but apparently through affecting different biochemical systems. However, through changing the activity or function of some respective systems they affect, dietary boron and omega-3 PUFA each influenced the effect of the other on vertebral trabecular physical characteristics, brightness discrimination, plus maze activity, and brain composition. The basis for the relationship between boron and fatty acid bioactivity is still unclear. However, the findings to date do not eliminate the possibility that alterations of the beneficial actions of boron and omega-3 PUFA by each other occur through changes at the cell membrane level and may involve changes in redox metabolism.

Acknowledgements

The authors thank Rhonda Poellot for performing the biochemical determinations, Kim Baurichter for the bone breaking and physical measurements, Denice Shafer and staff for animal care, Jim Lindlauf for animal diet preparation, Craig Lacher and staff for mineral element analyses, Sheila

Bichler and LuAnn Johnson for statistical analysis assistance, and Christine Bogenreif for secretarial assistance.

References

Armstrong T.A., Spears J.W., Crenshaw T.D., and Nielsen F.H., 2000 Boron supplementation of a semipurified diet for weanling pigs improves feed efficiency and bone strength characteristics and alters plasma lipid metabolites. J. Nutr. **130**: 2575–2581.

Becker C.C. and Kyle D.J., 2001 The importance of DHA in optimal cognitive function in rodents. In Fatty acids: physiological and behavioral functions (D. Mostofsky, S. Yehuda and N. Salem Jr., eds.) Humana Press, Totowa, NJ, pp. 357–375.

Eckhert C.D., and Rowe R.I., 1999 Embryonic dysplasia and adult retinal dystrophy in boron-deficient zebrafish. J. Trace Elem. Exp. Med. **12**: 213–219.

Fort D.J., Rogers R.L., McLaughlin D.W., Sellers C.M., and Schlekat C.L., 2002 Impact of boron deficiency on Xenopus laevis. A summary of biological effects and potential biochemical roles. Biol. Trace. Elem. Res. **90**: 117–142.

Hoffman D.R., 2000 Fatty acids and visual dysfunction. In Fatty acids in foods and their health implications (C.K. Chow, ed.) Marcel Dekker, New York, NY, pp. 817–841.

Hunt C.D., 1996 Biochemical effects of physiological amounts of dietary boron. J. Trace Elem. Exp. Med. **9**: 185–213.

Hunt C.D. and Idso J.P., 1999 Dietary boron as a physiological regulator of the normal inflammatory response: a review and current research progress. J. Trace Elem. Exp. Med. **12**: 221–233.

Hunt C.D. and Nielsen F.H., 1981 Interaction between boron and cholecalciferol in the chick. In Trace element metabolism in man and animals (TEMA-4) (J. McC Howell, J.M. Gawthorne and C.L. White, eds.) Australian Academy of Science, Canberra, Australia, pp. 597–600.

Hunt C.D. and Shuler T.R., 1990 Open vessel, wet ash, low temperature digestion of biological materials for inductively coupled argon plasma spectroscopy analysis of boron and other elements. J. Micronutr. Anal. **6**: 161–174.

Johnson W.T., 2005 Copper and brain function. In Nutritional neuroscience (H.R. Lieberman, R.B. Kanarek and C. Prasad, eds.) Taylor & Francis, Boca Raton, pp. 289–305.

Miret S., McKie A.T., Sáiz M.P., Bomford A., and Mitjavila M.T., 2003 IRP1 activity and expression are increased in the liver and the spleen of rats fed fish oil-rich diets and are related to oxidative stress. J. Nutr. **133**: 999–1003.

Moriguchi T. and Salem N. Jr., Recovery of brain docosahexaenoate leads to recovery of spatial task performance. J. Neurochem. **87**: 297–309.

Nielsen F.H., 1996 Evidence for the nutritional essentiality of boron. J. Trace Elem. Exp. Med. **9**: 215–229.

Nielsen F.H., 2004 Dietary fat composition modifies the effect of boron on bone characteristics and plasma lipids in rats. BioFactors **20**: 161–171.

Nielsen F.H., 2005 Fish oil instead of safflower oil as the dietary fat source modifies the oxidative stress response to boron deficiency in rats. FASEB J. **19**: A1705.

Nielsen F.H., Penland J.G., and Newman S.M. Jr., 2004 Dietary boron modifies the effect of changing dietary fatty acid composition on rat behavior and eye mitochondrial morphology. FASEB J. **18**: A491.

Okuyama H., Fujii Y., and Ikemoto A., 2001 Brightness-discrimination learning behavior and retinal function affected by long-term α-linolenic acid deficiency in rat. In Fatty acids: physiological and behavioral functions (D. Mostofsky, S. Yehuda and N. Salem, Jr., eds.) Humana Press, Totowa, NJ, pp. 219–235.

Park M., Li Q., Shcheynikov N., Zeng W., and Muallem S., 2004 NaBC1 is a ubiquitous electrogenic Na^+-coupled borate transporter essential for cellular boron homeostasis and cell growth and proliferation. Mol. Cell **16**: 331–341.

Pellow S., Chopin P., File S.E., and Briley M., 1985 Validation of open:closed arm entries in an elevated plus-maze as a measure of anxiety in the rat. J. Neurosci. Methods **14**: 149–167.

Penland J.G., 1998 The importance of boron nutrition for brain and psychological function. Biol. Trace Elem. Res. **66**: 299–317.

Penland J.G. and Prohaska J.R., 2004 Abnormal motor function persists following recovery from perinatal copper deficiency in rats. J. Nutr. **134**:1984–1988.

Rico H., Crespo E., Hernandez E.R., Seco C., and Crespo R., 2002 Influence of boron supplementation on vertebral and femoral bone mass in rats on strenuous treadmill exercise: a morphometric, densitometric, and histomorphometric study. J. Clin. Densitometry **5**: 187–192.

Rossi A.F., Miles R.D., Damron B.L., and Flunker L.K., 1993 Effects of dietary boron supplementation on broilers. Poult. Sci. **72**: 2124–2130.

Sakaguchi K., Morita I., and Murota S., 1994 Eicosapentaenoic acid inhibits bone loss due to ovariectomy in rats. Prostaglandins Leukot. Essent. Fatty Acids **50**: 81–84.

Sandstead H.H., Frederickson C.J., and Penland J.G., 2000 History of zinc as related to brain function. J. Nutr. **130**: 496S–502S.

Schlemmer C.K., Coetzer H., Claassen N., and Kruger M.C., 1999 Oestrogen and essential fatty acid supplementation corrects bone loss due to ovariectomy in the female Sprague Dawley rat. Prostaglandins Leukot. Essent. Fatty Acids **61**:381–390.

Sheng M.H.-C., Taper L.J., Veit H., Qian H., Ritchey S.J., and Lau K.-H.W., 2001 Dietary boron supplementation enhanced the action of estrogen, but not that of parathyroid hormone, to improve trabecular bone quality in ovariectomized rats. Biol. Trace Elem. Res. **82**:109–123.

Siegel R.C., 1978 Lysyl oxidase. Int. Rev. Connect. Tiss. Res. **8**: 73–118.

Tang A.H. and Ho P.M., 1988 Both competitive and non-competitive antagonists of N-methyl-D-aspartic acid disrupt brightness discrimination in rats. Eur. J. Pharmacol. **22**: 143–146.

Takashi T., Fukumoto Y., and Harada E., 2002 Influence of a dietary n-3 fatty acid deficiency on the cerebral catecholamine contents, EEG and learning ability in rat. Behav. Brain Res. **131**:193–203.

Tuderman L., Myllylo R., and Kivirikko K.I., 1977 Mechanism of the prolyl hydroxylase reaction. I. Role of co-substrates. Eur. J. Biochem. **80**: 341–348.

Wallach S., 1990 Effects of magnesium on skeletal metabolism. Magnesium Trace Elem. **9**:1–14.

Yavin E., Brand A., and Green P., 2002 Docosahexaenoic acid abundance in the brain: a biodevice to combat oxidative stress. Nutr. Neurosci. **5**: 149–157.

Yuan Y.V. and Kitts D.D., 2003 Dietary (n-3) fat and cholesterol alter tissue antioxidant enzymes and susceptibility to oxidation in SHR and WKY rats. J. Nutr. **133**: 679–688.

Boron and Prostate Cancer a Model for Understanding Boron Biology

Curtis Eckhert*, Wade Barranco and Danny Kim

Department of Environmental Health Sciences, University of California, Los Angeles, CA 90095-1772 USA E-mail: *ceckhert@ucla.edu*

Introduction

The aim of this paper is describe how boron was identified as an anticancer agent for prostate cancer and how research using human prostate cells identified a biochemical mechanism that may explain many of the diverse effects reported for the element in plants and animals. At physiological pH, boric acid binds to NAD^+, a coenzyme in metabolism and substrate for an enzyme in the NAD^+ cADPR system of calcium signaling. The chemical structure that results from the boration of NAD^+ is a poor substrate for ADP-ribosyl cyclase leading to a reduction in its product, cyclic ADP-ribose (cADPR). cADPR is intracellular signaling molecule in plants and animals that stimulates the release of internal Ca^{2+} stores. In plants, cADPR is the central mediator of abscisic acid, a phytohormone involved in integrating responses to the environment and stress (Grill and Himmelbach, 1998). In animals, cADPR is a mediator of many aspects of cellular function in embryonic, pancreatic, bone, muscle, nerve, hemopoietic and immune cells (De Flora et al. 2004).

Prostate cancer

The prostate is a secretory gland in human males that provides the major portion of the ejaculate fluid. The fluid originates as a secretion of the epithelial cells, which line the ducts of 20 to 30 arborized prostatic gland structures, that are surrounded by smooth muscle and empty into the urethra. The epithelium undergoes continual renewal with cells first originating from stem cells in the small apical ducts and then migrating down the branches and into the large proximal ducts where they undergo apoptosis. A human prostate with attached seminal vesicles is shown in Figure 1. About 70% of prostate cancers occur in the lower peripheral zone of the prostate. Unlike breast cancer, prostate cancer cell growth is diffuse and does not form large tumors. There are no reliable biochemical markers for prostate cancer and even the widely used measurement of prostate serum antigen (PSA) has been only been shown to be an indicator of benign prostatic hyperplasia (Stamey et al. 2004).

F. Xu et al. (eds.), Advances in Plant and Animal Boron Nutrition, 291–297.

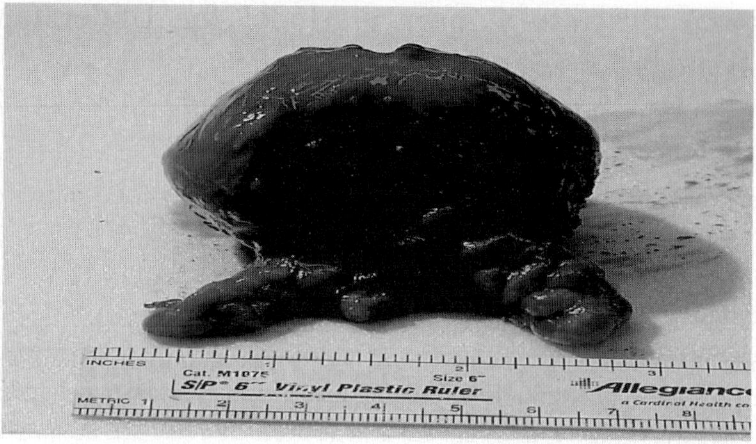

Figure 1. Human prostate seminal vesicles entering the top of the gland

Prostate cancer is the most common non-skin cancer in men living in western countries (Hing and Devesa, 2001). The National Cancer Institute now estimates that only 5% to 10% of new cases are attributable to genetics alone. The importance of the environment in the etiology of the disease is readily apparent by looking at its global distribution. There is a 100 difference in incidence and mortality between North American and Northern European men, and men living in China and Japan. In a search for B related human disease, we divided the U.S. National Heath and Nutrition Examination Survey (NHANES III) into four levels of boron intake to screen for cancers that responded to dietary boron. Our analysis showed that the risk of the disease was inversely proportional to dietary boron intake (table 1). Boron decreased the risk of prostate cancer in a dose dependent manner with a 54% difference in the adjusted odds ratio between the lowest and highest dietary intake (Zhang et al. 2001; Cui et al. 2003).

Table 1. The Risk of Prostate Cancer is Reduced by Foods Containing Boron

Diet Intake (mg B/d) of Cases	Adjusted Odds Ratio	95% Confidence Intervals
≤ 0.52	1.0	
≤ 0.86	0.61	0.32 – 1.16
≤ 1.36	0.59	0.30 –1.16
>1.36	0.46	0.21- 0.98

Modified from Oncology Reports 11:887-892, 2004.

Biological plausability

The results of this epidemiological screen prompted us to explore the biological plausability that boron was an anticancer agent. For this we investigated the impact of boric acid on the growth of cultured prostate cells (Barranco and Eckhert, 2004). Researchers at the National Institute of Environmental Health (NIEHS) also examined the effect of boric acid on PSA and tumor growth (Gallardo-Williams et al. 2004). The NIEHS researchers injected immunocompromised nude mice with the androgen dependent human prostate cancer cell line LNCaP and found that boron supplementation (1.7 mgB/kg/d) decreased tumor growth 38% and PSA, a clinical indicator of prostate growth, by 89%. We showed that boric acid decreased the proliferation of androgen dependent LNCaP and androgen independent cells (DU-145) in a dose dependent manner (Figure 2). This occurred in the absence of apoptosis and without a major shift in the cell cycle (Barranco and Eckhert, 2004).

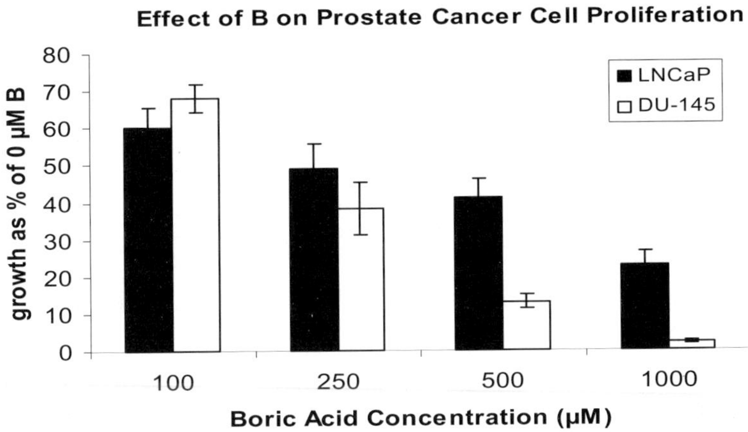

Figure 2. Boric acid inhibits the proliferation of human prostate cell lines LNCaP and DU-145. The androgen receptor and PSA are expressed by LNCaP, but not DU-145 cells. Thus the inhibitory effect of boric acid is not dependent on the androgen receptor or PSA

Boric acid binding molecules

Boric acid binds to cis-diol groups particularly those on five carbon sugars. The relative binding affinity of boric acid with nucleotides present in the cytosol of animal cells was determined using mass spectrometry binding analysis (*Kim* et al. 2003, 2004). This afforded direct measurements that could be used to determine the association constants (K_A) of boric acid for different target molecules. Dinucleotides had the strongest association constant with NAD^+ >

NADH > $NADP^+$ > NADPH. The dinucleotide with the highest affinity, NAD^+, was therefore deemed the most physiologically relevant. The association constants (K_A) to nucleotides were found to decrease with increasing phosphorylation, for example AMP>ADP>ATP (Kim et al. 2004)

NAD^+ as a substrate

NAD^+ serves as both a coenzyme in intermediary metabolism and as a substrate for the formation of cADPR and NAADP. The formation of cADPR is shown in Figure 3. NAD^+ is converted by ADP-ribosyl cylase to cADPR a signaling molecule that releases Ca^{2+} from the endoplasmic reticulum into the cytoplasm to raise intracellular concentrations. In animal cells, ADP-ribosyl cyclase activity is associated with one of the active sites of the multifunctional enzyme CD38 located on the plasma membrane (De Flora et al. 2004).

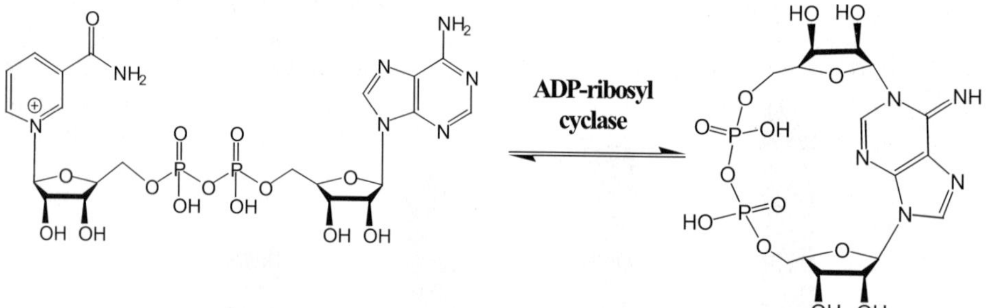

NAD^+ **cADPR**

Figure 3. NAD^+ is a substrate for the formation of cyclic ADP-ribose (cADPR), a signaling molecule for Ca^{2+} release from internal stores in plant and animal cells. cADPR has both autocrine and paracrine properties

Mechanism underlying boron's antiproliferative effect

Cell proliferation and apoptosis are regulated by Ca^{2+} signaling. Since NAD^+ is the substrate for cADPR we examined boric acid's affect on NAD^+ stimulated release of Ca^{2+} stores in the boric acid sensitive DU-145 cell line. The results are shown in figure 5. Ca^{2+} signaling had not been studied before in DU-145 cells so it had to be established that NAD^+ was active in releasing stores. This was accomplished quantifying Ca^{2+} in cells using the indicator dye Fluo-4 which emits fluorescence when it binds to Ca^{2+}. Fluo-4 fluorescence was excited with a 488 nm argon laser line and emissions were collected with a 505 LP filter using

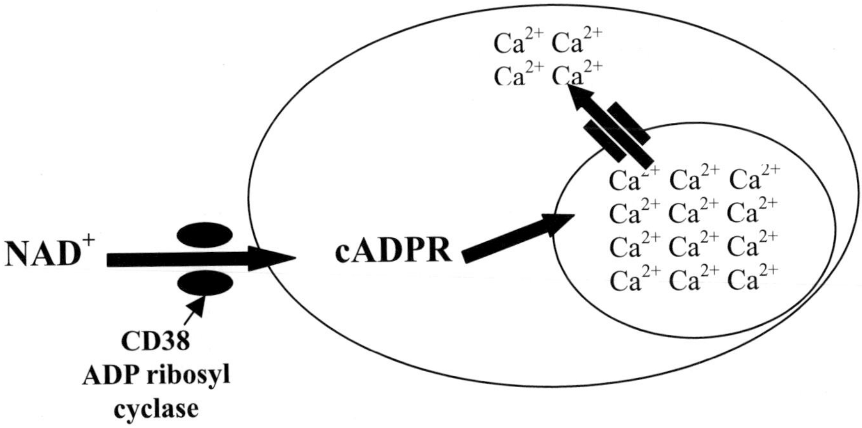

Figure 4. NAD^+ is converted by ADP-ribosyl cylase to cADPR a signaling molecule that releases Ca^{2+} from the endoplasmic reticulum into the cytoplasm to raise intracellular concentrations. In animal cells the multifunctional enzyme CD38 has ADP-ribosyl cylase activity. NAD^+ is transported to the active site of the cyclase of CD38 by connexin 43 (Cs43). cADPR is transported by nucleoside transporters to ryandine receptors which are required for the release of Ca^{2+} from the endoplasmic reticulum

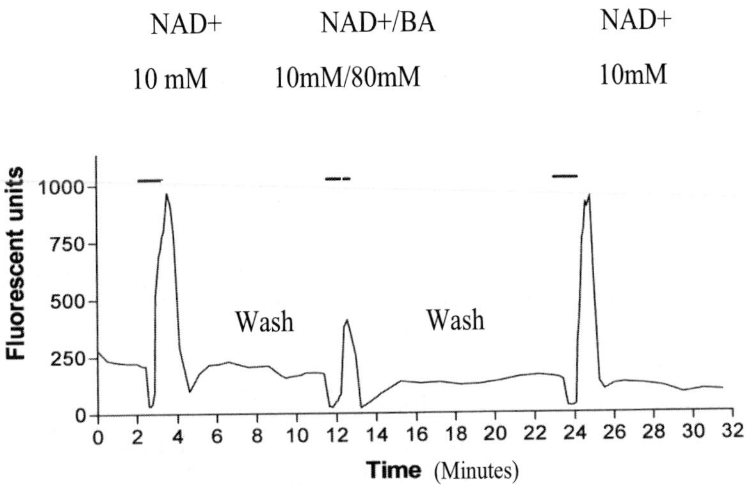

Figure 5. Boric acid inhibition of NAD^+ stimulated release of internal stores of Ca^{2+}. DU-145 were stimulated using NAD^+ to release Ca^{2+} from internal stores (peak on left). The release of Ca^{2+} was greatly diminished when boric acid (BA) was added with NAD^+ (center peak). Boric acid inhibition was reversible since after a wash the response to NAD^+ returns to normal (right). Ca^{2+} was measured using Fluo-4 and a Zeiss LSM 5 Pascal confocal microscope

a Zeiss LSM 5 Pascal confocal microscope coupled to an upright fixed stage microscope) Axioskop 2 FS mot), and equipped with a Axoplan 63X (NA 0.95) water-immersion objective. NAD^+ stimulated release of Ca^{2+} was greatly diminished in the presence of boric acid. The inhibition was reversible as shown by the return to a normal response after boric acid was removed. Subsequent studies have determined that boric acid inhibits the ability of ADP-ribosyl cyclase to form cADPR by substrate inhibition (Kim et al. submitted).

Discussion of findings

The observation that boric acid inhibits the release of Ca^{2+} stores by the NAD^+ cADPR system identifies a general mechanism that explains many of the effects of boron on the prostate as well as other biological systems. Basic intracellular calcium functions are well conserved between animals and plants (Nagata et al. 2004). The NAD^+ gradient is steep with the intracellular concentration of NAD^+ in the mM range and the plasma and extracellular fluid at 50-100 nM (De Flora et al. 2004). NAD^+ is transported to the active site of the cyclase of CD38 by connexin 43 (Cs43). The cyclase product, cADPR, is then transported by nucleoside transporters to ryandine receptors that function in the release of Ca^{2+} from the endoplasmic reticulum. In plants, cADPR is the central mediator of abscisic acid, a phytohormone involved in the expression of over 1000 genes and a plethora of intracellular messengers (Himmelbach et al. 2003). A field trial of boron and zinc supplementation to balsam pear grown in boron and zinc deficient soils has shown that boron supplementation reduces abscisic acid content in leaves (Shi and Cheng 2004). This result is consistent with what we have observed in animals and supports our hypothesis that many of the physiological effects of boron are a result of its ability to reduce the formation of cADPR and its subsequent intracellular Ca^{2+} signal. The list of abscisic acid functions is long and includes cell cycle inhibition, elongation, RNA processing, stomatal closure, seed formation, germination, and environmental responses to desiccation, salt and cold (Grill and Himmelbach, 1998; Himmelbach et al. 2003). In animals, cADPR is a mediator of many aspects of cellular function including: fertilization, proliferation, bone remodeling, insulin release, smooth muscle contraction, and the activity of neurons, hemopoietic and immune cells (De Flora et al. 2004). The long list of functions regulated by the NAD^+ cADPR system in plants and animals brings some clarity as to why it has been such a challenge to identify specific mechanisms that explain the its broad range of physiological effects. The regulation of cellular functions by Ca^{2+} signaling is an expanding and very complex area of cellular physiology. The road to understanding the importance of boron in modulating Ca^{2+} signaling is sure to be filled with exciting opportunities for discovery and new challenges.

References

Barranco W.T. and Eckhert C.D., 2004 Boric acid inhibits human prostate cancer cell proliferation. Cancer Lett. **216**: 21–9.

Cui Y., Winton M.I., Zhang Z.F., Rainey C., Marshall J., De Kernion J.B., and Eckhert C.D., 2004 Dietary boron intake and prostate cancer risk. Oncology Reports. **11**, 887–892.

De Flora A., Zocchi E., Guida L., Franco L., and Bruzzone S., 2004 Autocrine and Paracrine calcium signaling by the CD38/NAD+/cyclic ADP-ribose system. Ann N Y Acad. Sci. **1028**, 176–191.

Gallardo-Williams M.T., Chapin R.E., King P.E., Moser G.J., Goldsworthy T.L., Morrison J.P., and Maronpot R.R., 2004 Boron supplementation inhibits the growth and local expression of IGF-1 in human prostate adenocarcinoma (LNCaP) tumors in nude mice. Tox. Path. **32**, 73–78.

Grill E. and Himmelbach A., 1998 ABA signal transduction. Current Opinion in Plant Biology. **1**, 412–418.

Himmelbach A., Yang Y., and Grill E., 2003 Relay and control of abscisic acid signaling. Curr. Opin. Plant Biol. **6**, 470–479.

Hing A.W. and Devesa S.S., 2001 Trends and patterns of prostate cancer. What do they suggest? Epidem. Rev. **23**, 3–13.

Kim D.H., Marbois B.N., Faull K.F., and Eckhert C.D., 2003 Esterification of borate with NAD^+ and NADH as studied by electrospray ionization mass spectrometry and ^{11}B NMR spectroscopy. J. Mass. Spect. **38**, 632–640.

Kim D.H., Faull K.F., Norris A.J., and Eckhert C.D., 2004 Borate-nuceotide complex formation depends on charge and phosphorylation state. J. Mass. Spect. **39**, 743–751.

Kim D.H., Faull K.F., Norris A.J., and Eckhert C.D. Submitted for publication.

Nagata T., Iizumi S., Satoh K., Ooka H., Kawai J., Carninci P., Hayashizaki Y., Otoma Y., Murakami K., Matsubara K., and Kikuchi S., 2004 Comparative analysis of plant and animal calcium signal transduction element using plant full-length cDNA data. Mol. Biol. Evol. **21**, 1855–1870.

Shi M. and Cheng R., 2004 Effects of zinc and boron nutrition on balsam pear (Momordica charantic) yield and quality, and polyamines, hormone, and senescence of its leaves. Ying Yong Sheng Tai Xue Bao. **15**, 77–80.

Stamey T.A., Caldwell M., McNeal J.E., Nolley R., Hemenez M., and Downs J., 2004 The prostate specific anigen era in the United States is over for prostate cancer: what happened in the last 29 years? J. Urol. **172**, 1297–1301.

Zhang Z.-F., Winton J.I., Rainey C., and Eckhert C.D., 2001 Boron is associated with decreased risk of human prostate cancer. FASEB J. **15**, A1089.

Boric Acid Inhibits Cell Growth in Breast and Prostate Cancer Cell Lines

Susan L. Meacham[1, 3], Kyler E. Elwell[2], Sarah Ziegler[2] and Stephen W. Carper[2,3]*

[1]Department of Nutrition Sciences, School of Health and Human Sciences, Division of Health Sciences

[2] Department of Chemistry, College of Sciences,

[3]UNLV Cancer Research Center, University of Nevada Las Vegas, PO Box 454003, Las Vegas, NV 89154-4003 *Corresponding author

Email: carpers@unlv.nevada.edu

Introduction

Boron, an element ubiquitous in the earth's crust can be found in most soil types as well as in fresh and salt water, consequently, boron is a natural constituent in the human diet (Nemodruk and Karalova, 1969, Meacham and Hunt, 1998). In a typical adult daily intakes of boron were measured to be approximately 1 mg/day (Meacham and Hunt, 1998, Rainey et al. 1999). Boron is considered an essential micronutrient for higher plants; however, the most recent US dietary recommendations concluded that insufficient evidence existed to consider boron an essential micronutrient for humans (Warrington, 1923, Loomis and Durst, 1992, IOM, 2001). However, evidence does exist that indicates boron plays a beneficial role in some physiological processes of various animal species and measurable responses to adjusted boron intakes in humans have been observed (Rowe and Echkert, 1999, Nielsen, 1998).

Boric acid, the most common form of boron in humans is a weak acid (pKa = 9.2) essentially unionized at physiological pH. Adults with daily adult boron intakes from food and water sources will typically have plasma boron concentrations of 60 μM (Ward, 1987). Boron appears to be homeostatically regulated with a rapid excretion (usually within 24 hrs) regardless of the method of administration, commonly as boric acid.

There is reason to believe that boron may demonstrate beneficial effects in cancer patients. Several recent publications have indicated that boron may be able to inhibit prostate cancer. An epidemiological study reported that the risk of prostate cancer was inversely proportional to dietary boron intake in a dose responsive manner (Cui et al. 2004). *In vitro*, boric acid inhibited growth of established human prostate cancer cell lines (Barranco and Eckhert, 2004). *In vivo* reduced tumor development was observed from human prostate cancer

F. Xu et al. (eds.), Advances in Plant and Animal Boron Nutrition, 299–306.

LNCaP cells implanted into nude mice receiving boric acid (Gallardo-Williams et al. 2004).

In the current study, boric acid inhibited the proliferation of a prostate cancer and breast cancer cell line. The mechanism of inhibition was investigated via cell cycle analysis determined by flow cytometry. Boric acid did not induce apoptosis in prostate cancer cells but did induce apoptosis in the breast cancer cell line.

Materials and methods

Cell lines were obtained from the ATCC (Manassas, VA) and cultured in T-25 plastic flasks (Invitrogen Corp., Carlsbad, CA). Breast cancer cell lines MDA-MB-231, MDA-MB-435 and MCF-7 were grown in minimal essential media (MEM) (Invitrogen Corp., Carlsbad, CA), supplemented with 10% fetal bovine serum, 25 mM HEPES buffer, 100 IU penicillin/mL, 100 μg/mL streptomycin sulfate (all supplements were from Invitrogen Corp., Carlsbad, CA). Prostate cancer cell lines DU-145, PC-3, LNCaP, T47-D, and breast cancer cell lines SK-BR-3 and ZR-75-1 were grown in RPMI 1640 media (ATCC, Manassas, VA) supplemented with 10% fetal bovine serum, 25 mM HEPES buffer, 100 IU penicillin/mL, 100 μg/mL streptomycin sulfate (all supplements were from Invitrogen Corp., Carlsbad, CA). All cell lines were maintained at 37°C in humidified incubators with 5%CO_2:95% air. Ultrapure water was used to prepare boric acid solutions. Boric acid was purchased from Sigma Chemical Co. (St. Louis, MO).

Growth curve experiments

Growth curve experiments were initiated by harvesting non-confluent flasks, counting cells using a Coulter Counter and plating between 80,000 to 100,000 cells in a T-25 flask containing 5.0 mL of media. Twenty-four hours after placing the cells in the flasks, boric acid was added to the flasks bringing the final concentration to 1 mM. Flasks were harvested and counted at the indicated times. All measures were done in triplicate.

Flow cytometry experiments

Flow cytometry was employed on cells harvested from growth curve experiments. Following counting, cells were centrifuged for 10 min at 1,500 X g, washed with 5 mL of phosphate buffered saline (PBS) (Invitrogen, Carlsbad, CA), re-spun and the pellet resuspended in 100 μL of PBS. Then 1.0 mL of -20°C 70% ethanol was added drop wise while vortexing the sample. The samples were stored at 4°C for at least three days prior to analysis by flow cytometry. Fixed cells had 1 mL of PBS added before being centrifuged for 10 min at 1,500 X g. The cell pellet was resuspended in 100 μl of 1.0% Triton

X-100 in PBS and 100 μl of RNase solution (1.0 mg/mL RNase S in PBS). Following a 15 minute incubation at room temperature, 200 μl of propidium iodide (PI) stain solution (100 μg/mL PI in PBS) was added and the sample vortexed gently. The samples were then incubated for 30 minutes in the dark at room temperature. Cytometry acquisition was done on a BD FACS Calibur with the argon laser set at 488 nm on the linear flow Channel 2 (FL-2) with Doublet Discriminatory Module (DDM) and Threshold set on FL-2. Flow data was analyzed for cell cycle distribution using Modfit 3.0 software.

Results

Boric acid inhibits the growth of DU-145 prostate cancer cells

In order to access the effects of boric acid on human prostate cancer cells, the growth rate of DU-145 cells was determined in the presence and absence of 1 mM boric acid. Under these conditions the DU-145 cells reached a plateau phase in five days but in the presence of boric acid no growth was observed (Figure 1).

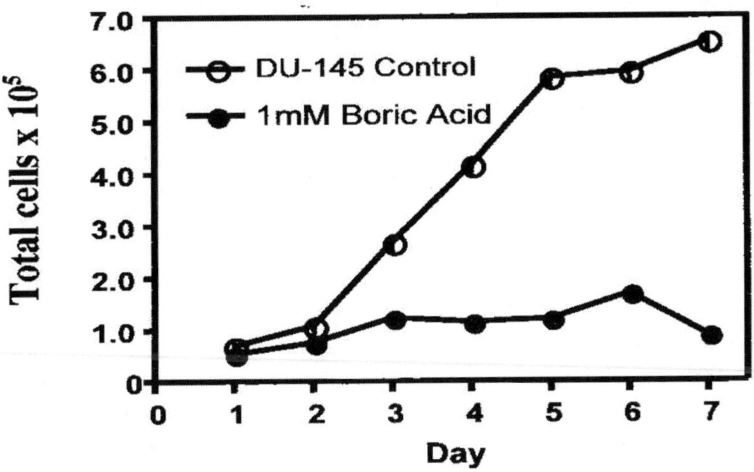

Figure 1. Growth curve for DU-145 cells. 1 mM boric acid was added to cells 24 hrs after seeding into T-25 flasks. A single typical result is shown. Error bars are smaller than the symbols used

We further characterized these cells using flow cytometry to determine their distribution in the cell cycle. Control cells displayed a steady increase in the G_1 phase during the seven days of growth while the percentage of cells in the G_2/M and S phases steadily decreased as was expected for cells reaching the plateau phase of growth (Table 1).

Table 1. Effects of boric acid on DU-145 cells in culture characterized by flow cytometry to identify cells in the phases of the cell cycle

Phase		Day 1	Day 2	Day 3	Day 4	Day 5	Day 6	Day 7
Control	G_1	48.35	50.16	57.46	62.76	61.11	69.68	77.95
	G_2/M	20.83	15.53	15.85	15.13	18.31	16.97	11.16
	S	30.82	34.31	26.70	22.12	20.58	13.35	10.90
	Apoptosis	2.17	4.93	0.20	2.30	2.30	2.08	7.65
Boric Acid(1 mM)	G_1	50.86	59.36	56.81	54.05	53.12	46.25	46.82
	G_2/M	13.66	16.07	18.15	16.76	16.40	19.29	18.91
	S	35.49	40.29	25.04	29.19	30.49	34.46	34.27
	Apoptosis	1.92	2.62	0.95	5.50	4.03	6.34	0.60

Apoptosis in control cells was below 3% except for the last day of growth. The cells treated with 1 mM boric acid maintained fairly consistent percentages for all phases of the cell cycle. Apoptosis values were 6% or less at all time points. These results indicate that in DU-145 cells boric acid acts as a cytostatic agent that does not appear to induce apoptosis. Also investigated were the effects of 1 mM boric acid on the growth rate of PC-3 and LNCaP human prostate cancer cell lines in which no growth inhibition was found (data not shown).

Boric acid inhibits the growth of ZR-75-1 breast cancer cells

Investigations continued observing the effects of boric acid on growth rates of estrogen receptor negative human breast cancer cell lines MDA-MB-231 and MDA-MB-435 (Figure 2).

Boric acid (1 mM) had no effect on the growth rate of either cell line. Additionally, the growth rate of estrogen receptor positive cell lines MCF-7 and T47-D (Figure 3) was observed in the presence and absence of 1 mM boric acid. Neither cell line showed any appreciable inhibition. It should be noted that the MDA-MB-231, MDA-MB-435, and MDF-7 cells were in MEM supplemented media while the T47-D cell line was in RPMI 1640 supplemented media.

Two human breast cancer cell lines were identified that did show growth inhibition in the presence of 1 mM boric acid. The estrogen receptor negative SK-BR-3 cell line display a 15% growth inhibition while the estrogen receptor positive cell line ZR-75-1 displayed 40% growth inhibition by day seven (Figure 4).

Both cell lines were cultured in RPMI 1640 supplemented media. To further characterize growth inhibition in ZR-75-1 cells, flow cytometric analysis confirmed the distribution of cells in the cell cycle (Table 2).

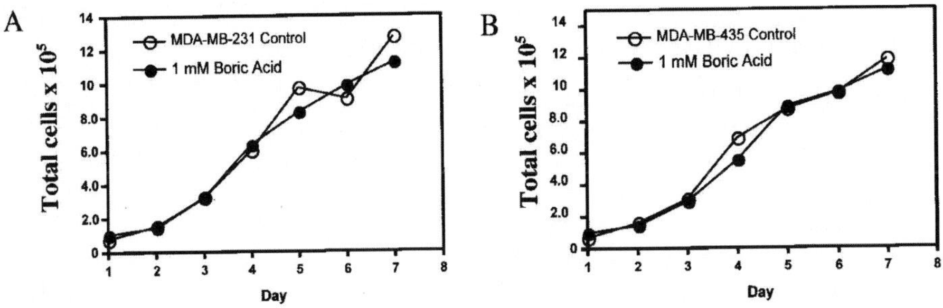

Figure 2. Growth rate for estrogen receptor negative human breast cancer cell lines MDA-MB-231 (panel A) and MDA-MB-435 (panel B). Cell lines were not inhibited by 1 mM boric acid. A single typical result is shown. Error bars are smaller than the symbols used

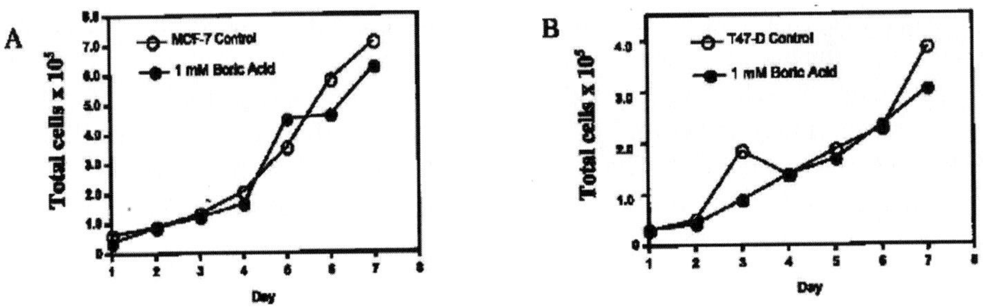

Figure 3. Growth rate for estrogen receptor positive human breast cancer cell lines. MCF-7 (panel A) and T47-D (panel B). Cell lines were not inhibited by 1 mM boric acid. A single typical result is shown. Error bars are smaller than the symbols used

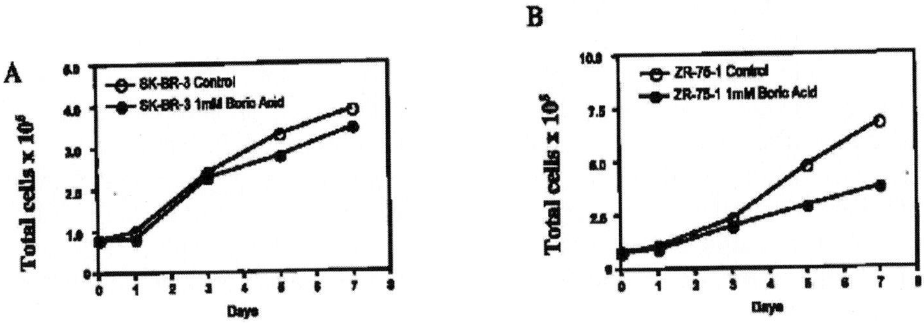

Figure 4. Growth inhibition in the presence of 1 mM boric acid. SK-BR-3 (panel A) and ZR-75-1 (panel B). Cell lines are growth inhibited in the presence of 1 mM boric acid. A single typical result is shown. Error bars are smaller than the symbols used

Table 2. Effects of boric Acid on ZR-75-1 cells in culture characterized by flow cytometry to identify cells in the phases of the cell cycle

Phase		Day 1	Day 3	Day 5	Day 7	Day 9	Day 11
Control	G_1	55.5	65.38	67.05	76.1	76.6	68.8
	G_2/M	22.8	14.0	13.9	8.9	8.0	15.76
	S	21.8	20.6	19.1	15.0	15.4	15.43
	Apoptosis	7.21	9.99	4.3	7.7	7.6	19.0
Boric Acid (1 mM)	G_1	59.3	66.05	67.4	74.4	71.8	66.8
	G_2/M	16.9	12.1	11.3	10.4	13.6	18.7
	S	23.6	21.3	21.3	15.2	14.6	14.7
	Apoptosis	7.4	7.4	6.5	13.9	16.4	11.3

Control treated cells displayed an increase in the percentage of cells in the G_1 phase of the cell cycle through the seven days of the experiment. G_2/M and S phases displayed decreasing percentages while apoptosis was below 10% except for the final day. Cells treated with 1 mM boric acid also showed an increase in the G_1 phase during the seven days. The percentage of cells in the G_2/M phase decreased slightly while those in the S phase did not change for the first five days after which they fell from 21% to 15%. The rate of apoptosis was similar to that seen in control cells for the first five days, but during the seventh day the rate of apoptosis doubled from 7% to 14%. This is the first indication that boric acid is able to induce apoptosis in a human breast cancer cell line.

Discussion

Data presented here indicate that boric acid is capable of inhibiting growth in selected prostate cancer and breast cancer established cell lines. At the concentration used (1 mM) the inhibition in DU-145 cells was complete while in the SK-BR-3 and ZR-75-1 cell lines there was only partial growth inhibition. Analysis of the cell cycle distribution of cells treated with boric acid suggested that boric acid is a cytostatic agent that does not cause a block in a specific phase of the cell cycle. Boric acid did not induce apoptosis in DU-145 cells but apparently did induce apoptosis in ZR-75-1 cells after seven days of exposure.

The results observed in the with DU-145 cell line (i.e., complete growth inhibition) are similar to those previously reported (Barranco and Eckhert, 2004). However, the finding that 1 mM boric acid did not inhibit growth in PC-3 and LNCaP cells differs from a previous report observing partial growth inhibition in these cell lines.

This is the first report that boric acid can inhibit the growth of human breast cancer cells. The breast cancer cell lines that demonstrated partial growth inhibition, as well as the DU-145 prostate cancer cell line, was cultured in RPMI 1640 media. Three of the breast cancer cell lines that did not respond to boric acid were cultured in MEM media while the other breast cell line, T47-D, that did not respond to boric acid was cultured in RPMI 1640 media. It does not appear that culture media played a role in the response to boric acid.

Since growth inhibition was observed in estrogen receptor negative SK-BR-3 cells and estrogen receptor positive ZR-75-1 cells, it would appear that estrogens are not involved in the response. A similar result observed in androgen receptor specific cells has been reported, again stating that gender specific hormones do not play a role in boric acid induced growth inhibition as reported for prostate cancer cell lines (Barranco and Eckhert, 2004). Therefore, it would appear that steroid hormones are not involved in the boric acid growth inhibition response.

The mechanism whereby boric acid can inhibit tumor growth is not understood. In a nude mouse tumor model of LNCaP cells, it was reported that boron inhibited the activity of the prostate specific antigen (PSA) which is a serine protease that can cleave insulin-like growth factor binding protein-3 providing increased local levels of IGF-1. Immunohistochemistry confirmed that expression of IGF-1 in tumors was markedly reduced by boron treatments (Gallardo-Williams et al. 2004). The mechanism for boron inhibition of the DU-145 cell line must be different since the DU-145 cells do not express PSA (Mickey et al. 1980). Boric acid has been determined to bind to the cis-diols present on the ribose moiety of nucleotides including NAD^+ (Ralston and Hunt, 2001, Kim et al. 2003). The authors proposed that boric acid's antiproliferative effect is through the formation of nucleotide-borate complexes that change the function or utilization of nucleotides (Barranco and Eckhert, 2004). From the results presented by this current study, the differences in responses to boric acid between DU-145 (complete growth response) and ZR-75-1 (partial growth inhibition and apoptosis) may reflect cell line specific targets or sensitivities to nucleotide-borate complexes.

Acknowledgments

This work was funded in part by a grant from the office of the Vice President for Research at UNLV. We thank Mr. John Adebay for performing flow cytometry at the BRIN core cancer lab, which is funded by the National Institutes of Health awards BRIN (1P20RR16464) and INBRE (2P20RR16464-04). We also thank Ms. Casey Hall and Ms. Shweta Tharkar for their excellent technical support.

References

Barranco W.T. and Eckhert C.D., 2004 Boric acid inhibits human prostate cancer cell proliferation. Cancer Lett. **216**: 21–29.

Cui Y., Winton M.I., Zhang Z.F., Rainey C., Marshall J., De Kernion J.B., and Eckhert C.D., 2004 Dietary boron intake and prostate cancer risk. Oncol. Rep. **11**(4): 887–92.

Gallardo-Williams M.T., Chapin R.E., King P.E., Moser G.J., Goldsworthy T.L., Morrison J.P. and Maronpot R.R., 2004 Boron supplementation inhibits the growth and local expression of IGF-1 in human prostate adenocarcinoma (LNCaP) tumors in nude mice. Toxicol. Pathol. **32**(1): 73–8.

Kim D.H., Marbois B.N., Faull K.F., and Eckhert C.D., 2003 Esterification of borate with NADC and NADH as studied by electrospray ionization mass spectrometry and 11B NMR spectroscopy, J. Mass Spectrom. **38**: 632–640.

Loomis W.D. and Durst R.W., 1992 Chemistry and biology of boron. Biofactors **3**: 229–239.

Meacham S.L. and Hunt C.D., 1998 Dietary boron intakes of selected populations in the United States. Biol. Trace Elem. Res. **66**(1–3): 65–78.

Nemodruk A.A. and Karalova Z.K., 1969 Analytical Chemistry of Boron. Ann Arbor-Humphrey Science Publishers, London.

Nielsen F.H., 1998 The justification for providing dietary guidance for the nutritional intake of boron. Biol Trace Elem Res. **66**(1–3): 319–30.

Rainey C.J., Nyquist, L.A., Christensen R.E., Strong P.L., Culver B.D., and Coughlin J.R., 1999 Daily boron intake from the American diet. J. Am. Diet. Assoc. **99**(3): 335–40.

Ralston N.V.C. and Hunt C.D., 2001 Diadenosine phosphates and S-adenosylmethione: novel boron binding biomolecules detected by capillary electrophoresis. Biochim. Biophys. Acta **1527**: 20–30.

Rowe R.I. and Eckhert C.D., 1999 Boron is required for zebrafish embryogenesis. J. Exp. Biolo. **202**: 1649–1654.

Ward N.L., 1987 The determination of boron in biological materials by neutron irradiation and prompt gamma-ray spectrometry. J. Radioanal. Nucl. Chem. **110**: 633–639.

Warrington K., 1923 The effect of boric acid and borax on the broad bean and certain other plants. Ann. Bot. **37**: 629–672.

Estimation of Dietary Boron and Silicon Intakes in China

Junquan Gao and Xiaowei Li

National Institute for Nutrition and Food Safety, Chinese Center for Disease Control and Prevention.29 Nan Wei Road, Beijing 100050, China.
e-mail: jqgao@vip.sina.com

Introduction

This report summarizes the recent research efforts to estimate the dietary intakes of both boron and silicon in the Chinese adult male population. In addition, the primary sources of boron in the Chinese diet were identified and compared to the previously identified primary dietary sources of boron in diets from the United States, Germany, Great Britain, Mexico, Kenya, and Egypt.

Materials and methods

The data reported here were generated from boron and silicon analyses of diet samples collected in the second Chinese Total Diet Study (CTDS-II). Food samples were collected in 1992-1993 from the same survey areas utilized in the CTDS-I survey conducted in 1990[(1)]. Sample preparation procedures and cooking methods were also the same as those used in CTDS-I. Concentrations of boron and silicon in the prepared food samples were determined by ICP-AES at the USDA, ARS Grand Forks Human Nutrition Research Center in Grand Forks, ND, USA. To estimate average boron and silicon dietary intakes, the analytical data were linked with estimates of intakes of representative foods identified in CTDS-II.

Results

Dietary boron intakes and major sources

The mean dietary boron intake estimates for adult Chinese males is 1.37 mg/d (Table 1). For this segment of the Chinese population, cereals are the major source of dietary boron intake (36%) (Table 2). Other significant sources of dietary boron are vegetables (24.3%), legumes (18.2%), and fruits (10.6%).

Dietary silicon intakes and major sources

The mean dietary silicon intake of Chinese adult males is estimated to be 43.15 mg/d (Table 3). The major contributors to dietary silicon intake are cereals (55.5%), vegetables (23.3%), legumes (5.9%), and potatoes (5.8%) (Table 3).

F. Xu et al. (eds.), Advances in Plant and Animal Boron Nutrition, 307–310.

Table 1. Estimated dietary boron intake of adult males living in China and compared to previous estimates of boron intake in other countries (mg/d)

Country	Mean	SD
Mexico	2.12	0.69
Kenya	1.95	0.57
Germany	1.72	0.91
China	1.37	
Egypt	1.31	0.50
Great Britain	1.30	0.63
United States	1.11	0.69

Discussion

The dietary boron intakes of adult males living in China are similar to those of adult males living in the United States, Great Britain, or Egypt (Table 1). However, adult males living in Mexico, Germany or Kenya consume more dietary boron than those living in China (Table 1). Grain-based foods are the top contributors of boron in the diets of adult males living in China (cereals: 36%), Mexico (tortillas: 56%), Kenya (maize: 35%), and Egypt (rural breads: 27%). In contrast, fruit-based beverages are the top dietary contributors of boron in the United States (coffee: 6.5%), Great Britain (wine: 14%), and Germany (wine: 15%). Cereal-based foods are the top contributors of dietary silicon in the Chinese diet. The dietary silicon intake data reported here for adult Chinese males are the only such data available from the Chinese population.

These dietary intake estimates for both boron and silicon provide information that will be useful for setting recommended daily intake levels of both element s when they are confirmed to be essential for humans.

Acknowledgments

The authors thank Forrest H. Nielsen and his colleagues at the USDA, ARS Grand Forks Human Nutrition Research Center, Grand Forks, ND, USA, for supplying analytical materials and technology.

Table 2. Main foods contributors (%) of dietary boron in China

Country	Cereals	Legumes	Potatoes	Meats	Eggs	Milk	Aquatic Foods	Vegetables	Fruits	Sugar	Beverage & Water	Alcohol	Total
China	36	18.2	6.9	1.8	0.2	0.1	0.9	24.3	10.6	0.02	0.8	0.1	100

Table 3. The Dietary silicon intake (mg/d) and main foods contributors (%) of dietary silicon in China

	Cereals	Legumes	Potatoes	Meats	Eggs	Milk	Aquatic Foods	Vegetables	Fruits	Sugar	Beverage & Water	Alcohol	Total
Silicon Intake	23.97	2.53	2.49	0.78	0.17	0.05	1.41	10.07	0.51	0.007	0.96	0.20	43.15
% of Si Intake	55.5	5.9	5.8	1.8	0.4	0.1	3.3	23.3	1.2	0.02	2.2	0.5	100

References

Chen J. and Gao J., 1993 The Chinese total diet study in 1990. Part I. Chemical Contaminants. J. AOAC Int. **76**, 1193–1205.

Rainey C., Nyquist L., Casterline J., and Herman D., 1999 Estimation of dietary boron intake in six countries: Egypt, Germany, Great Britain, Kenya, Mexico, and the United States. J. of Trace Elements in Experimental Medicine **12**: 263–270.

Part III

Boron is Soils

New Advances in Boron Soil Chemistry

Sabine Goldberg[1*] and Chunming Su[2]

[1]USDA-ARS George E. Brown, Jr., Salinity Laboratory, 450 W. Big Springs Road, Riverside, California 92507 USA

[2] USEPA National Risk Management Research Laboratory, 919 Kerr Research Drive, Ada, Oklahoma 74820 USA

*Corresponding Author: sgoldberg@ussl.ars.usda.gov

Introduction

Boron is an essential micronutrient element required for plant growth. Boron deficiency is wide-spread in crop plants throughout the world especially in coarse-textured soils in humid areas. Boron toxicity can also occur, especially in arid regions under irrigation. Plants respond directly to the B concentration in soil solution and only indirectly to the amount of B attached to soil surfaces (Keren et al. 1985). Therefore, the soil adsorption complex acts as both a source and a sink for dissolved B and can mitigate phytotoxic soil solution B concentrations.

The primary B adsorbing surfaces in soils are: aluminum and iron oxides, clay minerals, calcium carbonate, and organic matter (Goldberg, 1997). Boron adsorption on all of these surfaces increases with increasing solution pH, reaches an adsorption maximum near pH 8 to 9, and decreases with further increases in solution pH. The mechanism of B adsorption on these surfaces is considered to be ligand exchange with reactive surface hydroxyl groups leading to strong specific adsorption.

In addition to solution pH, other factors affecting the availability of B in soils are: soil texture, soil moisture, and soil temperature (Goldberg, 1997). Fine-textured soils usually contain more available B than coarse-textured soils because of their greater content of clay minerals (Gupta, 1968). Boron is generally less available in dry soil and increases with increasing temperature (Fleming, 1980).

Plant B response is affected by many aspects of B soil chemistry. In this review we will treat several of these topics: i) methodologies for determining the mechanisms of B attachment to soil particle surfaces; ii) kinetics of B adsorption reactions; iii) description and prediction of B adsorption reactions using chemical models; iv) use of B soil tests to predict plant response in field situations.

F. Xu et al. (ed.), Advances in Plant and Animal Boron Nutrition, 313–330.

Determining boron attachment to soil surfaces

To accurately describe the adsorption behavior of B in soils, the mode of attachment of B to soil particles must be known. Indirect and direct experimental procedures have been used to establish B adsorption mechanisms on oxide minerals. Indirect methods include point of zero charge (PZC) shifts and ionic strength dependence. Direct observation of ion adsorption mechanisms is provided by a variety of spectroscopic techniques.

Electrophoretic mobility measures that movement of charged particles in an applied electric field. Lack of mobility indicates the PZC of the particles. Inner-sphere surface complexes are strong, specific, and contain no water between the adsorbing ion and the point of surface attachment. Inner-sphere anion adsorption shifts the PZC to lower values and causes reversals of electrophoretic mobility with increasing anion concentration (Hunter, 1981). Adsorption of B lowered the PZC of an aluminum oxide mineral, gibbsite (see Figure 1) providing indirect evidence of strong, specific, inner-sphere adsorption.

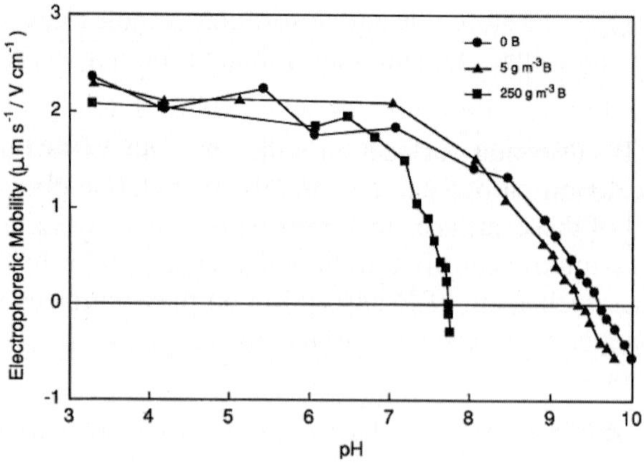

Figure 1. Electrophoretic mobility of the aluminum oxide, gibbsite as a function of total solution B concentration and pH. From Goldberg et al. (1993)

Ionic strength dependence of adsorption can be used to distinguish inner- and outer-sphere surface complexes. Outer-sphere surface complexes are weaker, nonspecific, and contain at least one water molecule between the adsorbing ion and the point of surface attachment. Decreasing adsorption with increasing solution ionic strength is considered evidence for outer-sphere surface complexation; while ions showing little ionic strength dependence or increasing adsorption with increasing ionic strength are considered to form inner-sphere surface complexes (McBride, 1997). Boron adsorption on an iron oxide mineral, goethite (see Figure 2) showed little ionic strength dependence, suggesting inner-sphere surface complex formation.

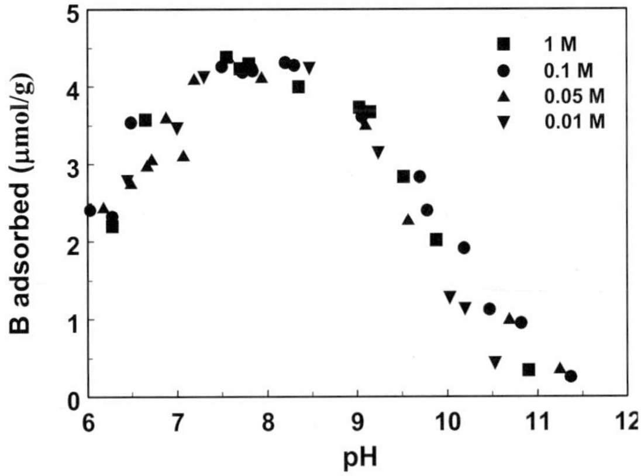

Figure 2. Boron adsorption on the iron oxide, goethite as a function of solution ionic strength and pH. Adapted from Goldberg et al. (1993)

Infrared spectroscopy is a technique that can provide insight into the type and number of bonds that an adsorbed ion forms with the solid surface. Attenuated total reflectance Fourier transform infrared (ATR-FTIR) spectroscopy can be used to observe the attachment of surface adsorbed ions in the presence of water. This is critical for applicability to natural systems such as soils.

Su and Suarez (1995) studied mechanisms of boron adsorption on amorphous aluminum and iron hydroxides using ATR-FTIR. Samples for ATR-FTIR spectroscopy consisted of either pure boric acid in 0.1 M NaCl or mineral pastes obtained from centrifugation of equilibrated suspensions. Four boric acid solutions (4.62, 9.25, 23.1, and 92.5 mmol L^{-1} in 0.1 M NaCl) were prepared from CO_2-free water and were adjusted to pH 7, 9, 10 and 11 with 1.0 M NaOH (CO_2-free). Mineral suspensions consisted of 0.25 g of either am-$Al(OH)_3$ or am-$Fe(OH)_3$ in 12.5 mL of 0.1 M NaCl adjusted to desired pH values near 7 and 10 and initial B concentrations (4.62, 9.25 and 23.1 mmol L^{-1}). Suspensions of am-$Fe(OH)_3$ were equilibrated for 20 h at 25°C and suspensions of am-$Al(OH)_3$ were equilibrated at pH 10.2 ± 0.1 and 5°C. It was found in a preliminary experiment that at a pH value greater than 8.6 at 25°C, am-$Al(OH)_3$ had been largely transformed to pseudoboehmite and bayerite via X-ray diffraction examination, but remained amorphous at 5°C for pH 3 to 11 after 20 h equilibration. Suspensions were centrifuged for 30 min at the corresponding temperature. The B concentration in the supernatant was determined using a Technicon AutoAnalyzer II and the azomethine-H method described by Bingham (1982).

Infrared spectroscopic analysis of boric acid solutions was performed with the reservoir module and analysis of mineral pastes with the pressure plate module of the horizontal ATR accessory using horizontal ATR and an internal reflectance element (IRE) of ZnSe crystal. Detailed experimental procedures can be found in Su and Suarez (1995). All ATR-FTIR spectra were recorded from 4000 to 700 cm^{-1} at 4 cm^{-1} over 1000 scans.

To obtain signals of B, the spectra of the pH-adjusted 0.1 M NaCl solution or mineral paste without B (reference) were subtracted from the spectra of the 0.1 M NaCl solution or paste with added B at the same pH (sample), both previously ratioed against the spectrum of the empty cell (background), respectively. A subtraction factor of 1.000 was used.

Boric acid is a neutral trigonal molecule. It is a very weak monobasic Lewis acid that accepts a hydroxyl ion to form the tetrahedral borate anion as indicated in Figure 3. The pK_a of boric acid is 9.24 (Bassett, 1980). Figure 4 shows the ATR-FTIR spectra of an aqueous B solution as a function of solution pH. The peak assignments are as follows: 1410 cm^{-1}, B-O asymmetric stretching of trigonal B, 955 cm^{-1}, asymmetric stretching of tetrahedral B, 1148-1170 cm^{-1}, mixture of B-OH bending of trigonal and tetrahedral B (Su and Suarez, 1995).

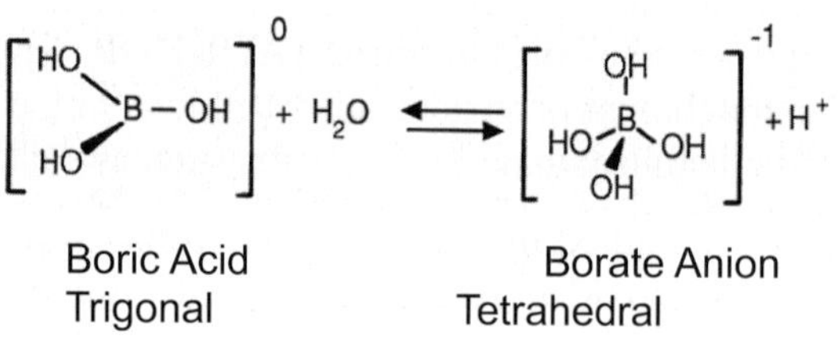

Figure 3. Dissociation of boric acid

ATR-FTIR difference spectra of am-$Al(OH)_3$ paste at pH 6.8 ± 0.2 with adsorbed B are shown in Figure 5a for three initial B concentrations of 4.62, 9.25 and 23.1 mmol L^{-1}. The equilibrium B concentrations are far lower than the initial concentrations and are plotted next to the spectra in Figure 5a. A broad band centered at 1420 cm^{-1} is assigned to trigonal boron asymmetric stretching mode and a band at 1280 cm^{-1} is assigned to trigonal boron B-OH bending mode. Up shift of both bands to higher frequencies in the mineral paste was evident compared to the pure boric acid solution (1410 and 1148 cm^{-1}) at pH 7. This is attributed to the strengthening of O-B and B-OH bonds in the surface complex -Al-O-$B(OH)_2$ when the boric acid molecule is complexed with surface functional groups such as -Al-OH. The intensity of these bands increased with increasing boron adsorption. Figure 5b shows the ATR-FTIR difference spectra of an am-$Al(OH)_3$ paste with and without added B at pH 10.2 ± 0.1 and 5℃. A band at 1412 cm^{-1} (B-O asymmetric stretching of trigonal B) was not significantly shifted compared to the 1410 cm^{-1} band for boric acid solution at pH 7; however, the B-OH bending of trigonal B band had a narrower width and was shifted to a higher frequency of 1266 cm^{-1} from 1148 cm^{-1} due to the formation of surface B complexes. Surface complexation is likely to make the B-OH bond stronger so as to increase the B-OH bending frequency.

Characterization of tetrahedral B at the mineral surface was not successful because of severe band interference in the range of 1000-900 cm^{-1} from the Al-O bond that shows strong absorbance centered at 969 cm^{-1}. Boric acid solution spectra show tetrahedral boron as the dominant species at pH 10.2 ± 0.1 (Figure 4c); however, tetrahedral boron may not necessarily be the dominant adsorbed species on am-$Al(OH)_3$ surfaces at pH 10.2. This may be explained by the change of surface charge of minerals as a function of pH. Am-$Al(OH)_3$ is expected to be negatively charged at pH 10.2 and the neutral $B(OH)_3$ species could be preferred due to its higher affinity for the negatively charged surface of am-$Al(OH)_3$ at high pH; whereas, the $B(OH)_{4-}$ ion would encounter charge repulsion. Another possibility is that polymerization of adsorbed B species occurred, resulting in both trigonally and tetrahedrally coordinated B.

Figure 6 shows the ATR-FTIR difference spectra of amorphous iron oxide with and without B (Su and Suarez, 1995). The asymmetric stretching of trigonal B was shifted downward while the B-OH binding was shifted upward. A weak tetrahedral B asymmetric stretching band was shifted upward and evident at 962 and 985 cm^{-1}. Shifts in frequency are indicative of surface complexation of B. A similar study of B adsorption onto amorphous iron oxide (Peak et al. 2003) assigned the 1395 cm^{-1} band to outer-sphere surface complexation and the 990, 1250, and 1330 cm^{-1} bands to inner-sphere surface complexation of trigonal B (Figure 7).

Based on both macroscopic and microscopic results, B forms strong inner-sphere complexes on the soil minerals: amorphous iron oxide, amorphous aluminum oxide, and allophane (Su and Suarez, 1997). Boron also forms weak outer-sphere complexes on amorphous iron oxide. This physically bound B could be readily leached and would be available for plant uptake.

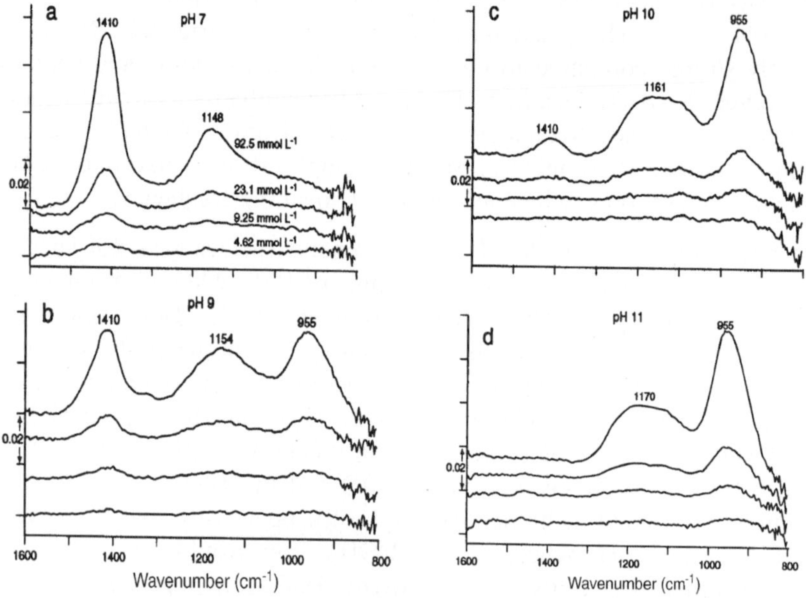

Figure 4. ATR-FTIR spectra of aqueous boric acid as a function of solution pH and total B concentration. From Su and Suarez (1995)

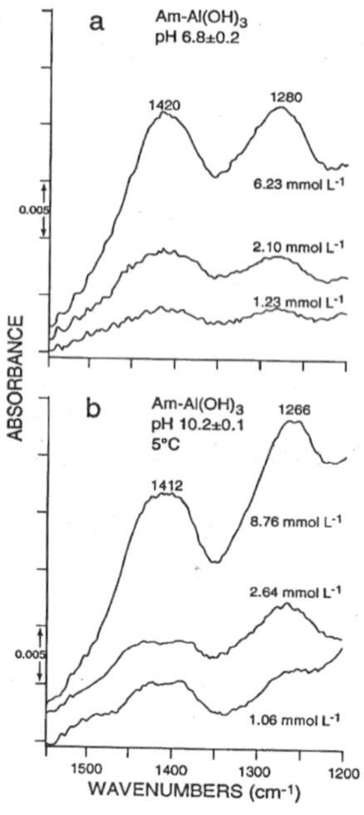

Figure 5. ATR-FTIR difference spectra of B adsorbed onto amorphous aluminum oxide as a function of pH and equilibrium boron concentration. From Su and Suarez (1995)

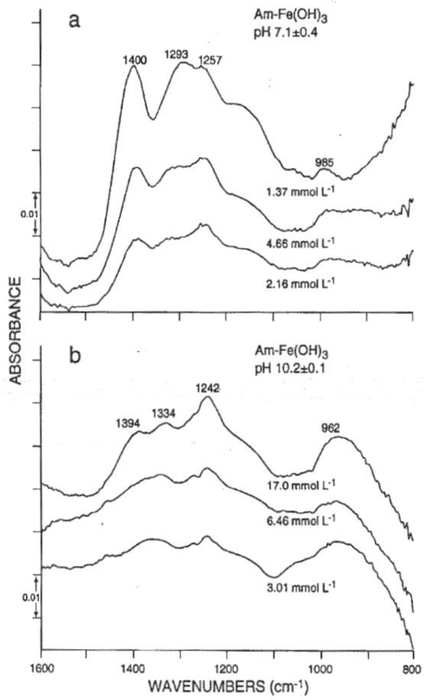

Figure 6. ATR-FTIR difference spectra of B adsorbed onto amorphous iron oxide as a function of pH and equilibrium boron concentration. From Su and Suarez (1995)

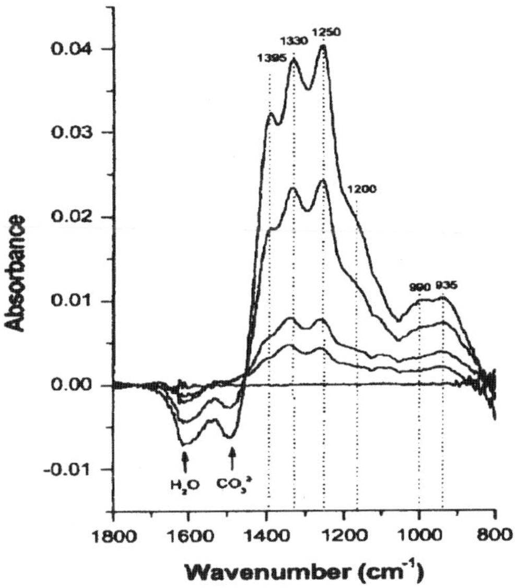

Figure 7. ATR-FTIR spectra of boron adsorbed on the hydrous iron oxide surface at pH 6.5 and I = 0.01 M as a function of initial boron concentrations (from bottom: 0, 50, 100, 500, and 100 μM). From Peak et al. (2003)

Kinetics of boron desorption reactions

Boron containing minerals generally do not control B solubility in soil solution because they are either too insoluble (tourmaline) or too soluble (hydrated B minerals). The amount of B in soil solution is usually controlled by B adsorption-desorption reactions (Goldberg, 1997). In many soils, B adsorption is readily reversible and the B desorption isotherm corresponds closely to the B adsorption isotherm. Other soils exhibit hysteresis, the lack of correspondance of the B desorption isotherm with the B adsorption isotherm. The apparent irreversibility of B sorption has been attributed to conversion of readily desorbable monodentate B surface complexes into less readily desorbable bidentate complexes, incorporation of B into tetrahedral sites of clay minerals, and B diffusion into particle interiors.

Excess soluble B in arid land soils has often been attributed to the weathering of B containing minerals. Su and Suarez (2004) studied the release of boron from representative minerals and soils. They included two specimen illites (Morris and Fithian), two shales (Salt Creek and Moreno Gulch), a fresh and a weathered serpentine (antigorite) from the Coastal Range of California, a Traver silt loam and a Twisselman clay loam both containing illite, chlorite, and palygorskite. The soils were collected from the San Joaquin Valley of California, USA. All samples were subjected to successive extraction (7 to 26 times) following each 12-h equilibration in 0.1 M (first three extraction) and 0.01 M $CaCl_2$ solution (subsequent extractions) until the supernatant B solutions were below the detection limit of 0.001 mmol B L^{-1}. Boron release from various minerals is depicted in Figure 8.

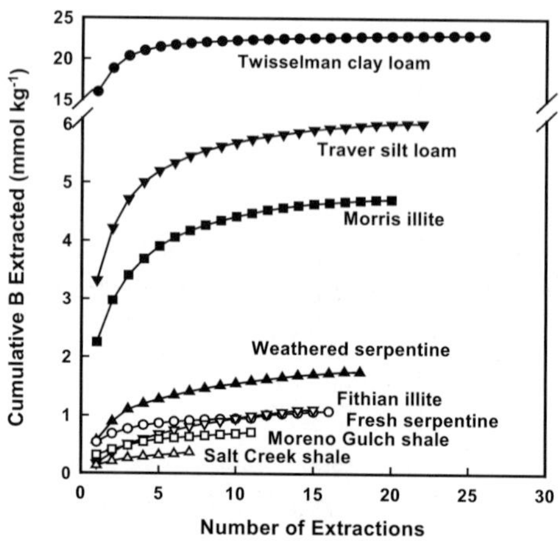

Figure 8. Cumulative B extracted as a function of sequential aqueous extraction for various minerals (12 h, 1:10 w/v with 0.1 M $CaCl_2$ for the first three extractions and 0.01 M $CaCl_2$ for subsequent extractions). From Su and Suarez (2004)

After the successive washing to remove surface-adsorbed boron, the < 2 μm and 2-20 μm size fractions were separated and reacted in deionized water at pH 5, 7, and 9 adjusted with HCl and NaOH for up to 180 days. Effect of particle size, solution pH, and time on B concentration and B release for the two San Joaquin Valley soils is shown in Figure 9. Boron release rates decreased with equilibration time and increasing pH. Reclamation of high B soils may be followed by continued release of sparingly soluble sources of B. Knowledge of long-term B leaching from high B irrigated soils is essential for prediction of future B contamination of drainage waters in such places as the San Joaquin Valley of California, USA. The presence of illite, chlorite, and palygorskite minerals was found to be responsible for the long-term release of B in these soils (Su and Suarez, 2004).

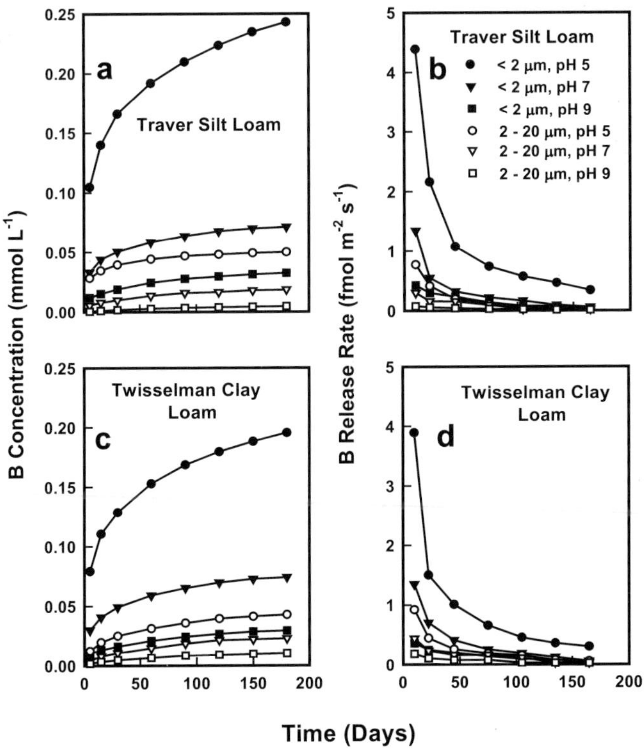

Figure 9. Effect of particle size, pH, and time on B concentration and B release from a soil from the San Joaquin Valley of California, USA. From Su and Suarez (2004)

Boron release rates were highest for the smallest particles and the lowest solution pH (Su and Suarez, 2004). The long-term release rates can be used to predict B concentration in the field. Weathering of such soils may produce toxic levels of B within a few years following reclamation unless continued leaching is maintained.

Prediction of boron adsorption using chemical models

Various modeling approaches have been used to describe B adsorption on soil minerals and soils. Empirical models such as the Freundlich and Langmuir adsorption isotherms produce parameters that are only valid for the conditions under which the experiment was conducted. The Keren model (Keren and Mezuman, 1981) is a phenomenological equation that can describe B adsorption on soil minerals and soils as a function of solution pH. Yet it is empirical in that no physical significance can be attributed to the parameter values.

Chemical surface complexation models such as the constant capacitance model and the triple layer model define surface species, chemical reactions, mass balances, and charge balance and contain molecular features that can be given thermodynamic significance. The triple layer model has been used to describe B adsorption by soils in only one study (Goldberg, 2005) because of the difficulty in determining meaningful values for its large set of adjustable parameters. The constant capacitance model contains a much smaller number of adjustable parameters than the triple layer model and is therefore more suitable for describing adsorption on complex natural materials such as soils.

The constant capacitance model assumes that B adsorption occurs via a ligand exchange mechanism with reactive surface hydroxyls forming inner-sphere surface complexes. The placement of ions in the constant capacitance model is depicted in Figure 10. A detailed discussion of the theory, assumptions, and equations of the constant capacitance model is provided in Goldberg (1992).

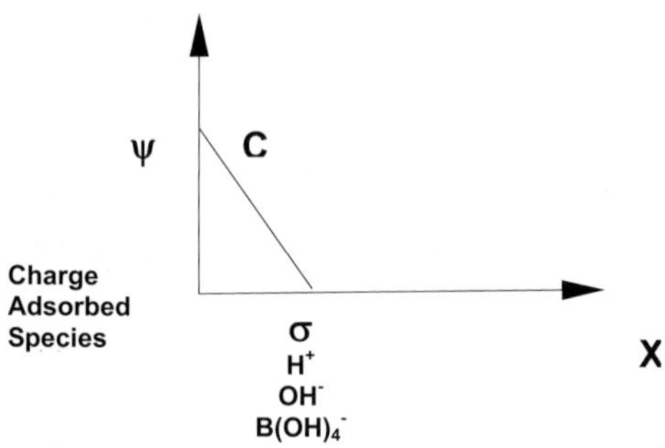

Figure 10. Placement of ions, capacitance, surface charge, σ, and surface potential, ψ for the constant capacitance model

In the application of the constant capacitance model to B adsorption the following surface complexation reactions for the surface functional group, SOH, are defined:

$$SOH + H^{+} \leftrightarrow SOH_2^{+} \quad (1)$$

$$SOH \leftrightarrow SO^{-} + H^{+} \quad (2)$$

$$SOH + H_3BO_3 \leftrightarrow SH_3BO_4^{-} + H^{+} \quad (3)$$

SOH represents a reactive surface hydroxyl group on an oxide mineral or an aluminol or silanol group on the edge of a clay mineral. The equilibrium constant expressions for the above reactions are:

$$K_{+}(\text{int}) = \frac{[SOH_2^{+}]}{[SOH][H^{+}]} \exp(F\psi / RT) \quad (4)$$

$$K_{-}(\text{int}) = \frac{[SO^{-}][H^{+}]}{[SOH]} \exp(-F\psi / RT) \quad (5)$$

$$K_{B-}(\text{int}) = \frac{[SH_3BO_4^{-}][H^{+}]}{[SOH][H_3BO_3]} \exp(-F\psi / RT) \quad (6)$$

where F is the Faraday constant (C mol_c^{-1}), ψ is the surface potential (V), *R* is the molar gas constant (J mol^{-1} K^{-1}), *T* is the absolute temperature (K) and square brackets indicate concentrations (mol L^{-1}). The electrostatic potential term, exp(Fψ/RT), can be considered as a solid phase activity coefficient that corrects for surface charge.

Boron adsorption experiments on soils were carried out in batch systems to determine adsorption isotherms (amount of B adsorbed as a function of equilibrium solution B concentration) and adsorption envelopes (amount of B adsorbed as a function of solution pH at a fixed total B concentration). The soils were chosen to provide a wide range of soil chemical characteristics.

Additional experimental details are provided in Goldberg et al. (2000). The constant capacitance model was able to describe B adsorption both as a function of solution B concentration and solution pH. The constant capacitance model was well able to fit B adsorption on 17 soils from the southwestern USA. Examples of model fit are shown in Figure 11 for two soils from California, USA.

A general regression model was developed to obtain soil surface complexation constants to allow prediction of B adsorption on additional soils using the constant capacitance model. The model surface complexation constants were obtained from easily measured soil chemical properties: surface area (SA), organic carbon content (OC), inorganic carbon content (IOC), and free aluminum oxide content (Al). These parameters are also soil properties that correlate with soil B adsorption capacity. Experimental details for these chemical measurements are provided in Goldberg et al. (2000). The prediction equations for obtaining the surface complexation constants are:

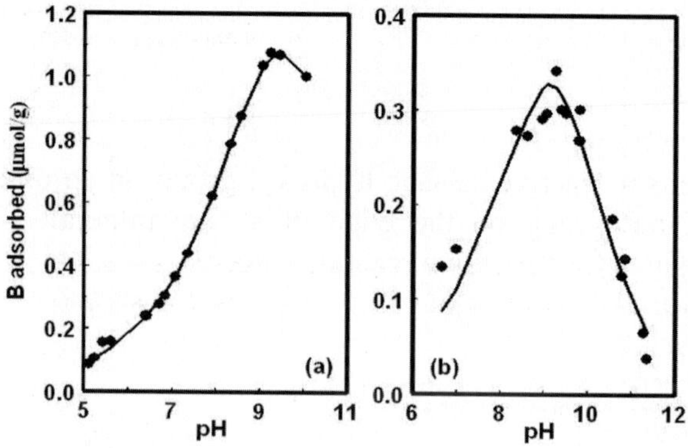

Figure 11. Fit of the constant capacitance model to B adsorption: (a) Diablo clay; (b) Fallbrook subsoil. Circles represent experimental data. Model fits are represented by solid lines. From Goldberg et al. (2000)

$$LogK_{B-} = -9.14 - 0.375\ln(SA) + 0.167\ln(OC) + 0.111\ln(IOC) + 0.466\ln(Al) \quad (7)$$

$$LogK_{+} = 7.85 - 0.102\ln(OC) - 0.198\ln(IOC) - 0.622\ln(Al) \quad (8)$$

$$LogK_{-} = -11.97 + 0.302\ln(OC) + 0.0584\ln(IOC) + 0.302\ln(Al) \quad (9)$$

Additional details on the statistical analysis are provided in Goldberg et al. (2000).

The prediction equations were used to predict surface complexation constants for other soils that had not been used to obtain the regression model. Using predicted constants, the constant capacitance model was well able to predict B adsorption behavior by 37 diverse soils from both the Southwestern (Goldberg et al. 2000, see Figure 12) and Midwestern (Goldberg et al. 2004, see Figure 13) parts of the USA. This approach represents a completely independent evaluation of the ability of the model to predict B adsorption using zero adjustable parameters. These results suggest widespread applicability of the prediction approach for describing B adsorption behavior in soils both as a function of solution B concentration and solution pH. The prediction equations have been incorporated into the UNSATCHEM speciation-transport computer program (Suarez and Simunek, 1997) to allow prediction of soil solution B concentrations for different soils and chemical conditions without the need for detailed characterization of B adsorption.

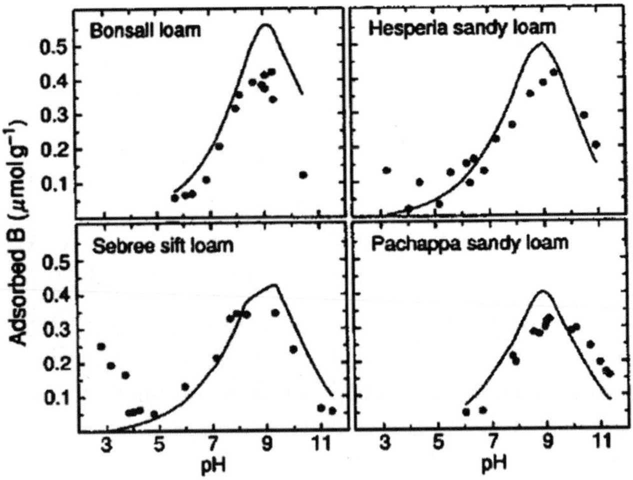

Figure 12. Constant capacitance model prediction of B adsorption by southwestern USA soils not used to obtain the prediction equations. Circles represent experimental data. Model predictions are represented by solid lines. Adapted from Goldberg et al. (2000)

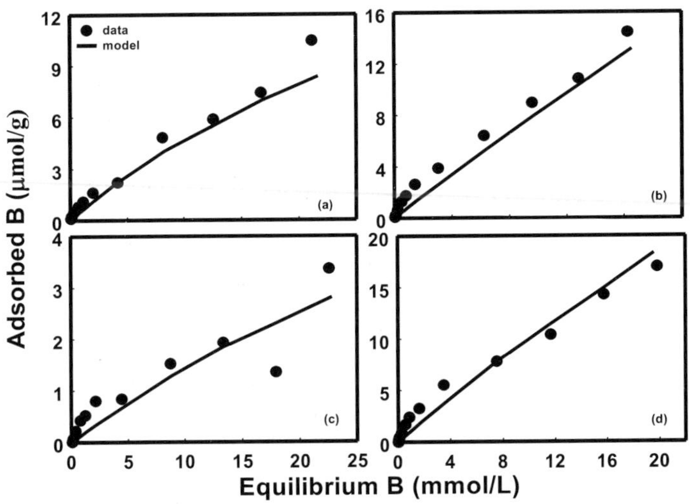

Figure 13. Constant capacitance model prediction of B adsorption by Midwestern USA soils not used to obtain the prediction equations: (a) Mansic soil; (b) Osage soil; (c) Pond Creek soil; (d) Summit soil. Circles represent experimental data. Model predictions are represented by solid lines. Adapted from Goldberg et al. (2004)

Use of boron soil tests to predict plant response

A wide variety of soil tests has been developed to predict plant B content. The most common soil test for estimating plant available B is hot-water-soluble (Berger and Truog, 1940). Calcium chloride-mannitol (Cartwright et al. 1983) and ammonium acetate (Gupta and Stewart, 1975) extracts have also been used extensively. A DTPA-sorbitol extractant has been recommended by the North American Proficiency Testing Program to estimate the potential soil bioavailability of Zn, Cu, Mn, Fe, and B (Miller et al. 2000). All of these soil tests have been found to be highly correlated (Goldberg et al. 2002).

Historically, B soil tests have been developed to predict B deficient soils and have not generally been evaluated for their ability to predict soil conditions that produce B toxicity effects in plants. Shallow groundwater usage by crops improves irrigation efficiency and is being tested in the San Joaquin Valley of California, USA. Such management systems have the potential to adversely affect crop yields due to build-up of soil solution salinity and B concentration. Several soil B extractants were used to determine B uptake by field and container grown plants. The extractants were evaluated for their ability to predict B content of both field-grown and container-grown plants under conditions of potential B toxicity (Goldberg et al. 2002, 2003).

For the container study, a heavy clay soil was collected from the Broadview Water District in the San Joaquin Valley of California, USA. Subsamples of soil were treated to attain seven different B levels replicated four times. Muskmelon (*Cucumis melo* L.) variety *Top Mark* was planted in outdoor containers in a randomized block design and fertilized and irrigated as needed. Plant leaves were sampled prior to fruit set and at harvest; stems and fruits were sampled at harvest only. Soils were sampled before the start of the experiment and after harvest. Additional experimental details are described in Goldberg et al. (2003).

For the field study, soil and plant samples were collected from the Broadview Water District. Soil samples at all sites were collected in the spring. For alfalfa, plants were sampled at ten sites at the time of soil sampling. Muskmelons were sampled at 28 sites prior to fruit set and at maturity. Cotton was sampled at 27 sites at flowering and two subsequent times during the growing season. Additional experimental details are described in Goldberg et al. (2002).

The soil samples were extracted with 1 M ammonium acetate, distilled water, and DTPA-sorbitol. The DTPA-sorbitol extractant consists of 0.005 M diethylenetriaminepentaacetic acid, 0.01 M $CaCl_2$, 0.1 M triethanolamine (TEA) adjusted to pH 7.3, and 0.2 M sorbitol. Boron concentrations in the extracts were determined using inductively coupled plasma (ICP) spectrometry.

In the container study, marginal chlorosis, characteristic of excess B, was found on the melon leaves for all B treatments. Deleterious effects of B on melon growth and development were found. At the highest two B treatments, the number of days to first flowering was significantly delayed, as seen in Figure 14.

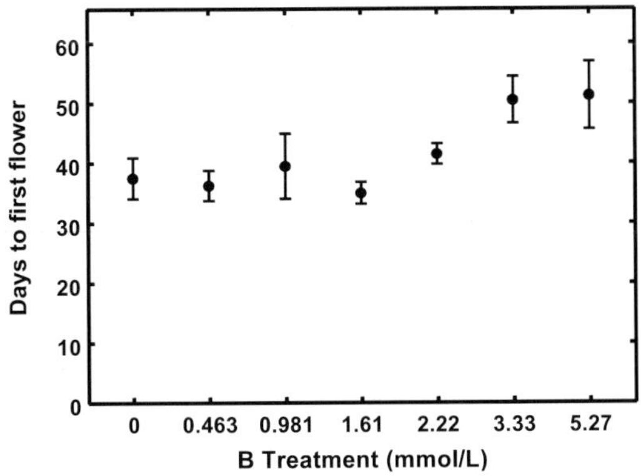

Figure 14. Delay in flowering of melons as a function of B treatment. Error bars represent one standard deviation from the mean of four replicates per treatment. From Goldberg et al. (2003)

At the highest B treatment, fruit set was completely inhibited (Goldberg et al. 2003). Similar effects have been observed in pear trees (Crandall et al. 1981) and are a much better indicator of B damage than reductions in dry matter since fruit is the marketable plant part for these crops.

Correlations between various initial soil B concentrations and B content of melon leaves, stems, and fruits at harvest were highly significant for all three extractants. The correlation with fruit B was especially high ($r > 0.99^{**}$). This is very useful since fruit is the marketable plant part for melons. The ability of the DTPA-sorbitol extractant to predict B content of melon leaves, stems, and fruits is indicated in Figure 15, since this extract showed the highest correlation coefficient.

In the field study, there were significant positive relationships between various extractable soil B concentrations and B content of muskmelon and cotton leaves for all three extractants. For melons, the best correlation was found for the sampling prior to fruit set, in agreement with the recommended sampling protocol for plant analysis. The relationship between melon B and DTPA-sorbitol extractable B is indicated in Figure 16. For cotton, the best correlation was found for the latest plant sampling. This result was surprising since the recommended sampling protocol for cotton is at flowering. The relationship between cotton B and DTPA-sorbitol extractable B is indicated in

Figure 17. Despite statistically significant relationships, the predictability of B content in field grown plants is poor using the B DTPA-sorbitol soil test that provided extremely high correlation with plant B content in the container study.

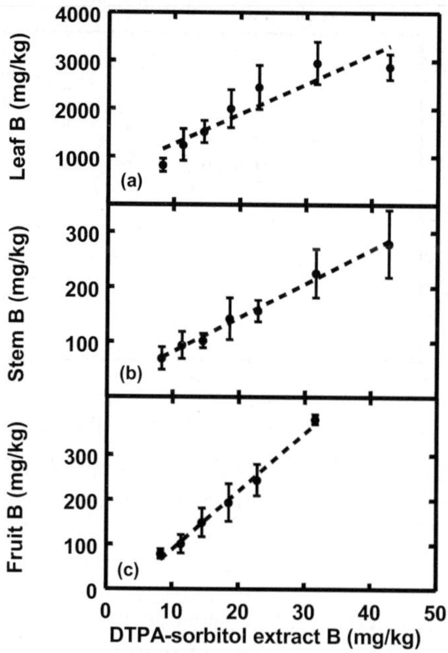

Figure 15. Ability of DTPA-sorbitol extractable B to predict B content of melons: (a) leaf B; (b) stem B; (c) fruit B. Error bars represent one standard deviation from the mean of eight plants per treatment. From Goldberg et al. (2003)

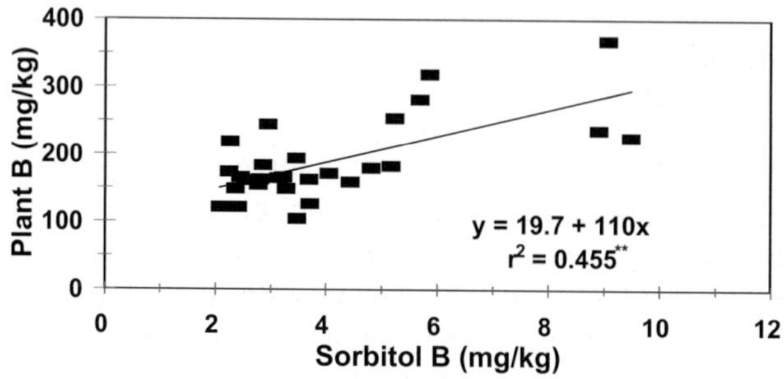

Figure 16. Boron content of melon plants prior to fruit set as a function of DTPA-sorbitol extractable soil B content depth averaged over the root zone (0-90 cm). From Goldberg et al. (2002)

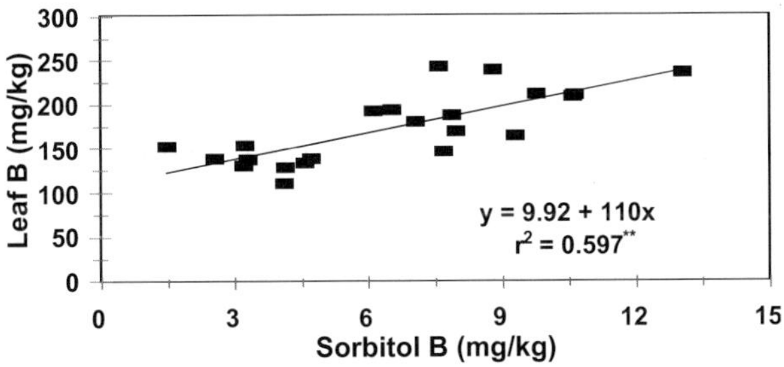

Figure 17. Boron content of cotton leaves at the end of the growing season as a function of DTPA-sorbitol extractable soil B content depth averaged over the root zone (0-180 cm). From Goldberg et al. (2002)

Historically, evaluations of the ability of B soil tests to predict plant B content have been conducted in the greenhouse. These conditions provide a much more controlled environment than the field, where clay and water content and root distribution vary considerably. The much lower, albeit significant, correlations between extractable soil B and plant B observed in the field study are attributed to sampling uncertainties and spatial and temporal fluctuations. These results draw into question the ability of soil B extractants to accurately predict soil solution B concentration experienced by field-grown plants. Soil B extractants that provide good correlation with plant B content in greenhouse studies should also be tested under field conditions.

References

Bassett R.L., 1980 A critical evaluation of the thermodynamic data for B ions, ion pairs, complexes, and polyanions in aqueous solution at 298.15 K and 1 bar, Geochim. Cosmochim. Acta **44**, 1151–1160.

Berger K.C. and Truog E., 1940 Boron deficiencies as revealed by plant and soil tests. J. Am. Soc. Agron. **32**, 297–301.

Bingham F.T., 1982 Boron In Methods of soil analysis (A.L. Page et al. eds.) Agronomy 9, Part 2, 2nd ed., American Society of Agronomy, Soil Science Society of America, Madison, pp. 431–447.

Cartwright B., Tiller K.B., Zarcinas B.A., and Spouncer L.R., 1983 The chemical assessment of the boron status of soils. Aust. J. Soil Res. **21**, 321–332.

Crandall P.C., Chamberlain J.D., and Garth J.K.I., 1981 Toxicity symptoms and tissue levels associated with excess boron in pear trees. Commun. Soil Sci. Plant Anal. **12**, 1047–1057.

Fleming G.A., 1980 Essential micronutrients. I: Boron and molybdenum. In Applied soil trace elements (B.E. Davies, ed.) John Wiley and Sons, New York, pp. 155–197.

Goldberg S., 1992 Use of surface complexation models in soil chemical systems. Adv. Agron. **47**, 233–239.

Goldberg S., 2005 Inconsistency in the triple layer model description of ionic strength dependent boron adsorption. J. Colloid Interface Sci. **285**, 509–517.

Goldberg S., 1997 Reactions of boron with soils. Plant Soil **193**, 35–48.

Goldberg S., Forster H.S., and Heick E.L., 1993 Boron adsorption mechanisms on oxides, clay minerals, and soils inferred from ionic strength effects. Soil Sci. Soc. Am. J. **57**, 704–708.

Goldberg S., Lesch S.M., and Suarez D.L., 2000 Predicting boron adsorption by soils using soil chemical parameters in the constant capacitance model. Soil Sci. Soc. Am. J. **64**, 1356–1363.

Goldberg S., Shouse P.J., Lesch S.M., Grieve C.M., Poss J.A., Forster H.S., and Suarez D.L., 2002 Soil boron extractions as indicators of boron content of field-grown crops. Soil Sci. **67**, 720–728.

Goldberg S., Shouse P.J., Lesch S.M., Grieve C.M., Poss J.A., Forster H.S., and Suarez D.L., 2003 Effect of high boron application on boron content and growth of melons. Plant Soil **256**, 403–411.

Goldberg S., Suarez D.L., Basta N.T., and Lesch S.M., 2004 Predicting boron adsorption isotherms by Midwestern soils using the constant capacitance model. Soil Sci. Soc. Am. J. **68**, 795–801.

Gupta S.K. and Stewart J.W.B., 1974 The extraction and determination of plant-available boron in soils. Schweiz. Landwirtsch. Forsch. **14**, 153–159.

Gupta U.C., 1968 Relationship of total and hot-water soluble boron, and fixation of added boron, to properties of Podzol soils. Soil Sci. Soc. Am. Proc. **32**, 45–48.

Hunter R.J., 1981 Zeta potential in colloid science. Principles and applications. Academic Press, London.

Keren R., Bingham F.T., and Rhoades J.D. 1985 Plant uptake of boron as affected by boron distribution between liquid and solid phases in soil. Soil Sci. Soc. Am. J. **49**, 297–302.

Keren R. and Mezuman U., 1981 Boron adsorption by clay minerals using a phenomenological equation. Clays Clay Miner. **29**, 198–203.

McBride M.B., 1997 A critique of diffuse double layer models applied to colloid and surface chemistry. Clays Clay Miner. **45**, 598–608.

Miller R.O., Vaughan B., and Kotuby-Amacher J., 2000 Extraction of soil boron with DTPA-sorbitol. Soil-Plant Anal. Spring, 4–10.

Peak D., Luther G.W., and Sparks D.L., 2003 ATR-FTIR spectroscopic studies of boric acid adsorption on hydrous ferric oxide. Geochim. Cosmochim. Acta **67**, 2551–2560.

Su C. and Suarez D.L., 1997 Coordination of adsorbed boron: A FTIR spectroscopic study. Environ. Sci. Technol. **29**, 302–311.

Su C. and Suarez D.L., 1997 Boron sorption and release by allophane. Soil Sci. Soc. Am. J. **61**, 69–77.

Su C. and Suarez D.L., 2004 Boron release from weathering of illites, serpentine, shales, and illitic/palygorskitic soils. Soil Sci. Soc. Am. J. **68**, 96–105.

Suarez D.L. and Simunek J., 1997 UNSATCHEM: Unsaturated water and solute transport model with equilibrium and kinetic chemistry. Soil Sci. Soc. Am. J. **61**, 1633–1646.

Boron Adsorption on Semiarid Soils of Tamil Nadu, India

R. Krishnasamy[1], U. Surendran[1], C. Sudhalakshmi[1] and M. Edward Raja[2]

[1]Department of Soil Science and Agricultural Chemistry, Tamil Nadu Agricultural University,Coimbatore –641 003.

[2] Indian Institute of Horticultural Research, Bangalore – 560080

E-mail: krishnasamydr@yahoo.com.

Introduction

Plants vary in their B requirement but the range between deficient and toxic soil solution concentration of B is very narrow than for any other nutrient element. In arid and semi arid regions, B toxicity results from high levels of B in soils and from addition of boron via irrigation water. However, B deficiency is of great concern in areas receiving heavy rainfall. Compared with other micronutrients, chemistry of B in soils is very simple. Boron doesn't undergo oxidation-reduction reactions or volatilization reaction in soils. Boron containing minerals are either insoluble (Tourmaline) or very soluble (Hydrated B minerals) and generally do not control the solubility of B in soil solution (Goldberg et al. 1993). Boron concentration in the soil solution is generally controlled by B adsorption reaction as is the amount of water soluble B available for plant uptake.

Boron adsorption is a decisive factor in boron nutrition of plants. It is because, the plant obtains its B from the soil solution and the equilibrium between B in solution and the fixed B partly determine its instantaneous supply to the plant. An additional important factor is the rate at which the B in solution is replenished by desorption from the soil or the effectiveness of a soil in controlling the concentration of available B in solution.

Boron present in soil solution or applied via irrigation water equilibrates with soil solid surfaces. The extent of this adsorption reaction is affected by the type and content of clay minerals, Al and Fe oxide content, organic matter, $CaCO_3$ and soil reaction (Keren and Bingham, 1985). Desorption rather than adsorption is essential as it describes the actual soil reaction, which controls the mobility of trace elements, the knowledge on both adsorption and desorption reactions is essential for the effective control over the concentration of these nutrients in soil solution (Krishnasamy, 2002).

Lime induced B deficiency in acid soils, presumably due to increased adsorption is a major phenomena occurring in Tamil Nadu. The role played by various soil physico-chemical properties in the mechanism of B adsorption is still

F. Xu et al. (eds.), Advances in Plant and Animal Boron Nutrition, 331–343.

ambiguous. Simple correlation inadequately explains the relationships since correlation does not ensure direct and indirect effect relationship (Wright, 1921). Path coefficient analysis that partitions correlations into direct and indirect effects has been used to investigate relationships between measured soil physico-chemical properties and adsorption parameters (Basta et al. 1993). Though adequate research pertinent to B adsorption in different type of soils was conducted, paucity of information regarding B adsorption as a function of soil pH with special reference to soils of Tamil Nadu necessitated the present study.

Hence, the present investigation was designed to derive the adsorption equation, which could be used to describe adsorption of B by soils from solution and to establish B release characteristics of these soils to constants derived from the adsorption equation.

Materials and methods

Surface soil samples (0 – 25 cm) collected from twelve sites covering all major soil series, representing various agro climatic zones of Tamil Nadu were air – dried and ground gently to pass through a 2 mm sieve .The homogenized soil samples were analyzed for pH, EC, CEC, organic carbon and particle size distribution by following the standard methods described by Piper (1966) and Jackson (1973).

Boron adsorption studies

Boron adsorption experiments were carried out in batch system to determine both adsorption isotherms (amount of B adsorbed as a function of equilibrium B concentration) and adsorption envelopes (amount of B adsorbed as a function of solution pH per fixed total B concentration. 10 mL solution containing varying levels of B (0 to 160 mg kg^{-1}) as borax ($Na_2B_4O_7$) was added in triplicate to 50 mL centrifuge tubes each containing 10 g of soil. The contents were equilibrated using a reciprocatory shaker at 27 $\pm$ 1° C for 22 h. The equilibrium period was fixed based on the previous study conducted by Krishnasamy et al. (1997). At the end of the equilibrium period, the tubes were centrifuged and the concentration of B in the equilibrium solution was determined by Azomethine –H spectrophotometric technique described by Bingham (1982) using HACH DR/2500 Spectrophotometer. Boron adsorbed by soil was calculated as the difference between the initial concentration and equilibrium B concentration.

Modeling

Boron adsorption was modeled with both the Langmuir and Freundlich adsorption isotherms. Langmuir adsorption isotherm was developed through the kinetic theory of gases to describe adsorption of gases on solids (Langmuir, 1918)

The Langmuir equation is given by

$$C/x/m = 1/Kb + C/b \qquad \text{------- (1)}$$

where, C is the equilibrium concentration of B (mg L^{-1}); x/m is the amount of B adsorbed (mg kg^{-1}); b is the adsorption maxima (mg kg^{-1}), while K is a constant related to bonding energy (L mg^{-1}).

Freundlich (1926) has given the following equation as suitable for describing the adsorption of ions from liquid on to a solid surface.

$$x/m = KC1/n \qquad \text{------- (2)}$$

where, x/m is the amount of B adsorbed per unit weight of soil (mg kg^{-1}); C is the equilibrium concentration of B (mg L^{-1}); while K and n are empirical constants. The constants K and n were obtained from the intercept and reciprocal of the slope of the straight line graph obtained by plotting the data on log-log scale.

Path coefficient analysis

Path coefficient analysis was made to determine whether any relationship exist between various soil properties and B adsorption parameters. It permits the partitioning of simple correlation coefficient between dependent variables (K values) and independent variables (soil properties) into direct and indirect effects. Path analysis results were obtained by following the procedure of Williams et al. (1990). Both direct and indirect effects of soil properties on adsorption maxima were delivered by using path analysis.

Sorption envelope as a function of pH

The isotherm parameters obtained using the adsorption isotherm equation through curve fitting procedure was valid only for the condition in which the experiment was conducted. Hence prediction of B adsorption behaviour under changing pH and solution B concentration was studied. For B adsorption envelopes, the equilibrating solution containing 20 mg B kg^{-1} had been adjusted to desired pH values using 4 M HCl or 4 M NaOH (3.0, 4.5, 6.0, 7.5 and 9.0 pH) (Goldberg and Forster, 1991). Equilibrating solution was prepared by dissolving the required quantity of borax. As per the procedure indicated in B adsorption studies, the quantity of B sorbed was calculated.

Boron desorption studies

The adsorption studies suggested that B may be held in different soil fractions or at different loci, each locus having unique retention energy. B sorption envelopes as a function of pH clearly proved that the retention was greatly influenced by soil reaction. Many earlier reports on B desorption reactions suggest that it was not studied in detail as that of adsorption (Elrashidi and O'Connor, 1982). The desorption study was conducted to know the rate at which it is desorbed from the soil by leaching, since this will decide the concentration of available boron in soil.

The desorption experiment was conducted in all the twelve soils which were used for adsorption studies. Desorption studies were initiated by adding 5 mL of B free 0.01 M $CaCl_2$ in triplicate (Jaishankar, 1994). This was vigorously agitated for 2 hrs on a reciprocating shaker and equilibrated for 22 hours at 27 $\pm$ 1°C. The suspension was centrifuged and the concentration of B in the equilibrium solution was determined. Desorbed B was calculated as the difference between equilibrium B concentration at the desorption step and B concentration at the adsorption step.

Results and discussion

The selected soils for B adsorption study represented a range of texture and amount of Boron (Table 1). The pH and EC of the soils varied from 4.03 to 8.13 and 0.037 to 0.838 dS m^{-1}, respectively. Organic carbon (2.7 to 12.6 g kg^{-1}), Fe_2O_3 (2.4 to 13.6 g kg^{-1}), Al_2O_3 (2.8 to 16.8 g kg^{-1}), $CaCO_3$ (1.5 to 29.2 g kg^{-1}) and CEC (8.4 to 36.2 cmol (p+) kg^{-1}) also varied extensively and it provided sufficient scope for the study of B adsorption.

Boron adsorption studies

The amount of B adsorbed increased and per cent adsorption decreased with increasing concentration of added B. It is important to understand the shape of the adsorption isotherm, because its classification depends on how the ion distributes itself in soil solution as well as in exchange phase (Sposito, 1981). When the amount of B adsorbed was plotted against the equilibrium concentration, the curve followed (L) type of adsorption isotherm given by Glassstone (1953) and Giles (1960) (Fig. 1).

This L curve isotherm indicated that the B has a reasonably high relative affinity for the exchange sites. The initial curvature of this L shaped isotherms depends on the rate of change of site availability with increase in the solute adsorbed. As more solute was taken up, there was usually less chance of bombarding solute molecules to find a suitable site on which it could be adsorbed; i.e., to cause adsorption of a given additional amount of solute, the external solution concentration must be raised by ever increasing amount.

Table 1. Physical and chemical properties of the selected soils

Soil No.	Location	pH 1 : 2.5	EC d Sm^{-1}	Org. C g kg^{-1}	CEC (cmol (p^+ kg^{-1})	$CaCO_3$	Fe_2O_3	Al_2O_3	Sand	Clay	Texture
						----	----	g kg^{-1} ----	----	----	
S1	Kattupudur	8.04	0.056	7.10	19.7	12.4	13.6	16.8	342	325	sil
S2	Arumanallur	8.00	0.040	9.60	8.4	8.2	4.10	6.80	784	136	scl
S3	Kanyakumari	7.80	0.037	12.60	36.2	8.6	4.92	5.04	432	464	cl
S4	Nagercoil	4.03	0.424	9.50	13.8	12.6	8.64	4.62	695	168	sl
S5	Ooty	5.29	0.113	6.20	13.2	4.8	6.88	5.81	711	217	sl
S6	Coonor	7.07	0.039	4.20	16.6	6.7	6.73	2.86	604	274	scl
S7	Irugur	7.71	0.053	8.60	14.9	13.6	4.80	9.12	510	295	sl
S8	Somayanur	7.82	0.049	9.30	24.1	18.4	3.86	8.16	599	274	scl
S9	Aravangadu	7.69	0.102	2.70	15.2	1.5	10.8	13.4	356	286	cl
S10	Bhavanisagar	8.13	0.838	10.4	21.5	9.8	7.41	9.68	303	325	sil
S11	TNAU Farm - I	7.93	0.681	6.20	32.5	2.8	2.40	4.80	355	369	cl
S12	TNAU Farm - II	5.47	0.730	5.10	27.1	1.6	6.80	3.20	446	344	l

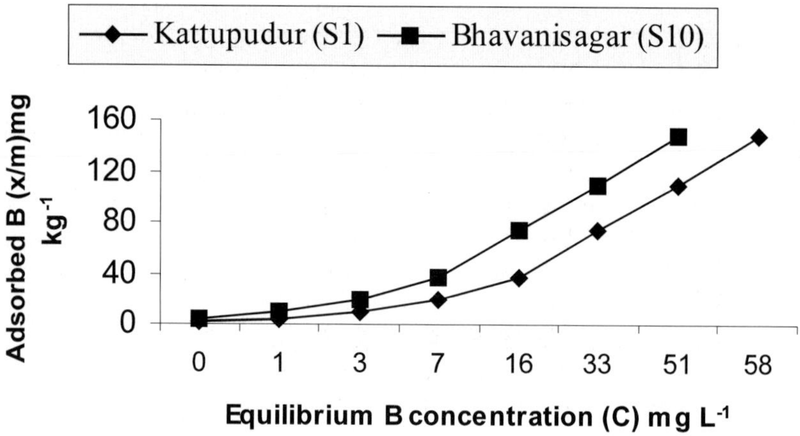

Figure 1. Amount of boron adsorbed and boron in equilibrium solution in selected soils

Though the B adsorption data conformed to both the Langmuir and Freundlich equations, the fit was excellent with the later. B adsorption on the 12 soil samples could be described by the Langmuir adsorption isotherm over a limited B concentration range, i.e., B adsorption can be expressed only approximately by this equation and that too only at low concentration. These findings confirm those of Hatcher and Bower (1958) and Krishnasamy et al. (1997). For most of the soils, the data deviated from the linear Langmuir behaviour. Lack of conformity to the Langmuir equation could be ascribed to the operation of other retention mechanisms in addition to adsorption or the presence of adsorption sites having variable bonding energies.

However, linear relationships were obtained between log (x/m) and log C for all the soils (Fig. 2).

B adsorption on all samples could be adequately described over the entire concentration range studied using Freundlich adsorption isotherm. The statistical analysis of the data for fit to this Freundlich equation by the method of least square gave correlations (r values) ranging from 0.92 to 0.99 (Table 2).

The values of K represented the amount of B adsorbed at unit equilibrium concentration and 1/n indicated the degree of linearity between solution equilibrium concentration and adsorption. The Freundlich isotherm was originally an empirical equation, which mathematically describes quantitative experimental

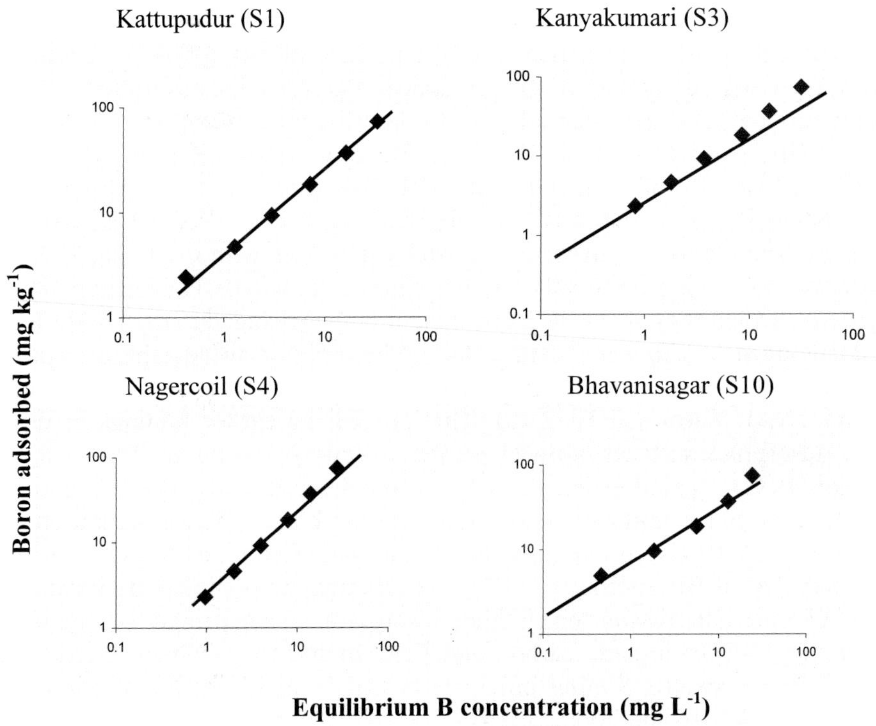

Figure 2. Freundlich adsorption isotherm constants (K, n) in soils of Tamil Nadu

Table 2. Freundlich Adsorption isotherm constants (K, n) in soils of Tamil Nadu

Soil No.	Location	K	n	r
S1	Kattupudur	0.42	1.31	0.991
S2	Arumanallur	0.11	1.36	0.997
S3	Kanyakumari	1.55	1.32	0.996
S4	Nagercoil	0.78	1.08	0.961
S5	Ooty	0.96	1.14	0.944
S6	Coonor	0.34	1.26	0.986
S7	Irugur	0.27	1.14	0.974
S8	Somayanur	1.04	1.28	0.924
S9	Aravangadu	0.55	1.12	0.987
S10	Bhavanisagar	1.12	1.96	0.946
S11	TNAU – Farm I	1.43	1.08	0.967
S12	TNAU – Farm II	1.09	1.15	0.998

results. However, this can be shown to confirm a model of adsorption in which the affinity term decreases exponentially as the amount of adsorption increases. By contrast, the Langmuir isotherm was developed to describe the adsorption of gases on solid surfaces, and has been adopted to describe the adsorption of many soil species (Harter and Baker, 1977). Freundlich 'K' values for the adsorption reaction of twelve soils ranged from 0.11 to 1.55. Coarse textured soils (Arumanallur S2 - 0.11) had less K values than fine textured soils (Kanyakumari S3 - 1.55). The maximum adsorption capacity increases with the fineness of the soil texture. Soils are seen to vary in their capacity to adsorb B as well as in the energy with which they retain it. Even though both Arvangadu and TNAU Farm soil I belonged to clayloam texture, the wide variation in adsorption capacity might be due to low content of organic carbon in TNAU Farm soil I (Arvangadu - 6.20 and TNAU Farm soil I - 2.70). This proved the earlier findings in which organic adsorptive surfaces showed greater affinity to retain B (Elrashidi and O'Connor, 1982). Ligand exchange is a possible mechanism by which mono and bi chelated (B-diol complexes) may form with the breakdown of organic matter products in soil. B adsorption capacity of Kattupudur soil was found to be low (0.42) than that of Bhavanisagar (1.12), even though the textural classification is similar. The possible reason for the high adsorption capacity might be due to the influence of $CaCO_3$, organic carbon and CEC of the soils, which override the effect of similar texture. Similar influence of $CaCO_3$ in adsorption of boron was recorded by Goldberg and Forster (1991).

Path coefficient analysis

Path coefficient analysis (direct effect) and correlation analysis (r values) showed that clay, CEC and OC are the most important soil properties affecting B adsorption by soils. Direct effect of soil properties on B adsorption are represented by single headed arrows while coefficients of intercorrelations between soil properties are represented by double headed arrows (Fig. 3).

Indirect effect of soil properties on metal adsorption are determined from the product of one two headed arrow and one single headed arrow. Path analysis results were obtained from the following equations (Williams et al. 1990)

$$r_{14} = P_{14} + r_{12}P_{24} + r_{13}P_{34} \qquad \text{------ (3)}$$

$$r_{24} = r_{12}P_{14} + P_{24} + r_{23}P_{34} \qquad \text{------ (4)}$$

$$r_{34} = r_{13}P_{14} + r_{23}P_{24} + P_{34} \qquad \text{------ (5)}$$

where, r_{14} – simple correlation; P_{14} -Direct effect and r_{24} P_{24} are the indirect effects. Subscript designations are : 1) Clay 2) OC 3) CEC 4) K values.

It could be seen that the direct effect contributed by clay (r = 0.919**) towards the K value was very high and revealed the significant relationship of B adsorption with that of fine mineral fractions (Basta et al. 1993). This was followed by significant effect of OC (r= 0.838*) and CEC (r= 0.612*). The formation of complexes between dihydroxy organic compounds and B can presumably explain the observed behaviour, showing the organic carbon plays an important role in B adsorption. Significant correlation values ('r') for these parameters were improved mainly due to the direct positive contributions (3.042, 2.212 and 1.889) (Table 3).

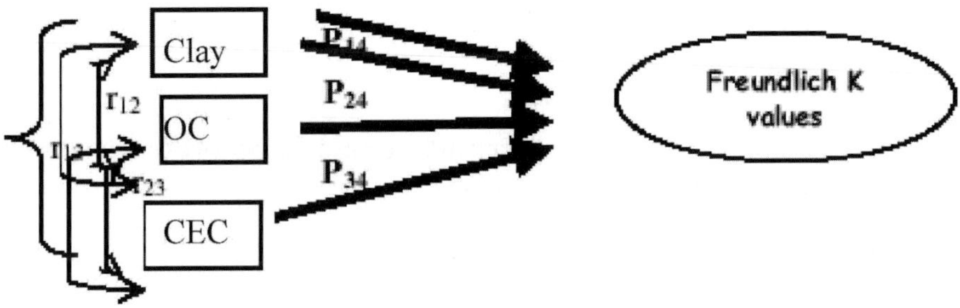

Figure 3 . Path diagram for the relationship between soil properties and Freundlich K values

Table 3. Path coefficient analysis using Freundlich K parameter and soil properties

	pH	R_2O_3	Org. C	CEC	$CaCO_3$	Clay	r	R^2	Residual effect (U2)
pH	**-.0957**	- 0.0211	- 0.0289	0.299	-.1075	0.156	0.428	0.899	0.3164
R_2O_3	1.204	**- 0.585**	-1.944	-0.379	3.804	-2.770	-0.699		
Org. C	1.385	- 3.474	**1.889**	0.550	0.034	0.448	0.838*		
CEC	0.314	- 4.031	0.714	**2.212**	-0.400	1.785	0.662*		
$CaCO_3$	0.279	0.040	-0.059	-0.003	**0.134**	-0.241	0.369		
Clay	-2.15	-1.22	0.003	1.560	0.065	**3.042**	0.919**		

Bold letters indicate direct effect : ** Significant at $P < 0.01$; * Significant at $P < 0.05$.

Path analysis exhibited the significant direct contribution made by clay towards CEC (1.785) in all the soils, thus revealing the relationship between CEC and adsorption of B. This is further evidenced by the large direct effect of CEC and the small indirect CEC effect via OC (0.714) and large indirect effect of OC via CEC (0.550) and the direct effect of OC with high significant correlation value (r= 0.838*) suggest that B adsorption was highly influenced by OC through not only by complexation sites but also by CEC sites. Though pH, R_2O_3 and $CaCO_3$ showed direct negative influence on 'K' due to indirect positive influences through clay, CEC and OC, its total 'r' value was not only improved but also become positive. Total "r" values for pH, $CaCO_3$ and R_2O_3 was not significant and the reason attributed may be due to the direct negative effect through other main soil parameters (clay, CEC and OC) which might have pulled down the effect of these parameters. Other factors contributed for this non significance are i) pH of the soils didn't show much variation among themselves except two soils ii) quality (fineness) of $CaCO_3$ present in the soil samples and limited range of $CaCO_3$ in the selected soil samples. Low uncorrelated residual effect value (U2 – 0.3164) and high significant coefficient determination (R^2- 0.899) value indicate that path coefficient analysis model explains most of the variations observed in B adsorption by soil.

Sorption envelope as a function of PH

Boron adsorption on soils was very dependent on solution pH (envelopes are shown in Fig. 4 for two selected soils samples).

Boron adsorption by soils increased as a function of solution pH in the range of 3 to 9. The shape of the adsorption envelopes were similar for almost all the samples, showing adsorption peaks in the pH range of 7.5 to 9.0. This corroborates with the findings of Bingham et al. 1971; Keren et al. 1985.

Mechanism of B adsorption includes ligand exchange, formation of bidentate surface complexes and incorporation into clay mineral lattices. Thus pH influenced ligand exchange mode of B adsorption is further confirmed (Goldberg et al. 1993). But in two soils, the adsorption of B increased up to pH 7.5 and decreased thereafter.

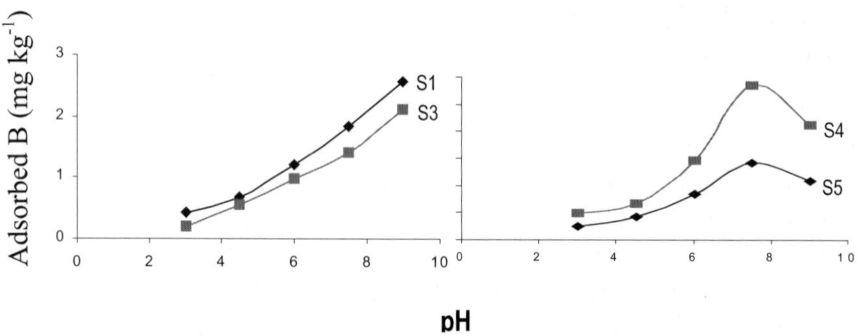

Figure 4. Variation in Boron adsorption as a function of pH in selected soils

Boron desorption studies

The amount of B desorbed from soil increased and percent desorption decreased with increasing concentrations of adsorbed B. This is in accordance with the findings of Rhoades et al. (1970) and they showed that a large fraction of B is initially removed by leaching but that the remainder persists even after large volumes of water have been applied. B desorption curves did not correspond to the

B adsorption curves, hence exhibit hysteresis. The possible reason for this difference may be attributed to the mechanisms of anion sorption, since anion desorption at constant pH exhibits varying degrees of irreversibility. For non-hysteric soils, the same adsorption parameters can be used to describe desorption but since the soils exhibit hysteresis, different sets of adsorption parameters are necessary. Hence the data obtained from desorption experiment was fitted both in Langmuir and Freundlich isotherms. The results clearly indicated that it did not obey the Langmuir isotherm but fitted well with the Freundlich isotherm as evidenced by the r values (0.94-0.99) obtained (Fig. 5).

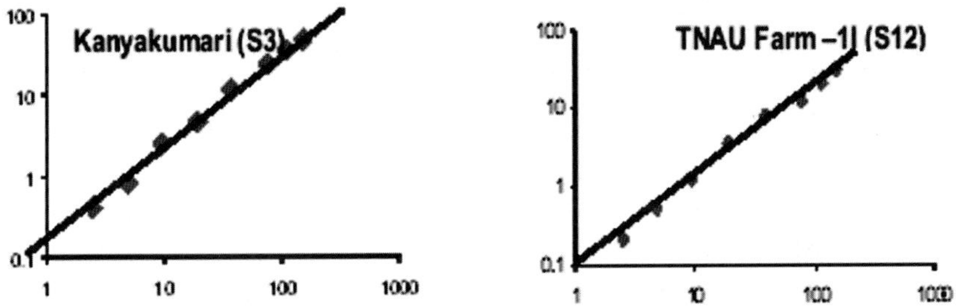

Figure 5. Freundlich desorption isotherms in two selected soils of Tamil Nadu

It is in consonance with Marzadori et al. 1991. The dersorbability K values ranged from 0.10 to 7.85 mg kg^{-1}. Significant positive correlation existed between soil texture and pH.

Conclusion

Boron adsorption capacity of the soils has been found to depend upon its texture, organic carbon and CEC. Langmuir adsorption equation was found to be valid over ranges of limited concentrations of the solutions, but when all the concentrations were taken into consideration, the relationship followed Freundlich adsorption isotherm. The studies of boron sorption as a function of pH showed an increase in boron adsorption with increase in pH range of 3.0 to 7.5 and thus exhibited a peak at the pH range of 7.5 to 9.0 in most of the soils. The data on B adsorption showed that the soils not only varied in their capacity to retain B but also in with which they adsorbed it. The findings will be useful to predict boron injury / deficiency in such soils types in Tamil Nadu.

References

Basta N.T., Pantare D.J. and Tabatabai M.A., 1993 Path analysis of heavy metal adsorption by soil. Agron. J. **85**, 1054–1057.

Bingham F.T., 1982 Boron. In Methods of soil analysis (A.L. Page et al. eds) Part 2. 2nd edition. Agronomy **9**, 431–447.

Bingham F.T., Page A.L., Coleman N.T. and Flach K., 1971 Boron adsorption characteristics of selected soils from Mexico and Hawaii. Soil Sci. Soc. Am. J. **35**: 546–550.

Elrashidi M.A. and O'Connor G.A., 1982 Boron sorption and desorption in soils. Soil Sci. Soc. Am. J., **46**: 27–31.

Freundlich H., 1926 Colloid and capillary-Translated from the third German edition by H. Stafford Hartfield. Methuen and Co, Ltd, London.

Giles C.H., Macewan T.H., Nakhwa S.N., and Smith D., 1960 Studies in Adsorption. Part XI. A system of classification of solution isotherms and its use in diagnosis of adsorption mechanisms and in measurement of specific surface area of solids. J. Chemical Soc, 3973–3993.

Glasstone S., 1953 Textbook of Physical Chemistry, pp. 1202. Macmillan & Co, Ltd, London.

Goldberg S. and Forster H.S., 1991 Boron sorption on calcareous soils and reference calcites. Soil Sci. **152**: 304–310.

Goldberg S., Forster H.S., and Heick E.L., 1993 Boron adsorption mechanisms on oxides, clay minerals and soils inferred from ionic strength effects. Soil Sci. Soc. Am. J. **57**: 704–708.

Harter R.D. and Baker D.E., 1977 Applications and misapplications of the Langmuir equation to soil adsorption phenomena. Soil Sci. Soc. Am. J. **41**: 1077–1080.

Hatcher J.T. and Bower C.A., 1958 Equilibria and dynamics of boron adsorption by soils. Soil Sci. **85**: 319–323.

Jackson M.L., 1973 Soil Chemical Analysis. Prentice Hall of India Private Limited, New Delhi.

Jaishankar J., 1994 Characteristics of Boron Adsorption-Desorption in major soil series of Tamil Nadu. M.Sc (Ag.) dissertation submitted to Tamil Nadu Agricultural University, Coimbatore, India.

Keren R. and Bingham F.T., 1985 Boron in water, soils and plants. Adv. Soil Sci. **1**: 229–276.

Keren R., Bingham F.T., and Rhoades J.D., 1985 Plant uptake of boron as affected by boron distribution between liquid and solid phases. Soil Sci. Soc. Am. J. **49**: 297–302.

Krishnasamy R., 2002 Sorption of Zinc and Boron in the soils of Tamil Nadu. Madras Agric. J. **89**: 198–206.

Krishnasamy R., Jaisankar J., and Suresh M., 1997 Characterization of boron adsorption in soils of Tamil Nadir. In boron in soils and plants (R. W. Bell and B. Rerkasem, eds.)

Kluwer Academic Publishers, Dordrecht, the Netherlands, pp 255–259.

Langmuir I., 1918 The adsorption of gases on plain surfaces of glass, mica and platinum. J. Am. Chemical Soc. **40**: 1361–1403.

Marzadori C., Vittori Anitsari L., Ciavatta C., and Sequi P., 1991 Soil organic matter influence on adsorption and desorption of boron. Soil Sci. Soc. Am. J. **55**: 1582–1585.

Piper C.S., 1966 Soil and Plant Analysis. Hans Publishers, Bombay.

Rhoades J.D., Ingvalson R.D., and Hatcher J.T., 1970 Laboratory determination of leachable soil boron. Soil Sci. Soc. Am. Proc. **34**: 871–875.

Sposito G., 1981 The Thermodynamics of Soil Solutions. Oxford University Press, Oxford, New York.

Williams W.A., Jones M.B., and Demment M.W., 1990 A concise table for path analysis statistics. Agron. J. **82**: 1022–1029.

Relationship Between Plant Availability of Boron and the Physico-chemical Properties of Boron in Soils

Duanwei Zhu[1*], Juan Wang[1], Shuijiao Liao[1, 2] and Wuding Liu[1]

[1]Microelement Research Center, Huazhong Agricultural University, Wuhan 430070, China;

[2]Department of Chemistry, College of Basic Sciences of Huazhong Agricultural University, Wuhan 430070, China. Corresponding author: zhudw@mail.hzau.edu.cn

Introduction

Boron is an essential micronutrient element required for the normal growth of plants. The range between boron deficiency and toxicity symptoms in plants is typically narrow, in the range of 0.028 to 0.093 mmol/L for sensitive crops and 0.37 to 1.39 mmol/L for tolerant crops (Goldberg, 1997). Much work has been done on plant availability of boron and physico-chemistry of boron in soil, but generally not in combination. Availability of boron to plants is affected by a variety of soil factors including soil solution pH, soil texture, soil moisture, temperature, oxide content, carbonate content, organic matter content, and clay mineralogy (Goldberg et al. 2000). However, these soil properties are functions of the physico-chemistry of soils, so that determination of relationships between plant availability of boron and the physico-chemical properties of boron in soils has still tremendous interest for us. This article summarizes results about plant availability of B, physico-chemical properties of soil B, chemical behavior of common B soil minerals and their relationships.

Availability of boron to plant

Hot-water soluble boron (HWSB) is a standard method to measure the plant availability of soil boron in China. Availability of B differs in different soils. In acid soils, B availability is always low. For example, HWSB is only 0.1~0.2 mg/kg on yellow-brown soil and red soil in southern China, obviously lower than on alluvial soils developed from the calcareous sediments in the Jianghan Plain beside the middle reaches of Changjiang River. The B concentration in the soil solution is generally significantly related to the amount of boron taken by plants, as plants respond only to the B activity in soil solution. It was reported that 80% of boron was transferred into unavailable forms during a cropping season, and only 20-60% of B could be extracted by hot water (Zhu, et al. 1994). Availability of boron for plant growth was strongly related to B capacity, boron supplying intensity, sorption and desorption behavior of boron in soils.

F. Xu et al. (eds.), Advances in Plant and Animal Boron Nutrition, 345–354.

Adsorption of B on soil mineral surfaces is important for managing B toxicity or deficiency because adsorbed B is not perceived as toxic by plants.

Capacity of B in soil and its availability to plants

Capacity of B is the sum of B in soil extracted for several times in succession, and is also called the chemical pool of soil boron (Zhu et al. 1995). The bigger the capacity, the more boron the plant absorbs. Table 1 shows the relationship between capacity of B in soils and the accumulated amount of B in plants. The higher the capacity, the more B was absorbed by the plant. There was a significant correlation between soil pH and the logarithm of boron accumulated by the plant. This reflects the close relationship between boron capacity in soil and availability of B to the plant (Zhu et al. 1999a). Table 1 also shows that B capacity was affected by soil pH. As pH increased, boron capacity increased also. This explains why in acid soils, boron capacity is always low and temporary B deficiency can be triggered by liming of acid soils because of increased B adsorption at higher soil pH. Boron becomes less available with increasing of solution pH. Soil pH was considered the most important factor for boron capacity (Keren et al. 1983). Different types of B extractants showed different activity (Jin et al. 1987). Hot water soluble boron is considered as an index of B supplying capacity of soil although soil boron extracted with $CaCl_2$, HCl or NH_2OH-HCl had greater correlation with B content or boron uptake by rape plants (Zhu et al. 1998a). Correlation coefficients between different B extractants and plant absorbed B are listed in Table 2. With the exception of mannitol-B, they were well correlated to plant B concentration and plant B uptake. NH_2OH • HCl-B was correlated to B absorbed by rape during the growing season. HWSB and boron extracted with hydrochloric acid had the best relationship with plant B concentration and plant uptake. This showed that these two kinds of boron extractants were more effective than the others and could represent the ability of the soil to supply B.

Table 3 shows coefficients of variation of two B extractants in the four soils after planting for several cuttings of crops. It was observed that the concentration of boron extracted by $CaCl_2$ was always low and its coefficient of variation was small. This pattern of B content did not change much as time went by and cutting time increased.

Due to this state, $CaCl_2$-B can represent the most stable pool of boron in soil. But, as cutting time increased, variability of hot water soluble boron was 16.8% to 36.8%, concentration of hot water soluble B declined and the coefficient of variation was larger than for $CaCl_2$-B. Furthermore, HWSB was not only affected by clay content in soil but also by seasonal temperature. Thus it can represent boron that is easily adsorbed or desorbed by soil solids. This kind of boron was more effective to the plant. Boron extracted by mannitol or $NH_2OH \cdot HCl$ was thought to be from fractions which were specifically adsorbed by soil solids.

Table 1. Relationship between repeated extractions of B by mannitol and absorbed amount of B by rape seedlings

Locality	Soil type	pH	First-order equation 100×k	$t_{1/2}$ (d)	r	Total amount of extracted B ($mg \cdot kg^{-1}$)	Accumulated amount of rape absorbed B ($\mu g \cdot pot^{-1}$)
Tianmen	Calcareous alluvial soil	8.1	10.7	6.27	0.998	7.9	261.61
Xinzhou	Alluvial soil	6.5	8.5	6.99	0.997	5.2	265.30
Xianning	Brown-red soil	5.5	8.3	8.04	0.992	4.4	72.11
Changsha	Red soil	5.2	7.7	8.4	0.997	3.6	46.68

Table 2. Correlation coefficients between soil extractable B and B content or B uptake of crops

Crop	Items	HWSB	HCl-B	Mannitol-B	$NH_2OH \cdot HCl$-B
Rape	Boron content	0.863**	0.813**	0.465	0.760**
	Boron uptake amount	0.885**	0.838**	0.474*	0.781**

Table 3. Variability of $CaCl_2$ - B and HWSB in soil as number of cutting times increases

Soil sample	Clay particle content (%)	$CaCl_2$ (mg/kg) Mean (×)	Coefficient of variation (CV%)	HWSB (mg/kg) Mean (×)	Coefficient of variation (CV%)
Calcareous alluvial soil	26.85	0.075	13.3	0.345	18.7
Alluvial soil	8.82	0.163	13.5	0.243	36.8
Yellow-brown soil	33.58	0.083	11.6	0.255	24.9
Brown-red soil	28.72	0.110	12.8	0.225	16.8

Dynamic properties of boron and its plant availability

Boron supplying intensity of soil is another factor which influences plant availability of boron. The dynamic process of soil boron described by the chemical kinetic equation can be used to reflect its plant availability (Zhu et al. 1997; 1999b). In Table 1, the k value of first order equation is the velocity constant of boron and $t_{1/2}$ is the half life of the extraction time. These constants are indicators which reflect the release velocity in boron desorption of the soils with mannitol. The bigger the value of k or the shorter $t_{1/2}$ was, the greater the amount of B that would be absorbed by rape seedlings (Zhu et al. 1999a).

Electro-ultrafiltration (EFU) equipment has been used frequently to measure available potassium. It was used to research dynamic properties of soil boron in a constant electric field when we collected different boron components at the anode and the cathode after incubation of added boron in acidic (brown-red soil) and alkaline soil (calcareous alluvial soil) (Zhu et al. 2000a). From the second to sixth fraction under conditions of 200V, boron was transported from the soil solid-phase to the soil solution. In order to describe the relationship between the dynamic processing of boron transported from the soil solid-phase to the soil solution and the B content of every soil fraction, the B content of each soil fraction was determined and dynamic equations were fitted to the experiment data. The equations were the zero-order equation $Q_i=k_0t+Q_0$, and the parabolic diffusion equation $Q_i=k_pt^{0.5}+Q_0$. As shown in Table 4, the desorption rate constants of the zero-order equation or the parabolic diffusion equation at the anode were smaller than those at the cathode under acidification and added boron incubation (ABI) for the brown-red soil, and that upon increasing the pH and added boron incubation (BBI), the desorption rate constants of the zero-order equation or the parabolic diffusion equation at the anode were bigger than those at the cathode in the same soil. On the contrary, in the case of calcareous alluvial soil, the rate constants in the ABI and BBI at the anode were all higher than those at the cathode. The results confirmed that cation-pairs of boron constituted an important portion of active boron under acidic condition, and that active boron existed mainly as borate ion in calcareous soil.

Physico-chemical characteristics of B in soil

Sorption and desorption behavior is the most important process which affects the plant availability of these nutrient elements. Several kinds of active sites exist on the surfaces of soil solids, especially on soil common minerals. Boron reacts with these sites and the extent of reaction determines its plant availability. Our research results showed that B sorption and desorption processes were not always reversible. On some soils, desorption hysteresis existed. We modeled our data using the Freundlich equation, obtained constant Ke and 1/n for the sorption and the desorption processes. Subsequently, we found that sum of △Ke and △

1/n, in which △Ke is the experimental constant Ke difference value and △1/n is another constant 1/n difference value of the desorption isotherm and adsorption isotherm of the test soils, had a significant and negative relationship with the B concentration of the plant (Zhu et al. 1998a). We named the value of △Ke+△1/n as boron hysteretic desorption coefficient. From Table 5, we can see that the lower the value of the hysteretic desorption coefficient, the stronger was the boron reversibility in the soil and the higher the B availability to the plant.

Table 4. Kinetic characteristics of boron desorption by EUF in two soils under various incubations

Equation	Velocity constant and regression coefficient	Pole	Brown-red soil		Calcareous alluvial soil	
			ABI	BBI	ABI	BBI
Zero-order	$k_0 \times 10^3$	anode	2.34	4.35	5.67	6.29
		cathode	2.68	1.51	4.04	1.45
	R^2	anode	0.995	0.969	0.959	0.932
		cathode	0.969	0.983	0.936	0.888
Parabolic diffusion	$k_P \times 10$	anode	1.57	2.93	3.86	4.30
		cathode	1.82	1.02	2.76	1.00
	R^2	anode	0.992	0.969	0.984	0.965
		cathode	0.986	0.995	0.966	0.932

Table 5. Relationship between B hysteretic coefficient and its plant concentration in soils

Sample	1	2	3	4	5	6	7	8	9	10	11
△Ke+△1/n	0.37	0.15	0.29	0.68	0.05	0.14	0.41	0.24	0.27	0.16	0.36
Plant B content	7.8	15.5	10.8	5.5	16.7	16.8	11.1	9.8	8.7	8.7	7.0

Table 6. Correlation analysis between soil components and boron hysteretic coefficient

Hysteretic coefficient	Correlation coefficient (n = 11)			
	O.M.	Amorphous Fe	Amorphous Al	$NH_2OH \cdot HCl$ Mn
□Ke+□(1/n)	0.041	0.255	0.790**	0.744**

**: Correlation is significant at the 0.01 level.

Research on B in acid soils was carried out later than that on neutral and alkali soils. In order to solve B deficiency problems on conifer soils, B chemistry in acid soil has been studied. It was reported that B capacity was low in acid soils because of the low B content of B containing minerals in acid soils. The reason was B sorption capacity decreased as the pH declined. The dominant B adsorbing surfaces in soil are oxides, clay minerals, calcite, and organic matter. Table 6 shows the relationship between soil solid materials and boron hysteretic coefficient in the soils (Zhu et al. 1998a). Results showed that amorphous aluminum and manganese oxides affected B sorption and desorption behavior in most in soils.

Soil pH is another important factor that affects B behavior in soil. It affects the solubility of minerals and B pools in soil solution. Many literature studies reported that B sorption amount increased with pH till 8 to 9, and then decline with pH in many soils. B deficiency in soil sometimes is related to temperature and soil moisture.

It was found that B sorption was exothermic on some soils; as temperature increased, adsorbed amount of B by soil or minerals increased also. Boron was harder to desorb when the temperature was high. This means that boron reversibility was enhanced by temperature (Zhu et al. 2000b, Cheng et al. 2002b). In order to measure the effect of heat on B adsorption for several minerals, a kind of two-thermometer method was used (Lei et al. 2004). From Table 7, we can see that the boron adsorption enthalpy was always negative, B adsorption on these soils were exothermic. But since this value was small, it indicated that B was specifically adsorbed by these soils. From Table 7 and Table 8, we can see B sorption maxima, sorption affinities, and hysteretic coefficients vary with temperature.

Relationships between physico-chemical properties of B and soil common minerals

Soil PZC declined when B was adsorbed. This process not only increased the surface negative charge but also increased the amount of protons on soil surfaces since proton is a kind of accompanying ion. Our research showed that potential amount of adsorbed protons without adding boron in the adsorption process (Q_1) was lower than that when adding boron in the adsorption process (Q_2). The result showed that adding boron into soil could increase the amount of proton adsorption (Table 9). This reaction suggested more protons would be adsorbed under conditions of adding trace boron concentrations into soils, especially into acidic soils such as brown-red soil or yellow-brown soil. The sorption molar ratio of boron to proton in the soils was almost 1:10 (Chen, et al. 2002). In boron desorption, the potential amount of adsorbed protons was still larger than that without boron treatment in acidic brown-red soil, namely, the difference of Q_3 and Q_1 of this soil was bigger than zero as shown in Table 9.

Table 7. Langmuir equation constants for B adsorption and heat of B adsorption in the soils

Sample	25℃			40℃			ΔHa
	K	Xm	R	K	Xm	r	(kJ/mol)
Brown-red soil	1.36	1.19	0.999	1.01	1.17	0.996	-15.4
Yellow-brown soil	1.80	0.89	0.997	1.33	0.84	0.984	-15.6
Calcareous alluvial soil	1.72	0.93	0.998	1.11	0.95	0.973	-22.6

Table 8. Boron adsorption and desorption constants and B hysteretic coefficient for two temperatures

Sample	25℃					40℃				
	Adsorption		Desorption		Hysteretic coefficient	Adsorption		Desorption		Hysteretic coefficient
	Ke	1/n	Ke	1/n	ΔKe+Δ1/n	Ke	1/n	Ke	1/n	ΔKe+Δ1/n
Brown-red soil	0.68	0.63	0.87	0.65	0.21	0.57	0.92	0.78	1.06	0.35
Yellow-brown soil	0.59	0.59	0.76	0.61	0.19	0.48	0.51	0.55	0.67	0.23
Calcareous alluvial soil	0.59	0.58	0.63	0.67	0.13	0.44	0.59	0.54	0.70	0.21

Table 9. Change in amount of potential adsorbed protons on soils under different PZCs with or without boron adsorption

Samples	ck-PZC	Q_1	ads-PZC	Q_2	des-PZC	Q_3
Brown-red soil	4.3	2.2	3.4	5.5	3.3	3.0
Yellow-brown soil	3.4	6.5	3.1	10	3.1	3.5
Calcareous alluvial soil	7.4	11	7.3	12	7.2	6.5

Boron was adsorbed predominantly by iron and aluminum oxides in soil. Research on the relationship between boron and soil common minerals can help us to understand the adsorption and desorption characteristics of B in soil (Zhu et al. 1998b, Cheng et al. 2002a, Wang et al. 2006). In Figure 1, a is B containing $Al(OH)_3$, b is $Al(OH)_3$ and d is the difference spectrum between a and b. We can see there were two bands at 1423 cm^{-1} and 1279 cm^{-1} which were inherent bands of borate, band intensity increased compared with c. These were the main differences from b. This indicates the presence of B on the $Al(OH)_3$ surface. Figure 2 shows the IR spectrum of manganite and its B containing samples (Cheng et al. 2002a). We can see that band intensity at 3000 cm^{-1} -3600 cm^{-1} weakened when manganite adsorbed or desorbed B. Changes in these bands indicate that B was adsorbed by substituting for hydroxyls of manganite. Band intensity also weakened at 1040.7 cm^{-1}, 1408.0 $cm^{-1}$and 1537.4 cm^{-1}. This indicated a structural change of manganite.

Figure 3 shows TEM pictures of goethite and two kinds of B containing ferric oxide (Liao et al. 2006). Figure 3a shows typical goethite crystal, Figure 3b shows goethite containing adsorbed B, Figure 3c shows ferric oxide occluded with B. It is obvious that crystal size in picture c was the smallest. We believe that boron entered into the lattice spaces of goethite and prevented crystal growth.

Figure 1 to Figure 3 summarize results about the effect of B on soil common minerals. It is obvious that B sorption and desorption behavior on soils was affected by several factors. Availability of B to plants is dependent on all of these factors.

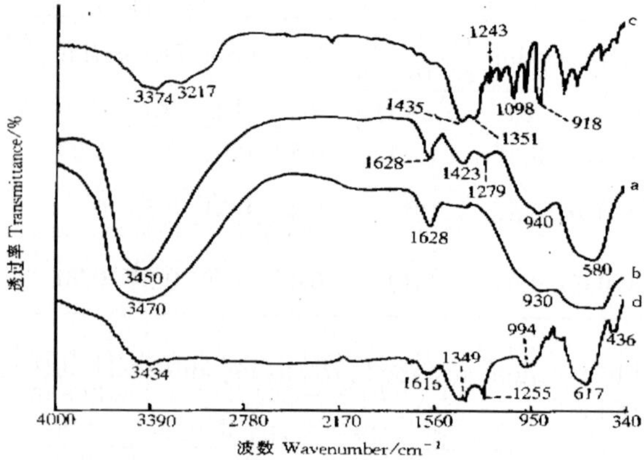

Figure 1. IR spectra of (a) boron-containing $Al(OH)_3$, (b) $Al(OH)_3$, and (c) $H3BO_3$ as well as (d) differential spectrum between boron-containing $Al(OH)_3$ and $Al(OH)_3$

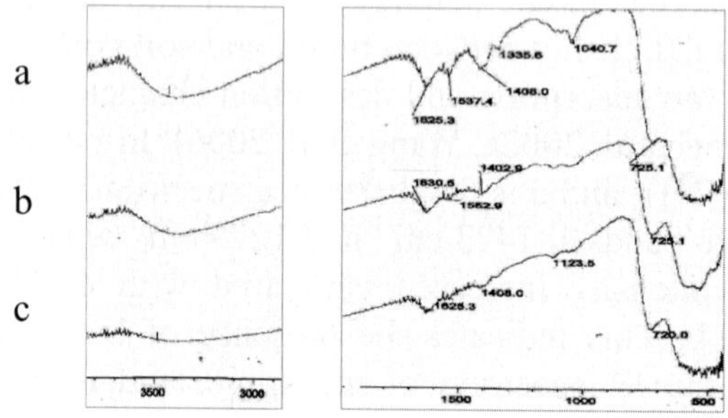

Figure 2. IR spectrum of manganite(a), manganite adsorbed B (b) and B-desorption (c) process

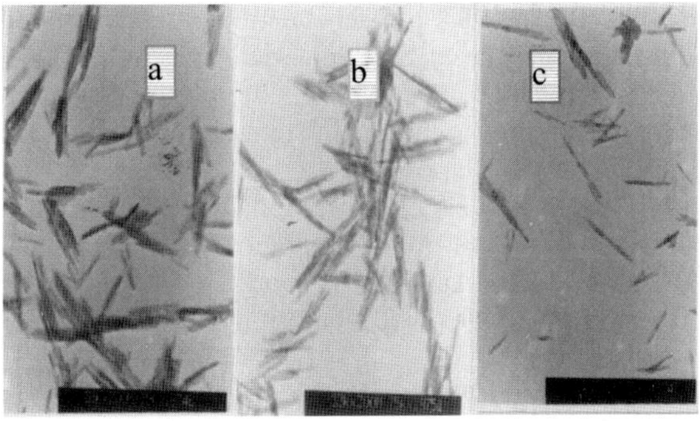

Figure 3. TEM pictures of Fe oxides (×33000): a-goethite, b-ad-B-goethite, -oc-B-goethite

Conclusions

B sorption and desorption research can guide us to proper B fertilization. It is necessary to investigate the relationship between B containing oxides and anions and cations in soil solution since we found that B sorption on soil surfaces could increase proton adsorption.

Furthermore, when boron was added to soil, it was not only utilized by plants, but also consumed by soil microorganisms, fixed or adsorbed by soil solids or lost with surface water and ground water. To study the amount and the pools of boron in each part of the soil is very helpful.

Acknowledgements

This work was and is being funded by The National Natural Science Foundation of China (39370415, 39670425, and 40371064).

References

Chen X., Zhu D., Cheng D., and Liu W., 2002 Effect of adsorption-desorption of boron on surface properties of soil. Acta Pedologica Sinica **39**(2), 145–151 (in Chinese).

Cheng D., Zhu D., and Liu W., 2002a Infrared spectral characteristics of reaction of boron with several minerals. Acta Pedologica Sinica **39**(5), 671–678 (in Chinese).

Cheng D., Zhu D., and Liu W., 2002b Effect of temperature on characteristics of boron adsorption-desorption by minerals. Acta Pedologica Sinica **39**(6), 822–829 (in Chinese).

Goldberg S., 1997 Reaction of boron with soils. Plant and Soil **193**, 35–48.

Goldberg S., Scott M.L., and Suarez D.L., 2000 Predicting boron adsorption by soils using soil chemical parameters in the constant capacitance model. Soil Sci. Soc. Am. J. **64**, 1356–1363.

Jin J., Martens D.C., and Zelazny L.W., 1987 Distribution and plant availability of soil boron fractions. Soil Sci. Soc. Am. J. **51**, 1228–1231.

Keren R. and Gast R.G., 1983 pH-dependent boron adsorption by montmorillonite hydroxyl-aluminum complexes. Soil Sci. Soc. Am. J. **47**, 1116–1121.

Lei H., Liu Y., Zhu D., Yin Y., and Hou D., 2004 The measurement on adsorption heat effect of several minerals to boron by two-thermometer method. Acta Pedologica Sinica **41**(2), 305–309 (in Chinese).

Liao S., Wang J., Zhu D., Ren L., Zhou W., and Ding J., 2006 Structure characteristics of goethite and its boron loaders. Acta Pedologica Sinica (in press) (in Chinese).

Wang J., Liao S., Zhu D., Ren L., Zhou W., and Ding J., 2006 Adsorption characteristics of Mn^{2+} on different oxides with boron. Acta Pedologica Sinica (in press) (in Chinese).

Zhu D., Pi M., and Liu W., 1994 Studies on property of different chemical pools of soil boron I. The plant availability of soil hot-water soluble boron. J. Huazhong Agricultural University **13**(3), 262–267 (in Chinese).

Zhu D., Pi M., and Liu W., 1995 Characteristics and intensity of chemical boron pool in soil. Proceedings of the Second Academic Conference of Young Scientists in China, Section of Agricultural Sciences., Chinese Sci. Tech. Press (Beijing), 322–325 (in Chinese).

Zhu D., Pi M., and Liu W., 1997 Relationship between soil boron adsorption kinetics and rape boron response. Phedosphere **7**(3), 269–274.

Zhu D., Pi M., and Liu W., 1998a Adsorption and desorption of boron in soils and its effect on boron uptake of plant. Acta Pedologica Sinica **35**(1), 69–75 (in Chinese).

Zhu D., Cheng D., Geng M., and Liu W., 1998b Analysis to structural regimes of boron-containing adsorption complex on the surface of aluminum hydroxide. J. Huazhong Agricultural University **17**(1), 40–44 (in Chinese).

Zhu D., Pi M., Liu W., Liu H., Chen X., and Cheng D., 1999a Study on the properties of different chemical boron pools in soil—Investigation on plant availability of soil boron. J. Huazhong Agricultural University, Supplement (**28**), 159–165 (in Chinese).

Zhu D., Shi L., and Liu W., 1999b Temperature effect on boron adsorption-desorption kinetics in soils. Pedosphere **9**(3), 243–250.

Zhu D., Chen X., Cheng D., and Liu W., 2000a Electrical characteristics and desorption kinetics of soil boron. Pedosphere **10**, 61–68.

Zhu D., Shi L., Chen X., and Liu W., 2000b. The studies on soil boron adsorption heat and temperature effect on properties of boron hysteretic desorption. Acta Pedologica Sinica **37**(2), 250–256 (in Chinese).

Screening B Sources for Suitability Under Different Rain Fall and Soil Condition for Mango in India

M. Edward Raja
Indian Institute of Horticultural Research,
Hessaraghatta, BANGALORE – 560 089, e-mail: medward@iihr.ernet.in

Introduction

Mango (Mangifera indica. L) is grown in 1.1 m ha in India in different climatic zones ranging from humid tropics receiving 2500 mm annual rainfall to semi arid tropics (600-900mm rainfall) and subtropics (800–1000mm rainfall). It is also grown in soils widely varying in its physico chemical properties. The pH of mango orchards range from a low of 4.5-5.2 in Konkan, Maharashtra to 7.2 to 8.1 in Uttar Pradesh and Chittor, Andhra Pradesh and accordingly it is affected by different types of macro and micronutrient disorders based on soil properties Boron deficiency is one of the important micronutrient disorders affecting mango both in high rainfall areas (Edward Raja et al 2005) and in semi arid tropics and average productivity is low at 6-7 t ha^{-1}. Besides mango is a B loving crop and needs very high level of B for its metabolic requirement (Agarawala et al 1988) and proper B nutrition is fundamental for its fruitset, yield and quality. Since B is highly immobile in plant and is continuously needed throughout the annual growth cycle, integrated nutrient management strategy involving both soil application and foliar spray with appropriate B sources is essential for efficient B deficiency correction. Since the range of sufficiency to toxicity is very narrow in Boron, extreme care is to be exercised in fixing the dose and selecting the sources of B for soil application. In the studies on correction of B deficiency in mango application of 250g of Borax/tree resulted in inadequate correction in Konkan due to leaching loss but resulted in mild B toxicity in Bangalore (Edward Raja et al 2005). Since B is lost by leaching, the right source of B which will resist leaching but at the same release just adequate B which will not result in toxicity is a challenging task. In Konkan, Maharashtra, about 2500mm rainfall is received in 4 months in an undulating topography resulting in heavy leaching loss and a B source which will resist leaching is needed. In the semi arid regions of peninsular India like Bangalore rainfall is low at 750 mm but the pH of the soil is high at 6.2 to 7.1 which reduces availability but with low humidity and a heavy subsoil which is

F. Xu et al. (eds.), Advances in Plant and Animal Boron Nutrition, 355–364.

favorable to B uptake. Hence a study was initiated to screen different B sources, which will resist leaching in high rainfall region but will be available in just adequate quantity in semi arid region.

Methods and material

To understand the soil characteristics, soil profiles in mango, orchards, one in high rainfall Konkan region and another in the semi arid region of Bangalore were studied. The soils were analyzed for physico-chemical properties (Black 1965) and B adsorption characteristics were studied adopting the method of Mondal et al 1993 and adsorption data were fitted to Langmur equations which describe the absorption: the adsorption maxima and bonding energy constants.

A laboratory incubation and leaching study was initiated to screen 3 different B sources: Boric acid (17 % B) sodium tetra borate penta hydrate (12% B) and Borax (11% B) with different water solubility characteristics and pH at 5mg kg^{-1} concentration with and without farm yard manure (FYM) compost with a no B control in 2 alfisols: Konkan (pH 4.9) and Bangalore (pH 6.2). The incubation study was carried out in 1liter plastic pots with one kg of processed mango orchard soil (<2mm) of the respective region. The completely randomized statistical design was followed. The soils were incubated at field capacity and after 30, 60 and 90 days were subject to leaching by water simulating the rainfall of each region 25 mm water for 24 hours for Konkan soil and 8 mm water for same duration in soil from Bangalore region. The hot water soluble B was estimated in the soil using Azomethine method of Parker and Gardner (1981). The results were statistically analyzed by ANOV and presented below.

Results

The study on evaluation of B sources for suitability to Mango in the humid tropics Konkan and semi arid Bangalore regions was carried out by first characterizing the representative soil profiles of the regions. The results of the characterizing the soil profile of mango orchards in Konkan presented in Table-1 indicates a deep, coarse textured surface and subsoil, with potential for heavy leaching. The acidic pH, low base status and high acidity due to exchangeable Al and low CEC indicate a heavy leaching already undergone by the soil. The low organic matter (0.2 to 0.57%) and clay status (7.5-20%) and low hot water soluble B throughout the profile (0.18-0.31$mgkg^{-1}$) indicate a poor fertility status especially very low B which is also the result of a heavy rainfall. The high free Al and its toxicity to roots are not favorable for nutrient uptake including B. The climate parameters also indicate a humid tropical climate with a heavy rainfall from June-October. (Fig. 1).

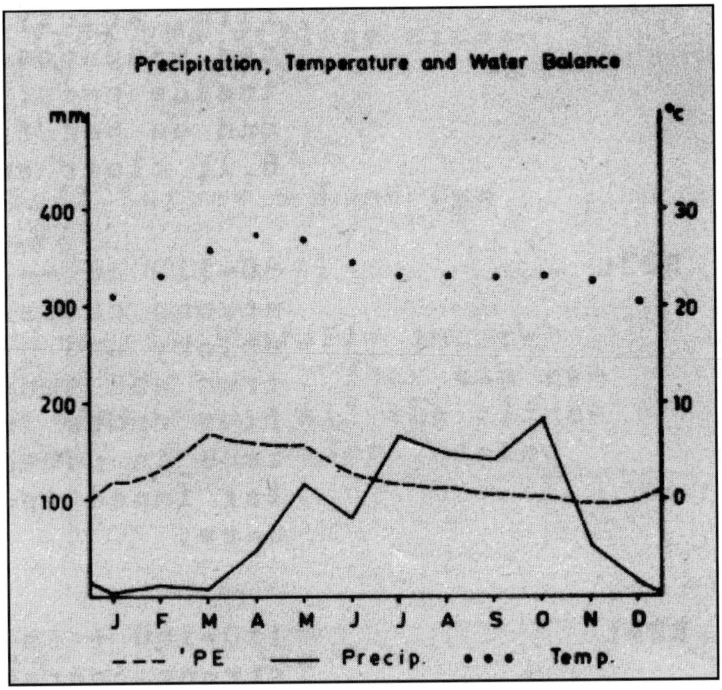

Figure 1. Weather parameters of Bangalore India

The soil and climate parameters of the mango orchards of Bangalore region are presented in Table 2, indicates a deep soil, loam in the surface with a heavy clay loam in the lower layers due to development of Bt horizon characteristic of alfisols. The soils are neutral with adequate base status (75-86%) though organic matter is low (0.5%OC) but the soil is adequate with available B (0.46 $mgkg^{-1}$) in the surface horizon but with a high status in the lower horizon (-0.62 $mgkg^{-1}$) which is totally different from the profiles of the Konkan region. The semi arid climate (Fig. 2) with a moderate rainfall more evenly distributed with transpiration potential more than precipitation in 8 out of 12 months are additional features in favor of a better B status in the soil as well as its availability, uptake and movement in the mango tree. B moves passively in transpiration stream and low humidity in the atmosphere and inadequate moisture by comparatively well distributed rainfall and moderate water holding capacity of the soil are very desirable features but the high bulk density is unfavorable for mass flow of B. These are the distinguishing features of soils of Bangalore over Konkan.

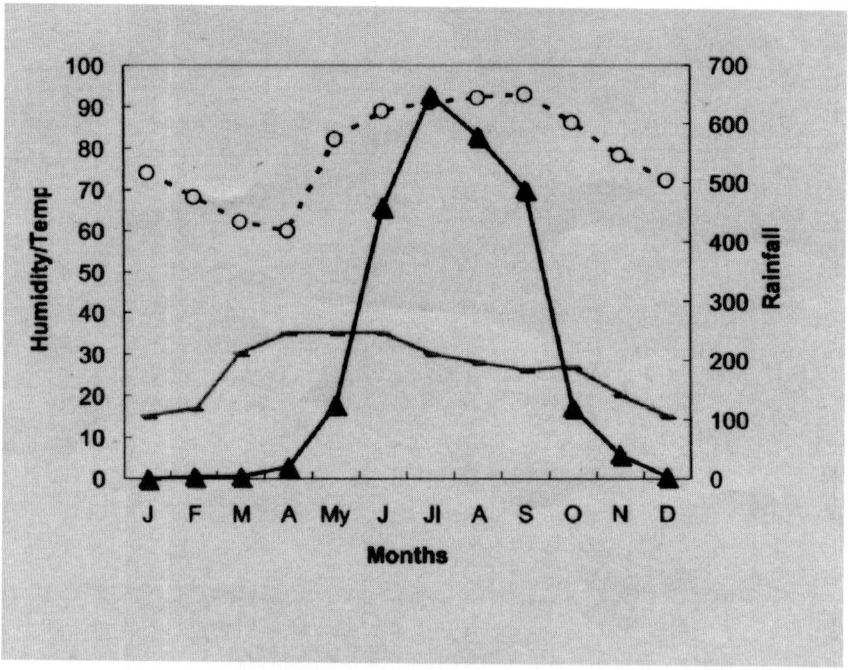

Figure 2. Mean weather changes of Konkan, Maharastra

Evaluation of B sources in Konkan soil

The evaluation of availability of three B sources: Boric acid, borax and sodium tetra borate with varying solubility and pH over 90 days of incubation in Konkan mango soil is presented in Table 3. It indicated out of 3 sources of B, boric acid (H_3BO_3 17%B) is superior to borax and sodium tetra borate (STB) in all incubation periods 30,60 and 90 days followed by STB and borax in the decreasing order. Studies on effect of duration indicated that available B is highest at 30 days followed by 60 and 90 days in decreasing order. But the significant aspect to be noted is in association with FYM compost the availability of B is much higher than without FYM in all sources very significantly in Boric acid. The highest available B has been recorded in Borax +FYM compost.

Evaluation of Boron sources in Bangalore soil

The similar evaluation of B sources in Bangalore soil (Table-4) indicated Borax is more efficient in maintaining high available B among three sources followed by STB and Boric acid. This is completely different from their behavior in Konkan soil. The positive effect of FYM compost on availability of all sources of B is maintained this soil also and the highest B is recorded in T_7 (Boric acid +FYM compost).

Table 1. Physico chemical properties of representative soil profile from B deficient mango orchard, Konkan

Depth (cm)	Horizon	Clay(%)	Texture	pH(1:2.5 (water)	Organic carbon%	Bases (cmol(+) Kg	CEC (cmol(+) Kg	Acidity Exch.H meq/100g soil	Exch.Al	Root abundance by number & Size (avrg.m^{-2})
0-15	Ap	16.3	SI	4.9	0.57	5.74	8.1	0.17	0.42	>1000, Fine
15-40	2A2	7.5	S	5.2	0.28	1.06	3.9	0.08	0.38	500-600, Medium
40-74	2BW1	20.2	SCI	5.8	0.27	1.24	2.8	0.07	0.17	10-15 coarse
74-108	2BW2	17.2	SI	5.6	0.26	2.33	4.2	0.09	0.23	Nil
108-145	2Bc	19	S	5.8	0.20	5.81	6.9	0.09	0.27	Nil

*Fine = 1-2, Medium = 2.5mm, Coarse = 2-10 mm

Table 2. Physico chemical properties of representative soil profile from B deficient mango orchard, Bangalore region

Horizon	Depth (cm)	Clay %(<0.002)	Bulk density g cc^{-1}	organic carbon%	pH (1:2.5) (1:2.5)H_20	E.C.(1:2.5) H_2O ds m^{-1}	CEC C mol (+) Kg-1	Base% Saturation (NH_4 OAC)	HWS Boron mgkg^{-1}	Root Abundance by number & size
AP	013	22.8	1.35	0.53	6.8	0.6	7.9	93	0.46	>800 fine
B21t	13-34	40.1	1.44	0.30	5.9	0.5	8	75	0.56	600-800 medium
B22t	34-80	43.3	1.50	0.21	6.1	0.46	8.7	80	0.62	300-400 medium
B23t	80-110	48.0	1.57	0.23	6.5	0.41	9.1	83	0.61	210-300 medium
B24t	110-160	49.7	1.53	0.22	6.5	0.42	9.4	86	0.54	80-100 medium

Table 3. Effect of sources of B on leachability in acid alfisols of mango orchards at Konkan Maharastra (HWS-B in mg kg^{-1}).

Treatments	Orchard I			Orchard II			Orchard III			Orchard IV			Orchard V			Orchard IV			Mean
T_1- NO B Controls	0.36	0.31	0.25	0.34	0.3	0.26	0.34	0.28	0.23	0.3	0.28	0.22	0.44	0.36	0.3	0.24	0.2	0.2	0..31
T_2-5 mg kg^{-1} B as Borax	2.2	2.2	1.8	2.2	2.3	1.8	2.2	2.1	1.3	2.8	2.9	1.9	2.5	2.6	2.3	2.5	2.3	2.0	2.42
T_3-5 mg kg^{-1} Borax as +FYM	3.5	2.6	2.0	3.6	2.7	2.2	3.5	2.6	2.1	3.4	2.9	2.2	3.3	2.9	2.8	3.4	2.7	2.4	2.82
T_4-5 mg kg^{-1} Sodium Tetra Borate	3.0	2.7	2.3	2.7	2.7	2.4	2.6	2.7	2.3	2.0	2.2	2.0	2.6	2.2	2.0	2.4	2.3	2.1	2.64
T_5-5 mg kg^{-1} Sodium Tetra Borate+FYM	3.0	2.9	2.0	3.0	3.3	2.8	3.0	2.6	2.1	3.0	3.6	3.0	3.7	3.4	2.9	3.0	3.2	2.9	2.96
T_6-5 mg kg^{-1} B as Boric Acid	3.7	3.3	2.8	3.3	3.0	2.8	3.6	2.9	2.8	3.1	3.0	2.9	3.2	3.0	2.8	3.1	3.1	2.8	3.06
T_7-5 mg kg^{-1} B as Boric Acid+FYM	3.4	3.2	2.3	3.4	3.4	3.0	3.4	3.2	3.0	3.4	3.3	3.2	3.6	3.4	3.0	3.5	3.4	3.0	3.22
Mean	2.95	2.45	1.92	2.93	2.58	2.18	3.0	2.34	1.97	2.86	2.74	2.2	3.02	2.6	2.3	2.83	2.62	2.2	
CD5%	0.18	0.18	-	0.14	0.18	-	0.21	0.18	0.19	0.18	0.19	0.13	0.14	0.13	0.14	0.14	0.14	0.13	

Table 4. Effect of sources of boron on leachability in neutral alfisol of mango orchard at Bangalore, Karnataka (HWS B in mg kg^{-1})

Treatments	Orchard I (Days after incubation)			Orchard II (days after Incubation)			Orchard III (days after Incubation)			Orchard IV (days after Incubation)			Orchard V (days after Incubation)			Orchard IV (days after Incubation)			
	30	60	90	30	60	90	30	60	90	30	60	90	30	60	90	30	60	90	Mean
T_1- Controls	0.5	0.4	0.3	0.6	0.5	0.4	0.5	0.4	0.4	0.6	0.5	0.4	0.4	0.5	0.4	0.4	0.5	0.5	0.47
T_2-5 mg kg^{-1} B as Borax	4.0	3.6	3.0	3.6	3.8	3.0	4.1	3.2	3.0	4.0	3.7	3.0	4.1	3.3	.3.0	4.5	3.8	3.2	3.55
T_3-5 mg kg^{-1} B as Borax +FYM	4.0	4.0	3.6	4.2	4.0	3.6	4.2	4.0	3.6	4.0	3.9	3.4	3.8	3.6	3.4	3.9	3.8	3.15	3.71
T_4-5 mg kg^{-1} B as Sodium Tetra Borate	3.1	3.6	3.0	4.0	3.3	3.0	3.0	3.8	.3.0	3.0	3.8	3.3	3.6	3.2	2.8	3.9	.3.0	2.2	3.1
T_5-5 mg kg^{-1} B as Sodium Tetra Borate+FYM	3.6	3.6	3.2	3.6	3.6	3.4	3.8	3.8	3.2	3.9	3.8	3.0	3.9	3.9	3.2	3.6	3.6	3.3	3.5
T_6-5 mg kg^{-1} B as Boric Acid	3.0	2.6	2.5	3.2	2.8	2.6	3.0	3.2	2.8	3.0	3.3	3.0	3.2	2.9	2.0	3.0	2.8	2.0	2.6
T_7-5 mg kg^{-1} B as Boric Acid+FYM	3.45	3.01	2.8	3.33	3.03	2.8	3.35	3.15	3.0	3.31	3.18	3.0	3.03	2.8	2.9	3.2	2.32	2.7	3.07
Mean	3.16	3.01	2.62	3.33	3.0	2.69	3.08	3.17	2.8	3.11	3 2	2.8	3.15	2.8	2.6	3.2	2.83	2.5	
CD* 5%	0.24	0.22	-	0.22	0.20	-	0.24	0.21	-	0.30	-	-	0.32	0.21	-	0.31	0.21	-	

Boron adsorption and behavior of B sources

The studies on adsorption behavior of the Konkan and Bangalore soil presented in (Table 5) indicated a totally different adsorption behavior in these two alfisols. In the Konkan soil, the mean adsorption maxima (b) and bonding energy constant (K) ranges from 0.40-0.124 with mean of 0.090 mLμg g^{-1} which is very low compared to the Bangalore soil with adsorption maxima 16.86-32.11 with a mean of 27.76 μg g^{-1} followed by a bonding energy constant of 0.143-0.42 with a mean of 0.213 mLμg g^{-1}. The B adsorption followed the adsorption isotherm.

Table 5. Boron adsorption characteristics of mango orchards soils of Konkan and Bangaolre.

Mango orchards in (K) different location	Adsorption Maxima(b) (ug g^{-1})		Bonding energy Constant (mL ug^{-1})	
	Konkan	Bangalore	Konkan	Bangalore
1	3.86	16.86	0.124	0.143
2	23.58	42.31	0.04	0.42
3	16.59	32.11	0.115	0.156
4	13.98	2.41	0.058	0.22
5	21.95	23.56	0.086	0.16
6	19.8	29.32	0.119	0.18
Mean	16.62	27.76	0.09	0.213

Discussion

The B deficiency in mango in India is very widespread and severe in the humid tropics of Konkan and moderate in semi arid tropics of peninsular India including Bangalore in Southern Karnataka. The B correction strategy followed earlier indicated foliar spray of 0.1% solubor was satisfactory but soil application of borax even up to 250 g $tree^{-1}$ was not efficient in the Konkan Maharashtra (Edward Raja et al 2005) whereas soil application was reaching toxicity and reduced yield drastically in semi arid region of Karnataka. Hence a study was initiated to identify a B source, which is suitable for mango cultivated in humid and semi arid tropics taking into consideration the heavy leaching in the former and reduced leaching but moisture stress in the latter. The undulating topography and the coarse texture of the entire profile plus heavy rain in June-Oct has resulted in increased B loss by leaching and this is the major reason in for widespread B deficiency. The Al toxicity also contributes to the B deficiency due to poor root health. The high humidity and long dry spell after the monsoon rain in June-Oct also are factors aggravating B deficiency in

Mango in Konkan. In the soils of Bangalore, though adequate soil B is noticed the hard setting alfisol (Muslim 1991) and heavy Bt horizon and inadequate moisture due to low rainfall are reason for B deficiency in mango. The lack of fine roots and moisture in B horizon makes the plants susceptible to B deficiency. Besides the high base situation compared to Konkan soil results in higher Ca/B ratio and raising metabolic B requirement. The higher soil pH also aggravates B deficiency compared to Konkan soil. In the present study Boric acid proved better in Konkan since among the three sources it has the lowest water solubility and hence can resist leaching. Its combination with FYM compost, which can form complexes, has proved best in this situation. In the less leaching situation of Bangalore, borax proved a good source of B and in combination with FYM compost, it has proved the best. The low clay content and low adsorption maxima and bonding energy constant has resulted in low affinity for Boron resulting in heavy loss in leaching even at 30 days after application of borax in Konkan. The higher B adsorption in B soil has prevented the B loss by leaching, in soils of Bangalore and reason for borax maintaining higher available B in the presence of enough (Goldberg, 1997) clay to retain it from loss by leaching.

Conclusion

B deficiency is widespread in Mango in India but due to its low yield and alternate bearing habit it is not considered as a profitable fruits crop. It is grown for its hardy nature (suited to rainfed and low input management) and hence any correction technology should take this into consideration since B is needed continuously since roots are the first affected in B deficiency. Adequate soil B is a must for any crop, more so for a B loving crop like mango. Due to variability in soil and climate Boric acid + FYM compost is ideal for the high rainfall region of Konkan maintaining enough B whereas for the low rainfall region of Bangalore Borax + FYM are the best methods of maintaining adequate B in soil. Soil application of B in organic matrix is needed for the humid tropical soils low in B.

Acknowledgements

The help given by Mr. Mazhar Jameel in conduct of the field experiment and Lab analysis by Usha.K in preparation of the manuscript is gratefully acknowledged. The technical and laboratory support given by Director IIHR Bangalore is also gratefully acknowledged.

References

Agarwala S.C., Nauriyal B.D., Chithralekha C., Sharma C.D., 1988. Manganese Zinc and Boron deficiency in Mango Sci. Hort. **35**: 99-107.

Black C.A., 1965. Methods of Soil Analysis – Part-II. Agronomy Monograph 9 American Society of Agronomy, Madison, Wisconsin, USA.

Edward Raja M., Anil Kumar S.C. and Raju S.Y., 2005. Boron deficiency in mango (*Mangifera indica* L.). A cause delineation study in acidic soils of Maharshtra, India. Soil Science and Plant Nutrition, **51**(4): 311-322.

Goldberg S., 1997. Reaction of boron in soils. Plant and Soil, **193**: 35-48.

Gupta V.C., 1979. Boron nutrition of crops. Advances in Agron., **31**: 273-307.

Jackson M.L., 1976. Soil chemical analysis. Prentice Hall of India Pvt. Ltd., New Delhi.

Mandal A.K., Mandal B., and Mandal L.N., 1993. Boron absorption characteristics of acidic alljuvial soils in relation to soil properties. Commun. Soil Sci. Plant Anal., **24**: 2553-2567.

Mullins C.E., Mcleoo D.A., Northcote K.H., Trisdale J.M., Young M., 1990. Hardsetting Soils. Behavior, Occurrence and Management. Adv.Soil Science, **11**: 37-108.

Parker D.R., Gaines E.H., 1981. The determination of hot water soluble boron in some acid Oregon soils using a modified Azomethine-H Procedure. Commun. Soil Sci. Plant Anal., **12**:1311-1322.

Widespread Boron Deficiency in Water-Eroded Soils of *Pothwar* Plateau in Pakistan: Identification, Establishment and Management

A. Rashid[1]*, E. Rafique[1], M. Mahmood-ul-Hassan[1] and M. Shafiq[2]

[1]Land Resources Research Program, National Agricultural Research Center (NARC), Islamabad-45500, Pakistan;

[2]Water Resources Research Institute, NARC, Islamabad, Pakistan, Corresponding author: abdul.rashid@comsats.net.pk

The Pothwar plateau

The *Pothwar* plateau in Pakistan (Fig. 1; latitude 32° 10' to 34° 9' N; longitude 71° 10' to 73° 55' E) is part of the great Indo-Gangetic synclinorium, separated from it, and elevated, at the end of the Tertiary Period. The plateau spans over 1.82 million hectare (Mha) and constitutes a major rainfed tract in the north of Punjab province, Pakistan (Fig. 2). It is situated in the north of Salt Range, in-between rivers Jehlum and Indus. Its climate is semi-arid to sub-humid continental. Rainfall pattern in the plateau is bi-model; the maximum being in late summer and winter-spring (Fig. 3). Because of torrential rains undulating topography and minimal vegetation cover, its soils suffer with water erosion of varying degrees. The plateau generally has a flat to gently undulating topography, locally cut-up by an intricate drainage network of gullies, stony lands and low hills/ranges. The gully formation is the consequence of geological erosion; the erosion process has been accelerated in recent times by human and animal activities. About 45% of the total area in the plateau is cultivated; the rest comprises of rough and *procal* land. Major features of the plateau are summarized in Table 1.

F. Xu et al. (eds.), Advances in Plant and Animal Boron Nutrition, 365–382.

Table 1. General characteristics of *Pothwar* plateau, Pakistan

Location	32 – 34° N, 71 – 74° E
Total area	1.82 Mha
Cultivated area	0.82 Mha
Rainfall	300–500 mm (~ 65% Monsoonal)
Parent material	Loess, alluvium, Tertiary rocks
Topography	Gently undulating plains (Gullied land)
Soil Orders	Alfisol, Entisol, Inceptisol
Major crops	Wheat, peanut, sorghum, corn, rapeseed-mustard

Source: Shafiq et al. (2005)

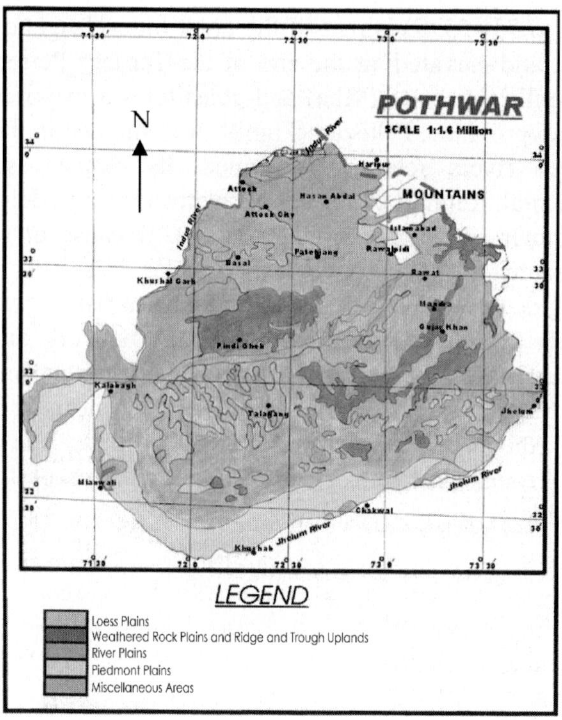

Source: Nizami et al. (2004)

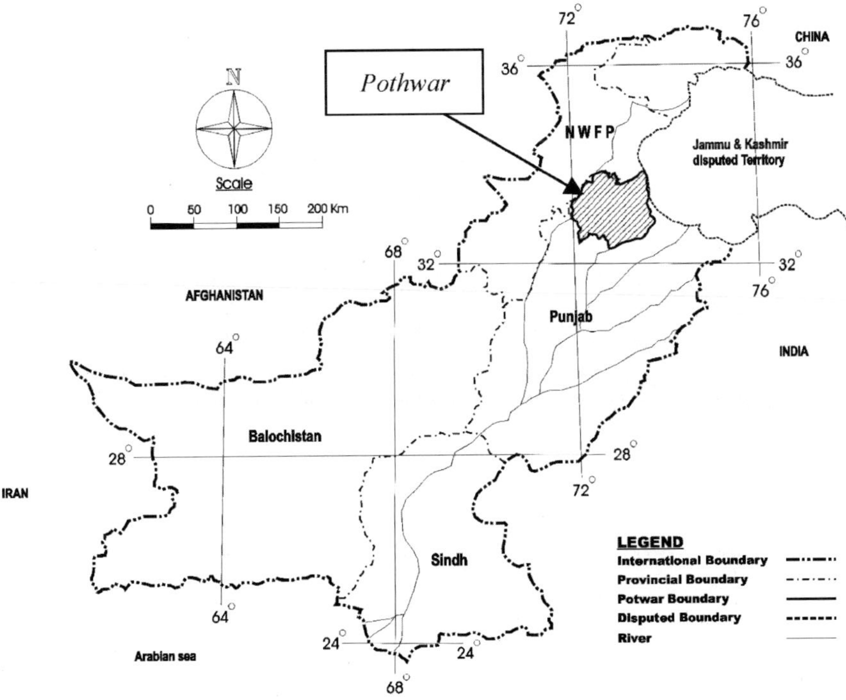

Figure 2. Location of Pothwar plateau in the north of Pakistan. Source: WRRI, NARC, Islamabad

Rainfall and temperature vary considerably over the plateau. Average annual rainfall is 300 mm in the south-west and 1500 mm in north-east of the plateau (Figs. 1, 3). Rainfall is highly erratic in terms of both space and time, with most of the rainfall occurring during monsoon period, i.e., July to September. Major crops grown there are wheat, rapeseed-mustard and chickpea during winter, and sorghum, maize and peanut during summer. Average cropping intensity is low, i.e. only 75%. Agriculture is rainfed, frequently suffering with drought, and, thus typical of a high-risk, low-input scenario. Highly erratic rainfalls, serious erosion hazard, nutrient depletion and inadequate soil depth are the major limitations adversely affecting crop productivity in the plateau. In view of uncertain productivity because of erratic rains, the resource poor, small land holders (average farm size 0.3 ha) are shy of using adequate farm inputs, including seed and fertilizers. Consequently, crop yields are very low.

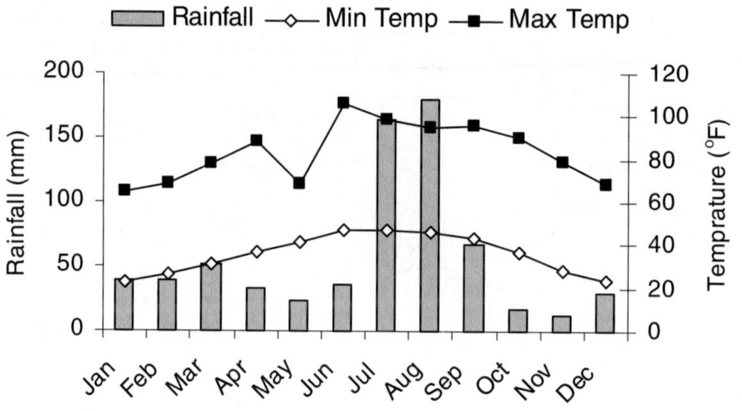

Figure 3. Mean monthly rainfall and temperature in *Pothwar* plateau. Source: adapted from Nizami et al. (2004)

The soils of Pothwar

The soils of this rainfed plateau in Pakistan have been developed from a wide range and kind of parent materials, i.e., loess, tertiary rocks like sandstone and shale, and alluvium. Loess plains are the characteristics landform of the region. The plains are comprised of thick deposit of calcareous, fine, wind–laid sediments that have undergone strong profile development, followed by erosion and a second cycle of pedogensis. Due to erosion, level plains, dissected loess plains, sub-recent valleys in the dissected loess plains, and redeposited loess plains have developed.

Major soils of cultivated areas in the plateau belong to *US Soil Taxonomy* Orders Inceptisol, Alfisol and Entisol. Although, most soils of the plateau are of Pleistocene age, progressive water erosion is responsible for existence of relatively young Soil Orders like Entisol and Inceptisol. The soils are mostly medium textured with a fair proportion of clayey soils and are alkaline, calcareous and low in organic matter (Table 2). Great variations occur in soil depth, slope and soil susceptibility to erosion.

Table 2. Generalized soil properties in Pothwar plateau

Property	Soil depth (cm)	Range	Mean
pH (1:1)	0 – 15	7.5 – 8.3	8.0
	30 – 45	7.6 – 8.4	8.1
$CaCO_3$ (%)	0 – 15	0.8 – 12.7	6.4
	30 – 45	1.2 – 15.9	6.6
Organic matter (%)	0 – 15	0.07 – 1.3	0.53
	30 – 45	0.05 – 1.0	0.41
Clay (%)	0 – 15	11 – 34	21
	30 – 45	12 – 36	21
Silt (%)	0 – 15	3 – 44	19
	30 – 45	3 – 37	20
AB-DTPA extractable:			
NO_3^- – N (mg kg^{-1})	0 – 15	2.9 – 22.4	9.3
	30 – 45	1.9 – 19.2	7.0
P (mg kg^{-1})	0 – 15	0.7 – 12.0	3.2
	30 – 45	0.5 – 6.2	2.1
K (mg kg^{-1})	0 – 15	32 – 332	97
	30 – 45	15 – 342	81
Zn (mg kg^{-1})	0 – 15	0.13 – 3	0.64
	30 – 45	0.12 – 1. 82	0.56

Source: Rashid et al. (2002b); Rashid and Qayyum (1990)

Nutrient depletion in *Pothwar* plateau

Loss of topsoils with water erosion, nutrient mining with centuries-old cropping, soil alkalinity (mean pH 8.0), soil calcareousness (mean $CaCO_3$ content 6.4%), no crop residue recycling and green manuring, and inadequate and imbalanced fertilizer use have led to minimum soil organic matter content and widespread multiple nutrient deficiencies of acute magnitude throughout plateau (Table 2). Organic manure use, at low rates (i.e., 0.7 t ha^{-1}), is confined to the fields in the surroundings of rural settlements. Despite widespread and severe deficiencies of nitrogen (N), phosphorus (P), zinc (Zn), boron (B), and iron (Fe), fertilizer use is restricted to bare minimal rates N and P (i.e., average rate for wheat are 12 kg N ha^{-1} and 4 kg P ha^{-1}).

Consequently, cereal, legume and oilseed crops grown there suffer with severe multiple nutrient deficiencies. Fertilizer use efficiency is low and varies drastically; e.g., fertilizer P uptake by current wheat crop is 7–43%, depending on soil type, fertilizer rate and fertilizer application method and moisture availability at critical crop growth stages.

Table 3. Extent of B deficiency in crops in *Pothwar* plateau

Species	Plant Part/ Soil depth	Range	Mean	Deficient sites
		------------(mg kg^{-1})------------		(%)
Wheat	Whole shoots	3.2 –13.7	6.18	64
	Soil (0-15)	0.04 – 1.20	0.45	72
	(30-45)	0.04 – 1.01	0.36	75
Sorghum	Whole shoots	1.5 – 20.5	6.83	50
	Soil (0-15)	0.10 – 0.96	0.38	77
	(30-45)	0.07 – 0.81	0.36	81
Rapeseed-mustard	Whole shoots	13.5 – 49.0	27.3	70
	Youngest mature leaves	17.4 –54.6	35.2	64
	Soil (0-15)	0.04 – 1.20	0.48	68
	(30-45)	0.04 – 1.03	0.37	73
Peanut	Shoot-terminals	12.0 – 97.9	34.9	51
	Soil (0-15)	0.10 – 1.08	0.47	60
	(30-45)	0.08 –1.00	0.43	63

Source: Rashid et al. (1997a, b; 2002a,b

Soil–crop boron deficiency: extent and severity

Soil conditions and agronomic practices in the *Pothwar* plateau are conducive to B deficiency in plants. Though the first ever report of B deficiency in Pakistan, pertaining to cotton dates back to 1970 Chaudhary and Hisiani 1970, a global study on micronutrients, carried out by FAO of the United Nation during late 70s to early 80s, postulated B adequacy or even B toxicity in soils of Pakistan (Sillanpaa, 1982). Thereafter, B research did not receive due attention in the country untill mid 1980s, primarily for the lack of lab facilities. However, the subsequent extensive research on B nutrition of crops has revealed B deficiency in a number of field and horticultural crops almost throughout the country (Rashid et al. 2002c). This paper reports on B status of crop plants and soils as well as crop response to B use in *Pothwar* plateau. The extensive research information was gained by virtue of: (i) monitoring B status of the area by nutrient indexing of farmer-grown crop plants and associated soils; (ii) mapping spatial variability of B status, within the target areas, using geostatistics and computer graphics; and (iii) field and greenhouse experimentation for studying crop responses to B fertilizer use and determination of internal B requirement of crop plants.

Extensive indexing in major crop plants and associated soils, conducted up to mid 1990s, has revealed that HWE B in these rainfed surface soils is 0.04–1.20 mg kg^{-1}, with mean B content of 0.4 mg kg^{-1} (Table 3). Boron deficiency existed in 70% rapeseed-mustard fields, 64% wheat fields, and 50% fields each of sorghum and peanut. However, a recent intensive B status investigation in two project areas within *Pothwar* has revealed a much greater magnitude of B deficient soils, i.e. 85% (Shafiq et al. 2005). Moreover, the deficiency is quite severe in most of the fields. In our experience, plant B concentrations were highly dependent on soil available B content [e.g., r = 0.79** between B concentration in rapeseed whole shoots and HWE B in associated topsoils]. Plant B contents were also related to certain soil properties, like pH sand, silt, organic matter and $CaCO_3$ content (Table 4). Boron status in subsoils is slightly lesser than topsoils, primarily because of lesser organic matter content and greater $CaCO_3$ content in the deeper soil layers. Root growth and crop residue recycling, though not appreciable in the rainfed *Pothwar*, also contributes to better nutrient availability in topsoil layers. However, a recent research investigation reveals lack of any specific pattern in soil B status down the profile depths (Shafiq et al. 2005). In fact, B status upto 150 cm soil depth had almost no change in a soil of Missa series, decreased with soil depth in Guliana series, and increased in deeper soil layers of Rajar series (Fig. 4). Our extensive nutrient indexing studies revealed that plant tissue B content was positively related with soil properties like soil B (r = 0.55 – 0.79; P <0.0.1), $CaCO_3$ content (r = -0.48; P <0.01), pH (r = -0.59; P <0.01), OM (r = 0.32; P <0.05), sand (r = -0.28; P <0.01), silt (r = 0.31; P <0.01). In our experience, incidence of B deficiency in soils and crop plants was also related to soil parent material. For example, B deficiency was observed in 64–68% fields each of loessal and alluvial soils, 74% of residual soils and 100% of redeposited loess–derived soils. However, as 65–70% the sampled fields of all the three Soil Orders suffered with B deficiency incidence of B deficiency was not related with the Soil Orders (Table 5).

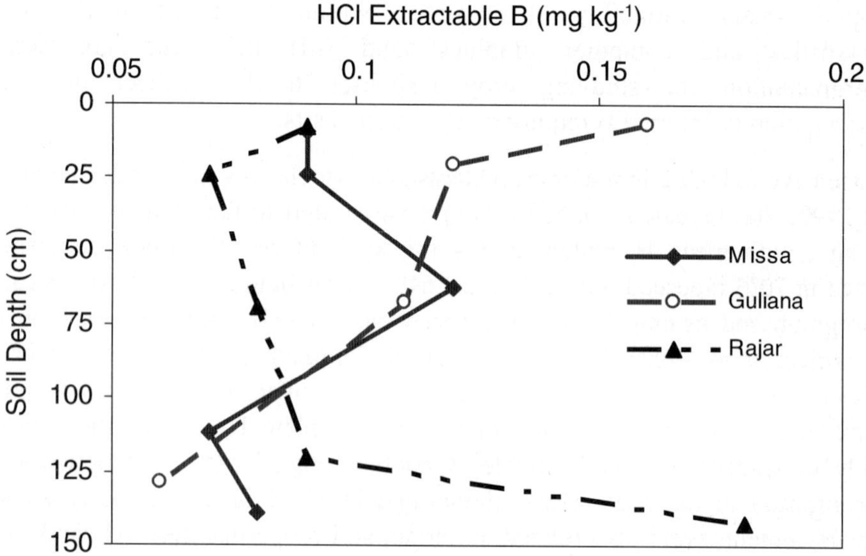

Figure 4. Boron status of three dominant soil profiles in *Pothwar* plateau. Source: adapted from Shafiq et al. (2005)

Table 4. Relationships between soil properties and boron content in rapeseed whole shoots and associated soils

Soil /Plant B	Topsoil characteristic				
	pH	Sand (%)	Silt (%)	OM (%)	$CaCO_3$ (%)
HWS B (mg kg $^{-1}$)	-0.59**			0.32**	-0.48**
Plant B (mg kg $^{-1}$)		-0.28**	0.31**		

**highly significant ($P \leq 0.01$) Source: Shafiq et al. (2005); Rashid (1993)

Table 5. Influence of soil parent material and soil classification on water soluble B content

Parameter	Mean HWE B in topsoils (mg kg^{-1})	% age of topsoil deficient in B
Parent material		
Alluvium	0.54	64
Loess	0.50	68
Rosiduum	0.40	74
Redeposited loess	0.26	100
Soil Order		
Alfisol	0.45	70
Inceptisol	0.46	67
Entisol	0.54	65

Source: Rashid et al. (2002b)

Spatial variability of boron in Pothwar plateau

Spatial variability in B status has been mapped for wheat, sorghum peanut and rapeseed–mustard crops grown in *Pothwar* plateau of Pakistan. These contour maps were prepared using B concentrations in diagnostic plant parts of farmer-grown crop plants and associated soils (0–15 and 30–45 cm depths) using geostatistical analysis techniques (Bhatti et al. 1991). Though spatial variability contour maps are available for a number of crop species, an example is presented for rapeseed–mustard (Fig. 5). In general, soil and plant B status maps are in good agreement and are quite effective in delineating B deficient areas. In the absence of effective soil advisory service, these maps can help identify areas needing B fertilizers and also help focus future research and development.

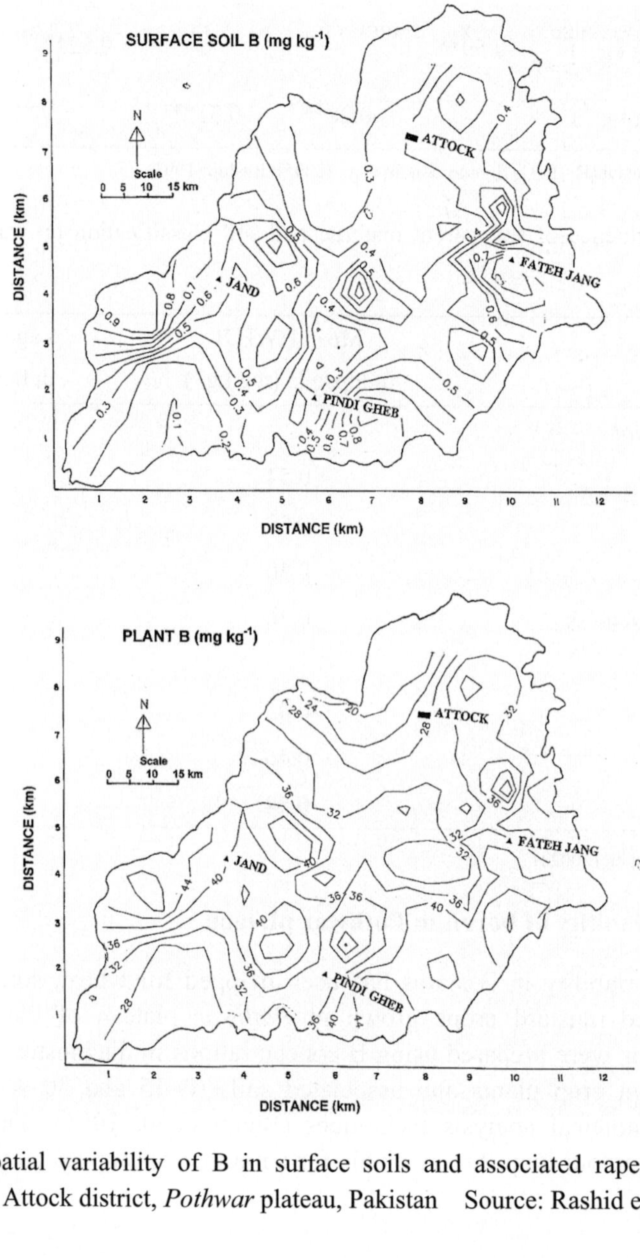

Figure 5. Spatial variability of B in surface soils and associated rapeseed–mustard leaves within Attock district, *Pothwar* plateau, Pakistan Source: Rashid et al. (1993)

Crop responses to boron

Appreciable yield increases with B use in the *Pothwar* soils have been observed for all the studied crops. Typical field-situation crop responses to graded levels of fertilizer B, applied to the *Pothwar* soils are given in Figs. 6, 7. Whereas B use upto 4 kg ha^{-1} increased wheat yield, were higher B rates proved detrimental because of toxic effect (Fig. 6). Fertilizer B requirement was much lesser for peanut crop, and rates more than 1 kg B ha^{-1} proved toxic (Fig. 7). On an average, yield increases with B use over control were 10–12% in wheat, maize and peanut and 20% in potato. Crop genotypes differ in susceptibility to B deficiency and response to B application. Land races are less susceptible to B deficiency than the introduced improved cultivars. Fertilizer B requirement, associated with near-maximum crop yield, ranges from 0.5 kg B ha^{-1} for peanut to 1.2 kg B ha^{-1} for wheat The economic impact of B use was highly attractive for farmers with value–cost ratio of 4:1 for wheat, 5:1 for maize, 11:1 for peanut, 13:1 for potato. Thus, B use, in soil deficient situations, is highly cost–effective. As fertilizer B use by the current crop is 3% of the applied B (Rashid 2006), subsequent crop(s) in the rotation would also benefit appreciably. Thus, actual value cost ratios for soil applied B are 2–4 times of those given in Table 6.

Table 6. Crop responses to B use and its economics

Species	HWE B (mg kg^{-1})	B requirement (kg ha^{-1})	Control yield (t ha^{-1})	Yield increase (%)	VCR
Wheat	0.4	1.2	2.32	11	4:1
Peanut	0.3	0.5	2.10	10	11:1
Maize	0.4		2.20	12	5:1
Potato	0.3		10.13	26	35:1

Source: Rashid et al. (1997a, 2002c)

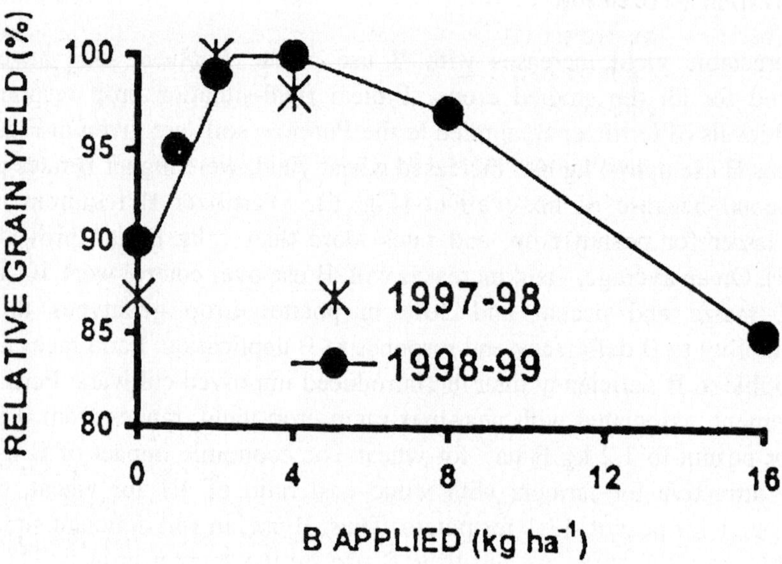

Figure 6. Relationship between B application rate and relative grain yield of wheat grown in a Typic Hapludalf of *Pothwar* plateau (maximum yield: 3.14 t ha^{-1}).Source: Rashid et al. (2002a)

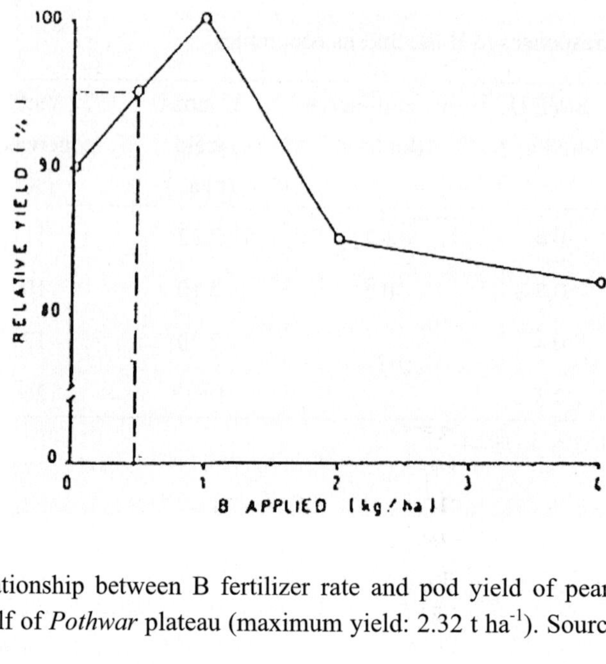

Figure 7. Relationship between B fertilizer rate and pod yield of peanut grown in a Typic Hapludalf of *Pothwar* plateau (maximum yield: 2.32 t ha^{-1}). Source: Rashid et al. (1997b)

We observed that in addition to yield reduction, soil B deficiency also delays crop flowering and maturity of certain crop species like rapeseed. Though B use could accelerate crop maturity by 7–10 days only, its practical impact in rainfed agriculture would be tremendous.

Though economics of B use is highly cost effective, fertilizer B use at farm level remains to be practiced by the resources-poor rainfed farming communities. Major constraints to B use are lake of awareness for the need to fertilizer and high-risk agriculture due to erratic rainfalls in the plateau.

Internal boron requirement of crops

The plant analysis diagnostic criteria for B are not only scarce but also frequently irrelevant to locally grown crop genotypes. Therefore, we determined internal B requirement of crop genotypes grown in rainfed *Pothwar* plateau. Salient examples of plant tissue B concentrations as affected by graded levels of fertilizer B are presented in Figs. 8, 9. The range of B rates used in our studies, for both crops, not only cured the deficiency but also induced toxicity; thus, yield depressions were observed at higher rates of fertilizer B. The estimated critical B concentration was 4 mg kg^{-1} in wheat compared with 28 mg kg^{-1} in sunflower leaves (Figs. 8, 9). Thus, the cited examples illustrate wide differences in internal B requirement of monocot and dicot plant species. Critical B level in diagnostic plant parts of local genotypes varied between species, varieties within the same species, and plant parts (Table 7). Whereas, critical range for whole shoots of wheat (i.e., 4-6 mg B kg^{-1}) are similar to the generally suggested values in the literature (e.g., Reuter et al. 1997) the determined critical levels for sorghum, rapeseed, mustard, chickpea, and sunflower are much greater (25–63 mg B kg^{-1}). The B deficiency critical concentration was greatest for sunflower and lowest for sorghum (Table 7).

Table 7. Boron deficiency critical concentrations/ranges

Species and cultivar	Critical level (mg B kg^{-1})	
	[1]Whole shoots	[2]Leaves
Wheat (cv. Pak-81)	5–7	4–6
Sorghum		
cv. Pothwar	17	31
cv. PARC-SS-1	18	25
Rapeseed (cv. Shiralee)	32	38
Mustard (cv. BARD-I)	41	49
Chickpea (cv. CM-72)		49
Sunflower (cv. Hysun 33)		46– 63

Source: Rashid et al. (1994, 1997b, 2002c); Rashid and Rafique (2005)

1 whole shoots ≤ 12 inches tall

2 leaves = youngest mature leaves at flower/ head initiation

Boron toxicity hazard

Contrary to the apprehension expressed by Sillanpaa (1982) about the possibility of B toxicity in soils of Pakistan, our comprehensive B indexing studies have rather identified B deficiency to be a widespread problem in soils of *Pothwar* plateau. Whereas the global study pertained to only a few field sites, we analyzed soils and associated crop plants from hundreds of random fields (Table 3). For investigating possible B accumulation in subsoils, with leaching in these soils possessing good hydraulic conductivity, we also analyzed subsoils (30–45 cm) of all the field sites, and observed that their HWE B was, infact, slightly lower (0.04–1.03 mg kg^{-1}; mean 0.38 mg kg^{-1}) than that of topsoils (0.04–1.20 mg kg^{-1}, mean 0.44). Moreover, B concentrations in diagnostic plant parts in farmer–grown crop plants were also in the deficient to adequate range, with no indication of B toxicity. However, we did observe B toxicity as a result of B over-application to B a deficient soil, i.e., @ > 2 kg B ha^{-1} in peanut (Rashid et al. 1997a) and @ > 4 kg B ha^{-1} in wheat (Rashid et al. 2002b). Whereas, the literature suggests high B in salt affected soils elsewhere

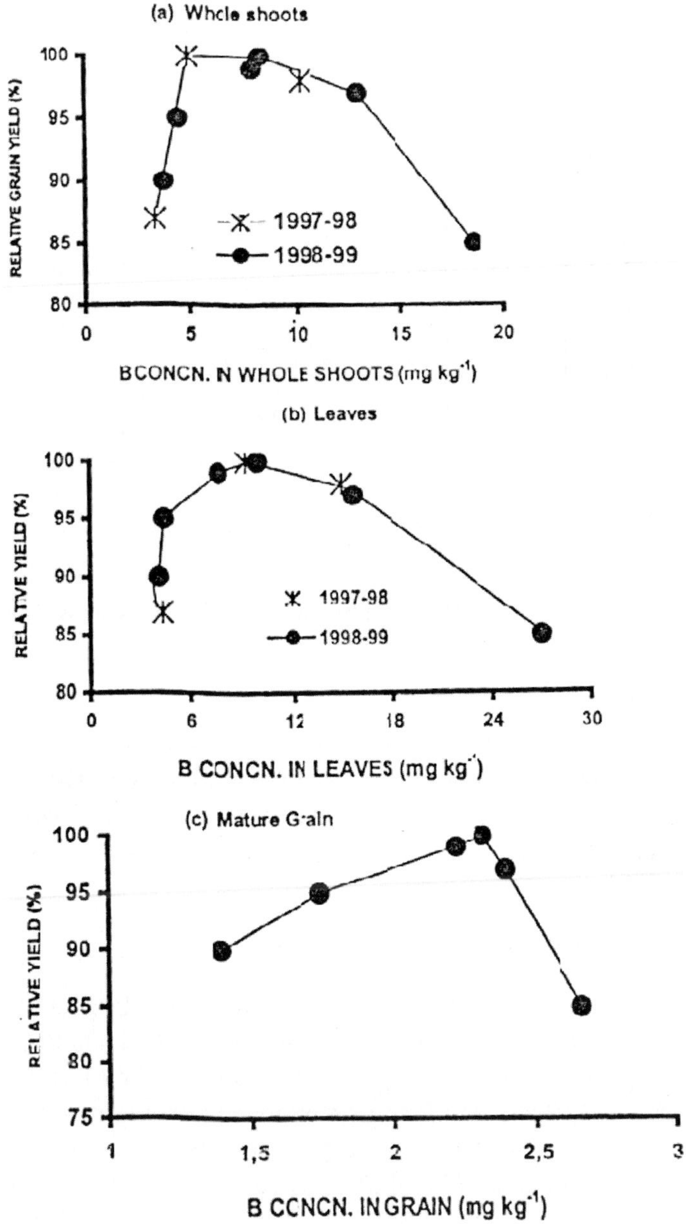

Figure 8. Relationship between B concentration (mg kg^{-1} dry matter) in diagnostic plant parts and relative grain yield of wheat: a) whole shoots, b) flag leaves, c) grains. Source: Rashid et al. (2002a)

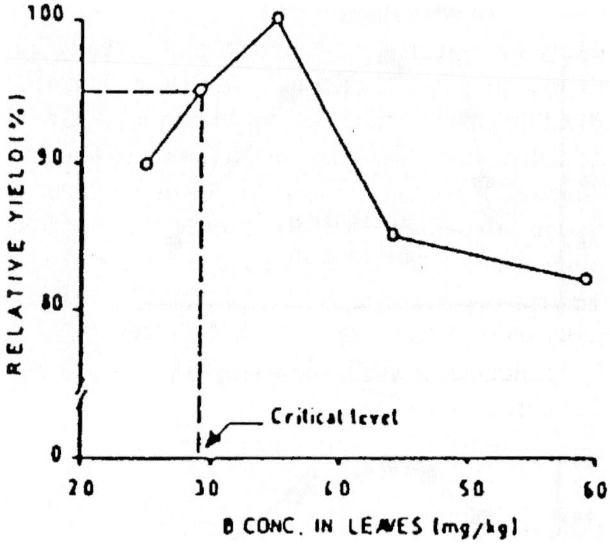

Figure 9. Relationship between B concentration in shoot terminals and pod yield of peanut. Source: Rashid et al. (1997b)

(Sillanpaa 1982), in our experience even the salt-affected soils of cotton belt in Pakistan suffered with B deficiency rather than B toxicity (Yasin et al. 2002). Therefore, there is no evidence of native B toxicity in soils of *Pothwar* plateau. However, great care is suggested in uniform field application of B fertilizer to avoid soil B build-up to toxic levels.

Managing boron deficiency in soils of Pothwar

Based on our research experience, suggested soil application rates for various crops are 0.75–1.0 kg B ha^{-1}. All standard sources of B, including borax and glanubor, are equally effective. Though foliar feeding and soil fertilization for B are equally effective, chances of B use adoption are rather remote for low–input and low–yield cropping systems in the plateau. This is primarily because of high risk agriculture in an erratic rainfall environment. However, foliar feeding of B, comprising of 2–3 sprays of 0.05% to 0.1% B solution, holds promise for irrigated vegetables and fruit orchards within *Pothwar* plateau.

Boron fertilizer use in Pothwar plateau

Non–availability of B sources as fertilizer until recently has been a major constraint to B use adoption in Pakistan. Moreover, numerous micronutrient products marketed by small vendors are not true-to-labeled-contents. However, it is encouraging that since 2005, two major fertilizer companies in Pakistan (i.e., Engro Chemical Pakistan Limited and Fauji Fertilizer Company) have launched marketing of good quality B fertilizer products. Additional constraint to B use adoption include lack of knowledge about the need to fertilizer and application problems (i.e., uniform field broadcast of a small B dose, thorough mixing with major nutrient fertilizers, spray solution preparation, etc.). Thus, in addition to increasing awareness, fertilizer industry is urged to provide B-fortified major nutrient fertilizers.

Acknowledgements

We thank Mr. Muhammad Irfan and Mr. Muhammad Jawad for word processing, graphics and composing of the manuscript.

References

Bhatti A.U., Mulla D.J., and Frazier B.E., 1991 Estimation of soil properties and wheat yields on complex hills using geostatistical and thematics mapper images. Remote Sensing Environ **37**: 181–189.

Chaudhary T.M. and Hisiani G.R., 1970 Effect of B on the yield of seed cotton. The Pakistan Cottons **1**: 13–15.

Nizami M.M.I., Shafiq M., Rashid A., and Aslam M., 2004 The soils and their agricultural development potential in Pothwar. National Agricultural Research Center, Islamabad, p. 158.

Rashid A., 1993 Nutritional disorder of rapeseed-mustard and wheat grown in Potohar area. Micronutrient 1991–92. National Agricultural Research Center, Islamabad. pp. 111.

Rashid A. and Qayyum F., 1990 Cooperative research program on micronutrient status of Pakistan soils and its role in crop production. Final Report 1983–90. National Agricultural Research Center, Islamabad, p. 84.

Rashid A. and Rafique E., 2005 Internal boron requirement of young sunflower plants: Proposed diagnostic criteria Commun in Soil Sci & Plant Anal **36**: 2113–2119.

Rashid A., Rafique E., and Ali N., 1997a. Micronutrient deficiencies in rainfed calcareous soils of Pakistan. II. Boron nutrition of peanut. Commun in Soil Sci Plant Anal **28**: 149–159.

Rashid A., Rafique E., and Bughio N., 1994 Diagnosing boron deficiency in rapeseed and mustard by plant analysis and soil testing Commun in Soil Sci Plant Anal **25**: 2883–2897.

Rashid A., Rafique E., and Bughio N., 1997b Micronutrient deficiencies in rainfed calcareous soils of Pakistan. III. Boron nutrition of sorghum. Commun in Soil Sci Plant Anal **28**: 441–454.

Rashid A., Rafique E., and Bughio N., 2002a Boron deficiency in rainfed alkaline soils of Pakistan. Incidence and boron requirement of wheat. In: HE Goldbach et al. (eds.), Boron in Plant and Animal Nutrition, Kluwer Academic/Plenum Publishers, New York, pp. 371–379.

Rashid A., Rafique E., Muhammad S., and Bughio N., 2002b Boron deficiency in rainfed alkaline soils of Pakistan. Incidence and genotypic variation in rapeseed-mustard. In: H.E. Goldbach et al. (eds.), Boron in Plant and Animal Nutrition, Kluwer Academic/Plenum Publishers, New York, pp. 363–370.

Rashid, A., Rafique E., and Ryan J., 2002c Establishment and management of boron deficiency in field crops in Pakistan: a country Report, In: Goldbach, et al. (ed.), Boron in Plant and Animal Nutrition. Kluwer Academic/Plenum Publishers, New York, pp. 339–348.

Shafiq M., Rashid A., Ghafoor A., Rafique E., and Aslam M., 2005 Nutrient status of water-eroded soils in Pothwar plateau. National Agricultural Research Center, Islamabad, p. 84.

Walsh L.M., and Beaton J.D., 1973 Soil testing and plant analysis. Soil Sci Soc Am Publication No. 368.

Westermann D.T., 1993 Fertility management in potato health management (ed R Rowe) American Psychopathological Society, St Paul, Minnesota, USA, pp. 77–86.

Yasin M., Rashid A., and Rafique E., 2002 Boron status of cotton soils: In: Abstract 9th Congress of Soil Science, 18–20 March, Faisalabad. Soil Science Society of Pakistan, Islamabad, p. 25.

Boron Environmental Geochemistry and Its Environmental Response

Enjing Zhu and Ende Wang

School of Resources and Civil Engineering, Northeastern University, Shenyang, China
E-mail: Zhuenjing2000@sina.com.cn

The geochemistry of boron

Form of boron

Boron is distributed in nature very widely. The average content of boron in the earth's crust is $1\times10^{-3\%}$. More than 150 boron minerals have been found, but only about ten of them are currently in use. The chemical composition of boron minerals is divided into three types: borosilicate minerals, boro-alumino-silicate minerals and borate minerals (including those containing water and anhydrous borate minerals). The borosilicate minerals are mainly datolite [$2CaO{\cdot}B_2O_3{\cdot}SiO_4{\cdot}H_2O$] and danburite [$CaB_2(SiO_4)_2$]. The boro-alumino-silicate minerals are mainly tourmaline [$Na_2O{\cdot}(Fe,Mg)O{\cdot}10Al_2O_3{\cdot}18SiO_2{\cdot}4B_2O_3{\cdot}5\ H_2O$] and axinite [$Ca(\ Mn,\ Fe)\ Al_2BSi_4O_{15}\ (OH)$]. Of these two kinds of boron minerals, only datolite has industrial value. Being the raw material used in industry, the third type of borate mineral plays an important role in boron processing. The characteristics of the primary boron minerals are presented in table 1.

The types of boron mines in China

There are two main types of boron mines in China, one is the continental salt lake boron deposit; the other is the skarn boron deposit. The former is the salt lake sedimentary deposit formed under the conditions of epigenesist. It belongs to volcanic sedimentary deposit according to the boron deposit classification of

F. Xu et al. (eds.), Advances in Plant and Animal Boron Nutrition, 383–391.

Table 1. Main characteristics of the borate minerals

Mineral name	molecular chemical formula	Content of B_2O_3%	Form of crystal	Luster	Color	Density (g/cm^3)	Mohs hardness scale	Solubility in water
Szaibelyite	$MgBO_2(OH)$	41.38	fibers or needle aggregates	Silky luster	White, gray, buff	2.62~2.75	3~4	imperceptible
Ludwigite	$(Mg,Fe)_2[BO_3]O_2$	17.83	needles, fibers, or pillar aggregates	pearl or diamond luster	Black and black green	3.6~4.7	5.5~6	imperceptible
Magnesioboracite	$Mg_2[BO_3]_4$	36.54	tiny particles	Glass luster	Colorless, white	3.03~3.10	6.5	imperceptible
Sassolite	$H_3[BO_3]$	56.40	squama, planks or stalactites	Glass or pearl luster	Colorless, white	1.46~1.52	1.0	soluble easily
Kernite	$Na_2(H_2O)_3[B_2B_2O_6(OH)_2]$	51.02	particles or fiber aggregates	Glass or sill luster	Colorless, white	1.90	2.5	soluble easily
Ulexite	$NaCa(H_2O)_6[B_3B_2O_7(OH)_4]$	42.95	tubercula or fiber nubby bodies	Glass or sill luster	Colorless, white	1.65~1.95	2.5	Sparingly soluble
Colemanite	$Ca(H_2O)[B_2BO_4(OH)_3]$	50.81	short pillars	Glass or diamond luster	Colorless, white	2.41~2.44	4.5~5	imperceptible
Priceite	$Ca_2(H_2O)[B_4BO_7(OH)_5]$	48.4	tubercles or compact masses	lusterless	White	2.26~2.48	3.0~4.5	imperceptible
Hydroboracite	$CaMg(H_2O)_3[B_2BO_4(OH)_3]_2$	50.53	actinomorphic forms or fiber aggregates	Glass luster	Colorless, white	1.90~2.17	2	Sparingly soluble
Chambersite	$Mn_3[B_3B_4O_{12}]OCl$	50.36	thin bean form aggregations	Grease luster	White with slight grey	3.49	7	imperceptible
Borax	$Na_2(OH)_8[B_4O_5(OH)_4]$	36.51	short pillars	Glass or grease luster	White, slight grey, buff	1.69~1 .72	2.0~2.5	insoluble
Suanite	$Mg_2[B_2O_5]$	46.34	bunches, actinomorphic or fibers	Glass or grease luster	White, hazel	2.91~2 .93	5.9	imperceptible
Pinnoite	$Mg[B_2O(OH)_6]$	42.46	pillars, short pillars or fibers	Glass luster	Colorless, white, offwhite	2.3	3.5	

the world. Examples are some large B lakes in northern Tibet, China, mainly containing borax and small amounts ulexite.

This deposit can be used easily. The skarn boron deposit results from endogenic mineralization, and is the borate deposit formed in the magnesio-skarn according to the boron deposit classification of the world. The types of boron ore used in China from endogenic mineralization deposits are listed in table 2.

Table 2. The type of boron ore found in active mines in China

The boron ore types	Typical mine area
Szaibelyite type	Zhuanmiao ditch, Houxian ditch of Liaoning Province
Szaibelyite — magnetite — serpentine type	Fengcheng County of Liaoning Province, Small east ditch of Jilin Province
Szaibelyite—carbonate type	Changning County of Hunan Province, Gaotai ditch of Jilin Province
Szaibelyite's ludwigite — magnetite type	
containingU ludwigite's sazibelyite—magnetite type	Wengquan ditch of Liaoning Province
Chambersite type	Ji County of Hebei Province

The distribution of boron resources in China

The boron resources of China are abundant. So far, about 55 boron mines have been discovered in the country. The explored reserves of China that are economically useful are the fifth largest compared with the reserves of the other countries of the world. The majority of boron deposits and reserves in China are found in the northeast and the southwest region, such as Liaoning , Jilin, Qinghai and Tibet province, and mainly found in three boron mineralization belts, i.e. the east of Liaoning — the south of Jilin boron mineralization belts, resulting from sediment — metamorphism, the Tibet plateau salt lake boron mineralization belts and Liuheye mountain of Jiangsu province—Zhong mountain yellow treasure the skarn boron mineralization belts of Guangxi province.

The boron reserves of the regions of Liaoning and Jilin province constitute 65% of all boron reserves of China. The ore types are mainly szaibelyite and ludwigite. The boron reserves of the regions of Qinghai and Tibet province constitute 29.8% of all boron reserves of China. The ore types are mainly ulexite, pinnoite, hydroboracite and kurnakovite types in Qinghai, and mainly borax, as well as, boracic lake mire and water in Tibet. There is still bittern water with boron in Sichuan province. Furthermore, szaibelyite has been found concurrently in some asbestos deposits in Jilin province. It has also been suggested that there are boron resources in some iron mines in Jiangsu province. Chambersite was found in Hebei province for the first time. There is also the middling scale skarn boron deposit in Hunan and Anhui province.

The distribution of boron in the environment

Boron is distributed widely and its content varies greatly in the environment. We will discuss the distribution of boron in the environment from the following aspects.

Boron concentration in air

Boron in air mainly results from the following sources. There is boron in the form of sassolite released into the air when volcanos erupt. Much of this B is captured by the oceans. At least 30% of the boron in coal from coal power plants is also in the form of sassolite released into the air. This is the main source of air pollution. Although the boron concentration in air is low, the atmosphere contains a great deal of sassolite vapor. In the active volcano districts, the average concentration of gaseous boron sprayed into the air is lower than 2.5 $\mu g/m^3$ ~31.4 $\mu g/m^3$. The particulate boron is lower than 4 $\mu g/m^3$. The average concentration of particulate boron is up to 8.5 $\mu g/ m^3$ in the smoke above the volcanic lake. The average concentration of gaseous boron is up to 16.1 $\mu g/ m^3$. The measured amount for the continental ocean shore and three distance ocean sites showed that the average concentrations of particulate boron in the atmosphere were 1.8 ng/m^3~12.2 ng/m^3, 2.4 ng/m^3~3.7 ng/m^3 and <0.5 ng/m^3~ 2.8 ng/ m^3.The average concentrations of gaseous boron were lower than 0.05 ng/m^3~20.7 ng/m^3,3.5 ng/m^3~82.8 $ng/m^3$and 0.6 ng/m^3~25 ng/m^3.

It can be estimated that 90% of boron occurs in gaseous forms, and only 10% of boron occurs in the form of the particulates in the atmosphere.

Boron concentration in natural water

Boron concentration is 5-10 μg/L in fresh waters, is above 1-3 mg/L in continental salt water, is 4450 μg/L in sea water and reaches 100-500 mg/L in some waters of salt lakes that can form borate crystals. The total transported quantity of the fluviowater is about 370,000 tons/ years. The water transfer coefficient is 8.3.

Boron in water commonly occurs in the form of borate, such as the BO^{2-}, BO_3^{3-}. The average content of boron in seawater is hundreds of times larger than in river water. The average concentration of boron is 4.5 mg/L in the ocean. In groundwater, boron mainly comes from rocks and soil (i.e. partial geology) which contain borates and borosilicates, which are eluviated by rain. The boron concentration changes in the groundwater, from <0.3 mg/L to>100 mg/L. Generally, the concentration of boron in the groundwater is influenced greatly by the boron minerals or boron containing rocks of the region. Below Kesterson Reservoir in California, United States, the surface water containing high boron concentrations has polluted the groundwater. The concentrations of boron, selenium, and arsenic are high in the reservoir, which received agricultural drainage water in the San Joaquin river valley. The boron content of surface water is controlled by the areal geochemistry and the drainage of industrial waste water and the municipal waste water characteristic of the region. The concentration of boron in European surface water is 0.001 mg/L~2 mg/L, the mean is below 0.6 mg/L. The boron concentration in waters of Pakistan, Russia and Turkey is similar to that in Europe, ranging from <0.01 mg/L to 7 mg/L, mostly below 0.05 mg/L. The boron concentration in Japanese surface water reaches 0.01 mg/L, and is about 0.3 mg/L in South Africa. The boron content of surface waters of the regions where boron is abundant, such as two rivers in South America rivers (Argentine Rio Arenales river and the Loa river of Chile) is 4 mg/L~26 mg/L. In other districts, such as the Rio Arenales river, the content of boron is lower than 0.3 mg/L. The boron concentration of surface waters of North America (Canada and the United States) ranges from 0.02 mg/L to 360 mg/L. This implies the existence of rich boron deposits. The median and the average concentrations of borate are 0.0031 mg/L and 0.0056 mg/L in rain and snow in the six regions of western Switzerland. In municipal sewage sludge, boron occurs mainly as sassolite. 50% of boron in waste water comes from the use of detergent products. The range of boron content in municipal sewage of 23 silts of the United States is from 7.1 mg/L to 53.3 mg/L (dry weight). The boron content is obviously high in the tailing waters in Kuandian, Liaoning province of China. In the water released from the boron mine, the average concentration of boron is 176 mg/L. The highest concentration is up to 1074 mg/L,and exceeds the boron concentration of other water bodies in other regions all over the world.

Boron concentration in soil

Boron is a naturally occurring dark brown/black substance found through out the environment. The content of boron in soil is about 10 mg/kg ~ 300 mg/kg (average 30 mg/kg), depending on the soil types, the soil organic matter content and the amount of rain. It is reported that in the soils of America, British, New Zealand, and Brazil, the background boron content is <20~300 mg/kg, 7~71 mg/kg, 2.5~47mg/kg and 31.3~54.0 mg/kg with a mean of 30 mg/kg, 33 mg/kg, 15.5 mg/kg and 42.9 mg/kg. The background content of boron in Kuandian, Liaoning province of China ranges from 39.2 ~ 82.4 mg/kg, and is higher than the above mentioned amount. The soil of Kuandian boron mineralization region contains boron ranging from 133 to 1195 mg/kg, exceeding the background boron concentration.

The boron content in a marsh soil and a forest soil is lower, 18.5×10^{-4}. In the deep ocean the average boron content is 230×10^{-4} in the lime silts and 55×10^{-4} in the clay silt. Boron in the ocean occurs in solution. The ocean biota absorb little boron. On the other hand, boron was easily adsorbed to the suspended detrietal particles accumulating in the clay silt.

The influence of boron in the ecosphere

Boron is an essential nutrient for plants and plays an important role in their sugar metabolism. Generally boron concentration in fronds is 2~100 parts per million (ppm), and it varies with the type and part of the plant. The ratio of the content of some element in plant ash to the mineral material of the lithosphere of the earth [parts per million (ppm)] is called the biological absorption coefficient. The biological absorption coefficient of boron is 50. Thus, it belongs to the strongly absorbing elements. Plants living in B deficient soil will be stunted, have restricted growth, and will have no buds, a decayed heart etc. The background content of boron in plants, such as soya, potato, corn, and kidney bean in Kuandian, Liaoning province of China is 34.7~47.9 mg/kg, 5.8~6.3 mg/kg, 24.4~25.5 mg/kg and 1.24~2.74 mg/kg (all are dry weight). The content of boron in boron mines area in Kuandian in the above fronds is 34.0 mg/kg, 12.0~12.8 mg/kg, 40.7 mg/kg and 2.48~2.68 mg/kg. No obvious relationship of boron content of soya and the soil can be found either in the background region or in the control area. There is an obvious relationship between the boron content of potato, corn, kidney bean and the soil of the control area. With higher content of boron in soil, the boron content of potato, corn, and kidney bean is higher. The boron content in the soil might be related to that in the fronds in Kuandian area. However, there are no toxicity symptoms in the fronds. More research is needed in this area.

Cows and sheep living in the areas of high boron content will have cow or sheep's bowel diseases, such as diarrhea and become emaciated etc. Many experiments have shown that excess absorption of boron by crops will cause boron toxicity. The half lethal amount of boron is 300 mg /kg of body weight for domestic rabbits and dogs. 1 mg/L of boron in the water will produce toxicity in the majority of crops. Boron is also an essential trace element for the human body. Just the right amount of boron can prevent osteoporosis, but excess boron will cause boron poisoning, such as diarrhea, vomiting, collapse, deep slumber etc. The higher the boron content in sea water, the higher the boron content in sea fronds. But some research has shown that there is no biomagnification in the food chain. In Kesterson National Wildlife Refuge in the San Joaquin river valley of California, United States, research has revealed that the bioaccumulated boron is very high. The boron content of seaweed ranges from 4.2 mg/kg to 14.9 mg/kg. Adams et al. carried out research on various hydrophytes of three main rivers (Delaware river, Susquehanna river and Allegheny river) in Pennsylvania, United States and measured the concentration of boron and 11 other possible polluting ions. The pollution sources of this region vary and include lumbering, open-air coal excavation, recreation, agricultural usage and the city industrial centers. The boron content of 21 kinds of vascular plants living underwater ranges from 26.3 mg/kg to 170 mg/kg. The boron content of 8 other kinds of vascular plants ranges from 11.3 mg/kg to 57 mg/kg. Tsui et al. once studied bioaccumulation of a few elements (including boron) in five kinds of fresh water fishes in the Cold River valley of the western region of Canada. Five fresh water fishes represented the different feeding habit and life styles. The northern dog fish and the lake trout are mainly carnivorous fishes, the lake herring is a fish that eats plankton, the lake white fish and the white mouth swallow fish are bottom feeders. In the spring and summer of 1978 Tsui et al. collected the fishes from 7 lakes of this district, and measured the boron content of their muscle tissue. The highest average boron content was 0.0627 mg/L in the lake water. The average content of boron of five fishes ranged from 3.23 mg/kg for the lake white fish to 12.44 mg/kg for the white mouth swallow fish. Wren et al. reported the boron content of fresh water fishes and clams of Precambrian Shield lake of Ontario, Canada. This lake is not under the direct influence of mankind. The B content in the musculature of the fishes is general lower than that of the fish of the Cold River valley of western of Canada. The average concentration (wet weight) of boron in the fish is 1.8 mg/kg ~ 2.9 mg/kg. But the content of boron in the parenchyma of the clams is 2.6 mg/kg.

The B content in the whole body sample of the blue gill fish and the carp of two branches of the San Joaquin river which received agricultural drainage water increased slightly. The highest boron content in the blue gill fish (dry weight) is 14 mg/kg (3.5 mg/kg, wet weight), and is 20 mg/kg (5 mg/kg , wet weight) in the carp. These concentrations are similar to those in the mosquito fish of the San Joaquin river valley. But according to a report, the higher boron concentration might be caused by the mining of sands and gravels in this area and the boron sediment of the neighboring soils. Someone analyzed the boron content in bird livers in the Grassland water area of California, United States in

1985 to 1988. The application of agricultural drainage water to the wetland, caused the composition of the food chain in this area to be polluted by trace elements. During the breeding and wintering periods, the boron content in the livers of the birds living in the northern grassland and the south were high ranging from 1.7 mg/kg to 40 mg/kg (dry weight).

The influence of boron mining and boron-sludge on the environment

Liaoning province is a main region for boron mining in China. In the region from Yingkou to kuandian there are many large boron mines. Along with the boron mining, a great deal of mullock and boron-sludge have appeared. Mullock is often the wall rock of the ore body, etc.. The boron-sludge is composed of the tailings which result from chemical milling. Presently, there are about ten million tons of boron-sludge accumulated in this region. Production is proceeding at nearly 1 million tons per year.

According to the analysis report, the boron-sludge in Liaoning province consists of B_2O_3=3.5-5%, MgO $\geqq$40%, Na_2CO_3<1%, $SiO_2 \leqq$20%, $Fe_2O_3 \leqq$10%, and 15-20% other constituents. No industrial process exists to make use of the resulting boron-sludge. Lean ore from boron mining has been discarded. Recycling uses for boron-sludge have not been found. It is a tremendous waste of boron resource. Because of the very thin granularity (100 μm or so) and the greater alkalinity of the boron-sludge, it has been piled up in the countryside in gutters in large quantity. No environmental protection exists for processing in the open air. This has resulted in pollution of the environment, and has become a big social pollution problem in the boride production area. The boron-sludge piles can also turn the soil into desert and raise the soil pH, destroy vegetation, prevent grass from growing.The mountain body is bare and the water body has been polluted resulting in death of life-forms, etc.. There are many bare rock bodies and the air is filled with process dust in the mine area. Some river waters have become red and fetid. The environment is getting worse and rainfall has declined year after year. Presently, the annual rainfall has already decreased from more than 1000 mm originally to 500-600 mm. There were even continuously dry seasons in the spring and autumn in 2000. That was seldom seen previously in history. Obviously, the boron-sludge pollution problem has become a restriction on the social economic development and must be urgently solved.

Boron influence on the environment

Existing life-forms have close geochemical contact with their environments. The basic principles of biogeochemistry are very important for understanding the enrichment of trace elements in plants. The increase of the content of any element in soil, water, and air will greatly increase the element content in the plant, i.e. the element content at the bottom of the food chain will increase. This will bring about a difference between the content of major elements and that of trace elements. Increasing amounts of the element entering the plant body will cause a variety of damage to the organism and even destroy its ability to function. The plant enrichment function to the chemical element has already become the basic means that can be used for seeking deposits in biogeochemistry. Through the analysis to the branch, trunk, leaf, root, flower bud, fruit of the plant etc., many kinds of mines have been discovered. Otherwise the plant which is unadapted to the high content of chemical element will grow abnormally and exhibit toxicity symptoms. As an element that is absorbed extensively by the plant (the absorption coefficient is above 50), boron can promote the activation of the enzyme in ribotide, thus affect the synthesis of protein, and cell division, producing a change in plant appearance.

References

Butterwick L., De Oude N., and Raymond K., 1989 Safety assessment of boron in aquatic and terrestrial environments. Ecotoxicology and Environmental Safety. **17**, 339–71.

Cöl M., Genc Y., and Saylı B.S., 2000 Osteoporosis prevalence in women living in a boron mineral region of Turkey, and combined factors. Archives of Complex Environmental Studies. 3-4(**12**): 30–41.

Howe P.D., 1998 A review of boron effects in the environment. Biol. Trace. Elem. Res. **66** (1-3): 153–166.

National Academy of Sciences (NAS) 1980 Boron In Mineral Tolerance of Domestic Animals. Natl. Acad. Sci. Natl. Res. Counc. Comm. Anim. Nutr. Washington, D C, pp. 71–83.

Ali O., Sezer J.I., and Brand Chemical vapor deposition of boron carbide Department of Chemical Engineering. The Uniuersity of Nebraska-Lincoln, **236** A6ery Laboratory. Lincoln. NE 68588–0126, USA.

Wang Chunli. and Xing X. et al. 2003 Analyses of boron distribution and pollution in soil and some crops in Kuandian Environmental Monitoring in China. vol. **19** No. 5.

Micronutrients Status in Citrus Orchards and Soils under these Orchards in Pakistan

M. Ibrahim [1], N. Ahmad, A. Niaz and T. Majeed

Soil Chemistry Section, Ayub Agricultural Research Institute, Faisalabad, Pakistan.

[1]Corresponding Author Email: soilchem@fsd.paknet.com.pk

Introduction

There were some reports containing information on micronutrients in citrus area (Kausar et al. 1979; Rashid et al. 1991). Upon looking the details, it was found that existing information were limited and unsystematic. The samples were collected disregarding the time and age of plant tissues during sampling which is very important aspect. To establish nutrients deficiency in fruit plants, the age of leaves taken from bearing and non-bearing branches are extremely important. These aspects were mostly lacking from the existing information. Keeping in view the previous work done in Pakistan, more appropriate efforts have been made in this study to investigate soil and plant micronutrients status of citrus growing area on high-pH calcareous soils.

Materials and methods

Before conducting the survey, the citrus area was divided into 334 equals units from where an orchard was selected for soil and plant samples collection. Soil and plant samples were collected during July-August. Out of 334 locations, 288 sites had the orchards. Soil and plant samples were collected from the same orchard. Keeping in view the calcareous nature of soil, AB-DTPA solution was used for soil extraction (Soltanpur and Workman, 1979). Boron was extracted with dilute 0.05 M HCl. A group of 20 trees with uniform canopy and growth were selected for leaves collection. The leaves from spring growth of 4-6 months age from non-bearing twigs were collected. The leaf tissues were digested and analyzed for Fe, Cu, Mn and Zn by Atomic-Absorption Spectrophotometer. Boron was determined calorimetrically after developing the color by Azomethane-H method

Results and discussion

Micronutrients in soils

The soil analysis results for Zn, Cu, Fe, Mn and B are presented in Table 1.

F. Xu et al. (eds.), Advances in Plant and Animal Boron Nutrition, 393–396.

Table 1. Micronutrients status of soils under citrus orchards

Nutrients	Limits ($\mu g\ g^{-1}$)	Total(samples	% of total	Mean
Zn				
Low	<1.0	266	80	0.5
Medium	1.0-1.5	38	11	1.2
Adequate	>1.5	30	9	2.4
Cu				
Low	<0.3	3	1	0.2
Medium	0.3-0.5	3	1	0.4
Adequate	>0.5	328	98	2.2
Fe				
Low	<3.0	24	7	1.5
Medium	3.0-5.0	57	17	3.1
Adequate	>5.0	253	76	8.6
Mn				
Low	<0.6	3	1	0.5
Medium	0.6-1.0	10	3	0.9
Adequate	>1.0	321	96	3.4
B				
Low	<0.55	288	86	0.3
Medium	0.55-1.0	42	13	0.7
Adequate	>1.0	4	1	1.2

Total soil samples = 334

Zinc (Zn)

The results show that deficiency of Zn is wide spread. Out of 334 sites 80% are deficient, 11% medium and 9% adequate. By taking deficient and medium ranges together, then almost 91% soils come under the range where deficiency of Zn could appear. The soils are high in pH (8.0 to 8.5) and are calcareous in nature that cause the deficiency of Zn.

Copper (Cu), iron (Fe) and manganese (Mn)

In most of the soils these nutrients are adequate in soils. The analysis showed that Cu is adequate in 98% soils. The Fe data showed that 76% sites fall in adequate, 17% in medium and 7% in low range. The extent of Mn deficiency is low and 96% sites are in adequate range, 1% deficient and 3% in medium range (Table 1).

Boron (B)

Out of 334 sites, 86% are deficient, 13% medium and 1% adequate (Table 1). Earlier researchers reported that B deficiency in soils under different cropping systems is of lesser extent, which do indicate general existence of B deficiency (Zia, 1993). The soil samples (177) taken by Sillanpaa (1982) from 20 different districts also showed deficiency in 49% samples. The area under this study is under citrus plantations having uniform cultural practices and climate condition than the areas investigated by earlier workers and that made the difference.

Micronutrients in citrus orchards

Zinc (Zn)

Like in soils, Zn deficiency is also wide spread in orchards. Out of 288 orchards 93% are deficient, 7% low and none in optimum range (Table 2). Earlier researchers reports are mostly based on limited number of leaf samples that showed the Zn deficiency occurrence in the range of 62-90% samples which support this investigation (Rashid et al. 1991 and Siddique, 1994). When soil and plant analyses data are seen together (Tables 1 and 2) it is observed that 100% orchards deficient in Zn are growing on deficient soils.

Copper (Cu), iron (Fe) and manganese (Mn)

Usually Cu deficiency symptoms are not seen in the field, which is confirmed by leaf analysis that showed 92% sites are in optimum range. It has been further observed that 98% orchards sufficient in Cu are growing on soils having sufficient quantities of this nutrients. Out of 288 orchards; Fe was optimum in 99%. Mn deficiency is not evident from leaf or from soil analysis. Out of 288 orchards 90% are in optimum and 3% in low range. Therefore, soils moderately alkaline in nature may not necessarily be deficient in Mn.

Boron (B)

Out of 288 orchards, B was deficient in 10%, low in 43% and optimum in 47% orchards (Table 2). If low range of nutrients is considered towards optimum range then 90% sites come under sufficient range because mean (29 $\mu g\ g^{-1}$) of low limit is more close to the mean (47 $\mu g\ g^{-1}$) of optimum limit. In this case soil test is not correlated with leave concentrations.

Conclusions

On the basis of these results, it is concluded that soils and orchards are deficient in Zn and its application is required either through soils or through foliage. In case of B about 53 orchards are in deficient to low range, its application should be based on plant analysis. Indiscriminate use may damage the orchards. The Cu, Fe, and Mn are not deficient in soils and orchards.

Table 2. Micronutrients status of citrus orchards

Nutrients	Limits*(μg g^{-1})	Total Samples	% of Total	Mean
Zn				
Deficient	<17	268	93	13
Low	18-24	20	7	19
Optimum	25-100	0	0	-
Cu				
Deficient	<3	0	0	-
Low	3- 4	11	4	5
Optimum	5-16	267	96	10
Fe				
Deficient	<35	0	0	-
Low	35-59	2	1	34
Optimum	60-120	21	99	103
Mn				
Deficient	<17	0	0	-
Low	18-24	8	3	23
Optimum	25-100	280	97	43
B				
Deficient	<20	29	10	17
Low	20-35	123	43	29
Optimum	36-100	136	47	47

* Hanlan et al. 1995. (Total plant samples = 288)

Acknowledgements

The authors are highly thankful to Pakistan Agricultural Research Council (PARC) for providing the funds for this study under Agricultural Linkages Program. Moreover, the cooperation of Project Staff at Head Quarter is duly acknowledged. We also acknowledge the cooperation of Mr. Mohsin Ali for typing the manuscript.

References

Kausar M.A., Alam S.M., Sharif M., and Pervaiz M.I. 1979 Pakistan J Sci and Indus Res **22**, 156–161.

Hanlan E.A., Obreza T.A., and Alva A.K., 1995 In Nutrition of Florida Citrus Trees SP 169, Univ.Florida Cooperative Extension Service, IFAS, S.A. Rashid A., Hussain F., Rashid A., and Din J., 1991 Pakistan J. Soil Sci. **6**, 25–28.

Sillanpaa M., 1982 A global study, Soils Bulletin **48**, FAO Rome.

Soltanpour P.N. and Workman S., 1979 Commun Soil Sci. Plant Anal. **10**, 1411–1420.

Zia M.S., 1993 Fertilizer use efficiency, Annual Report. 1992–93 NARC, Islamabd, Pakistan.

Author Index

Subject Index